THE ATCHAFALAYA RIVER BASIN

Protecting nature. Preserving life.™

This book was written
and published with
the support of
the Louisiana Chapter of
The Nature Conservancy.

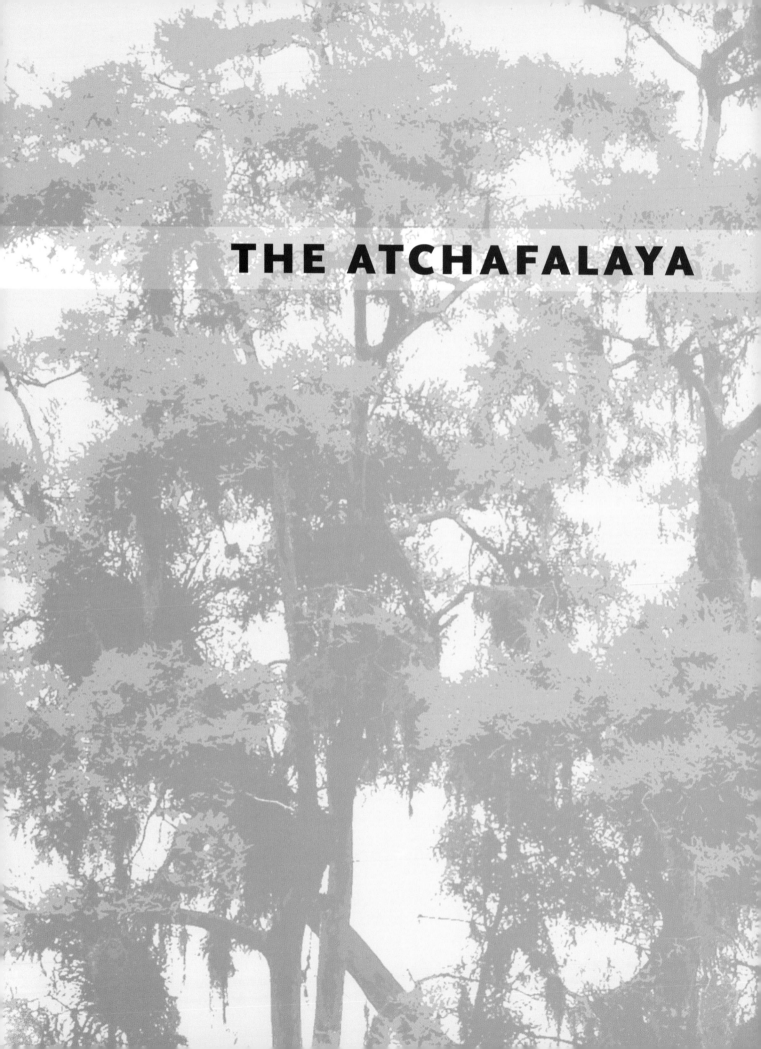

THE ATCHAFALAYA

RIVER BASIN

History and Ecology of an American Wetland

BRYAN P. PIAZZA

TEXAS A&M UNIVERSITY PRESS
College Station

Manufactured in China

This paper meets the requirements of ANSI/NISO Z39.48–1992 (Permanence of Paper).

Binding materials have been chosen for durability.

Library of Congress Cataloging-in-Publication Data

Piazza, Bryan P. (Bryan Patrick), 1970– author.
    The Atchafalaya River Basin : history and ecology of an American wetland /
Bryan P. Piazza. — First edition.
        pages cm —
    Includes bibliographical references and index.
    ISBN 978-1-62349-039-3 (cloth : alk. paper) —
    ISBN 978-1-62349-141-3 (e-book)
 1. Atchafalaya River Watershed (La.)   I. Title.
    F377.A78P53 2014
    976.3—dc23
                                    2013030344

For Sarai, Kade, and Reed

CONTENTS

STILL REMEMBER RECEIVING that telephone call. I was working up at Delta Water-fowl Research Station in the summer of 1993. I had just graduated from college at the University of Wisconsin—Stevens Point; a new wildlife biologist on the loose, want-ing to further my education in Louisiana's vast coastal wetlands. The call was news that I had been accepted into the wildlife program at LSU to pursue my master's degree, and came with a simple question: would I like to live on the Atchafalaya River Delta and study colonial wading birds? After nearly jumping out of my skin with excitement, I calmly accepted, loaded my Ford Ranger, and within days, I was on the road, pointed south, about to begin my journey into one of the largest wetlands in the world.

I immediately fell in love with the Atchafalaya and Louisiana's wetlands and culture. Whether I was helping my buddies on their research projects in the Basin, harvesting crawfish, hunting and fishing, or pursuing my own research, I was never far from the marsh or swamp. And since that time, I have continued to pursue a deeper knowledge of wetlands and have worked to conserve and restore them. My research has taken me into countless areas of the coast, but wherever I was, the Atchafalaya was not far from my thoughts (or my muddy boots). So, when the opportunity arose to write this book, I jumped at the chance.

This book tells the story of the Atchafalaya River Basin through the lens of science. When I was hired at The Nature Conservancy (TNC), we were just beginning our work in the Atchafalaya Basin. Our goal is to help conserve and restore this magnificent landscape so that it can continue to sustain biodiversity and provide livelihoods and enjoyment for generations to come. My first task was to assess the current ecological state of the Atchafalaya River Basin. I quickly realized that, while there were a multi-tude of good scientific studies that had been done in this great landscape, nobody had ever synthesized them into a comprehensive volume to tell its story from a scientific perspective. So, that's what I did, and this book is the result.

This book reviews hundreds of studies done in the Atchafalaya River Basin across multiple ecological disciplines—geology, hydrology, forestry, fisheries, wildlife science, social science, and economics. I did not review cultural studies, as a number of good books already exist on the Cajun culture and its close ties to the Atchafalaya Basin.

I want this book to aid in the conservation and restoration of this American wetland by providing instant access to a wealth of scientific information and an objective assessment of the system's current status and trends. It is also my desire that this book will serve as a model for other natural systems worldwide where conservation and restoration are needed. I hope that readers come away with an appreciation of the importance of the Atchafalaya Basin and river-floodplain systems, not only locally and regionally, but nationally and internationally. Further, I wish readers to come away with a sense of the interconnectedness of ecology and economy and the importance of keeping natural systems intact, because a good environment is critical for a good economy. Lastly, I hope that readers gain an appreciation for the abundance of great scientists who live and work in Louisiana and the importance of all of our state's wetland science and monitoring programs (agency, NGO, and academic). These scientists, many of whom are my colleagues and friends, do great work and provide a tremendous service to our state, the nation, and the world.

The Atchafalaya and all of Louisiana's wetlands are important to me. Not only do I make my living from them, but they have given me many of the things that are dear to me—my family and friends, wonderful food, and the opportunity to live in a culture that is unique, vibrant, and extremely close to the land. Writing this book has been a wonderful journey. I have had the privilege of reading many great studies done in these magnificent wetlands and wetlands worldwide, and I have learned much in the process. While I have done my best to represent each author's work as it was intended, I take all responsibility for any errors in this book.

This book would not have been possible without a team effort. I am indebted to many people for helping me through this process. My colleagues at TNC provided constant mentorship, support, and encouragement throughout this process. Dr. Keith Ouchley and Dr. James Bergan provided me the leadership and space to get this done. Dr. Steve Haase and Richard Martin provided multiple, thoughtful reviews. Nicole Love, Joe Baustian, and Kacy King helped proofread citations, the bibliography and page proofs. Seth Blitch always picked up my slack when I needed help. David P. Harlan created the final maps in the book. Don McDowell, Harriet Pooler, and Jennifer Browning made sure we could make this a quality publication. Seth Blitch, Richard Martin, Don McDowell, and Matt Pardue provided a number of photographs. Kacy King coordinated with photographers to help me obtain images and usage rights.

A number of terrific photographers provided images for free or at a greatly reduced rate, because they valued the mission of this book and The Nature Conservancy: Brittany App, Patti Ardoin, Charles Bush, Eric Engbretson, Christina Evans, Zeb Hogan, Alison M. Jones, C. C. Lockwood, Bob Marshall, Joel Sartore, and Todd Stailey. The final graphs and diagrams were produced by STUN Design in Baton Rouge, Louisiana.

A number of scientists and researchers also provided figures or images from their presentations or publications so that I could incorporate them into the manuscript: Dr. Chris Bonvillain, Dr. Richard Condrey, Alaric Haag, Dr. Guerry Holm, Daniel E. Kroes, Cassidy Lejeune, Gary W. Peterson, Chet Pilley, Dr. Nancy Rabalais, John Troutman, and Dr. Nan Walker. Micah Bennett made sure my species lists were correct, updated, and complete. Dr. Megan La Peyre provided valuable input and support whenever I needed her. Yvonne Allen provided expertise and friendship on multiple occasions. Whether providing data, data analysis, photos, figures, maps, advice, multiple reviews, editorial assistance, or her honest opinion, Yvonne always made time for me, and this book would not be possible without her.

Lastly, I thank my family. My wife Sarai provided expertise, support, and tolerance of late nights, multiple revisions, and piles of scientific papers all over the house. My sons Kade and Reed inspire me by constantly reaffirming my drive for conservation and restoration and my desire to make the world better for them.

This book was created with funding from the Walton Family Foundation, the Charles Lamar Family Foundation, and Canal Barge, Inc. It was produced by the Louisiana Chapter of The Nature Conservancy with support from The Nature Conservancy's Great Rivers Partnership.

The following is a list of peer reviewers who invested their time and effort and made this book better. One reviewer wished to remain anonymous:

Yvonne C. Allen, formerly with the US Army Engineer Research and Development Center, Baton Rouge, Louisiana, now with the US Fish and Wildlife Service, Baton Rouge.

Dr. James F. Bergan, The Nature Conservancy, Baton Rouge, Louisiana

Michael R. Carloss, Louisiana Department of Wildlife and Fisheries, Baton Rouge, Louisiana.

Stephen Chustz, Atchafalaya Basin Program, Louisiana Department of Natural Resources, Baton Rouge, Louisiana.

Stephen P. Faulkner, US Geological Survey, Leetown Science Center, Kearneysville, West Virginia.

Angelina Freeman, formerly with the Environmental Defense Fund, Washington, DC, now with Louisiana Coastal Protection and Restoration Authority, Baton Rouge.

David Fruge, Louisiana Department of Natural Resources, Baton Rouge, Louisiana.

C. Stephen Haase, The Nature Conservancy, Rockwood, Tennessee.

Guerry O. Holm Jr., CH2M Hill, Baton Rouge, Louisiana.

Richard F. Keim, School of Renewable Natural Resources, Louisiana State University Agricultural Center, Baton Rouge, Louisiana.

William E. Kelso, School of Renewable Natural Resources, Louisiana State University Agricultural Center, Baton Rouge, Louisiana.

G. Paul Kemp, formerly with the National Audubon Society, Baton Rouge, Louisiana, now with G. Paul Kemp and Associates, LLC, Baton Rouge.

Daniel E. Kroes, US Geological Survey, Louisiana Water Science Center, Baton Rouge, Louisiana.

Richard Martin, The Nature Conservancy, Baton Rouge, Louisiana.

Lawrence P. Rozas, NOAA Estuarine Habitats and Coastal Fisheries Center, Southeast Fisheries Science Center, Lafayette, Louisiana.

Harry H. Roberts, School of the Coast and Environment, Louisiana State University, Baton Rouge, Louisiana.

Mike Walker, Louisiana Department of Wildlife and Fisheries, New Iberia, Louisiana.

Dean Wilson, Atchafalaya Basinkeeper, Water Keeper Alliance, Bayou Sorrell, Louisiana.

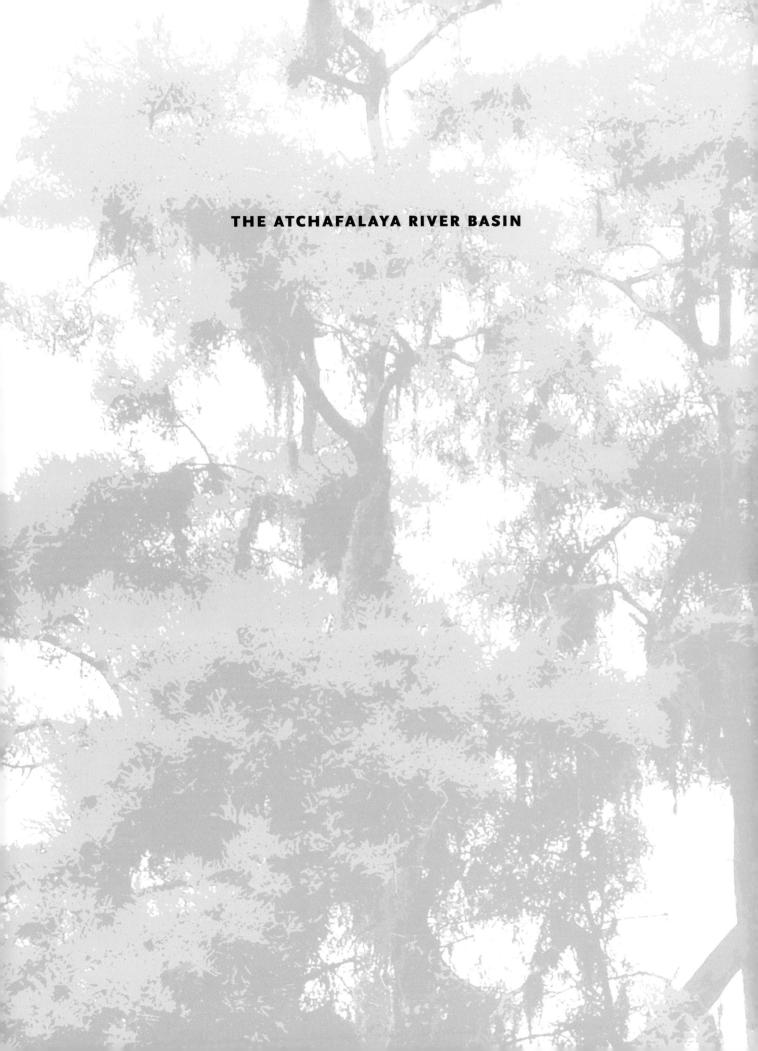

THE ATCHAFALAYA RIVER BASIN

The Atchafalaya River Basin (ARB) is a large, shallow interdistributary basin located within the Mississippi River delta plain in south Louisiana (Fig. 1.1; Fisk 1944, 1952, Tye and Coleman 1989a, b). The historic extent of the ARB was approximately 200 km long and encompassed an area of 8,345 km² (3,222 mi², 2,062,000 acres; Lambou 1990, Sabo et al. 1999a, b). The ARB begins at the confluence of the Mississippi, Red, and Atchafalaya Rivers (Three Rivers) near Simmesport, Louisiana. Its natural ecological boundaries were the Bayou Teche ridge on the west, the Mississippi River on the east, and the Gulf of Mexico to the south (Fig. 1.2).

The boundaries of the ARB were changed after the Great Mississippi River Flood of 1927, when large-scale loss of property and human life spurred a national effort to control the Mississippi River and eliminate catastrophic flood loss (Fig. 1.3). After receiving designation as the principal floodway in the Mississippi River and Tributaries Project in 1928, flood protection levees were erected on its east and west sides, constraining the basin into a system of floodways (West Atchafalaya Floodway, Morganza Floodway, Atchafalaya Basin Floodway System; hereafter ARB floodway) approximately 25 km (15 mi) across and designed to divert Mississippi River floodwaters quickly to the Gulf of Mexico (Fig. 1.2). The ARB floodway, now constrained to approximately half of the basin's original size, encompasses 339,100 ha (838,000 acres) of forested wetlands, bayous, and lakes from Three Rivers to Morgan City (Louisiana Department of Natural Resources 2010). The East and West Atchafalaya Protection Levees also severed the connection of two large areas of swamp forest (Lake Fausse Pointe on the west, Verret Swamps on the east) to the Atchafalaya River.

South of Morgan City, the Atchafalaya and Wax Lake Deltas contain an additional 57,060 ha (141,000 acres) of coastal delta habitat, making the total area of the modern ARB (ARB floodway and river deltas) almost 400,000 ha (396,200 ha, 979,000 acres; Fig. 1.2). Floodway designation also prompted the 1941 construction of the Wax Lake Outlet in the lower basin to more quickly divert floodwaters to the Gulf of Mexico (Reuss 1982). Because the Atchafalaya River also offers a more efficient flow path for the Mississippi River, a water control structure complex (Old River Control Structure)

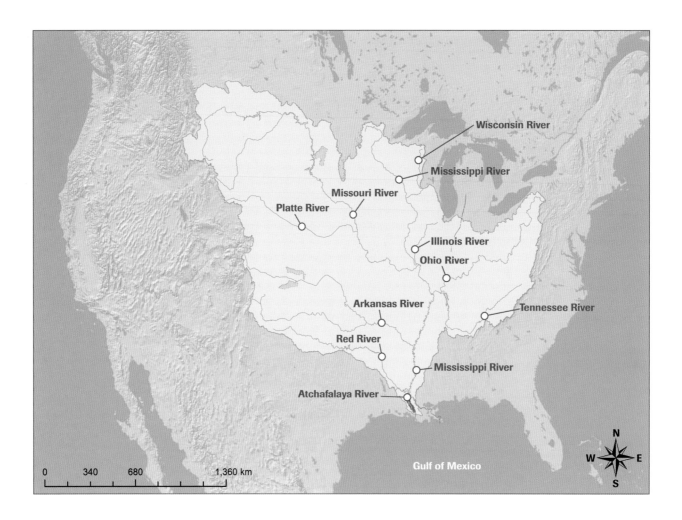

FIG. 1.1. Map showing the Mississippi River drainage basin, including major tributaries and the Atchafalaya River, its major distributary. Also highlighted is the Atchafalaya River Basin Floodway System. Map courtesy of Yvonne C. Allen, US Army Engineer Research and Development Center, Baton Rouge, Louisiana.

was constructed at its origin in 1963 to halt the stream capture process by regulating the amount of Mississippi flow that is allowed to enter the Atchafalaya.

The ARB is an extremely dynamic system that has been the site of at least four large-scale deltaic deposition events during the Holocene epoch and, since the late 1500s, the Atchafalaya River has been progressively capturing the flow of the Mississippi River (Tye and Coleman 1989a, b, Coleman et al. 1998, Roberts et al. 2003). Consequently, the ARB contains a range of habitats from bottomland hardwood forest to a growing marine delta ecosystem (Fig. 1.4). However, the ARB is likely best known for its expansive relict backswamp environments, dominated by baldcypress (*Taxodium distichum*) and water tupelo (*Nyssa aquatica*; Fig. 1.5).

Like many large river systems worldwide, the ARB supports a disproportionately high level of biodiversity (Calhoun 1999). It provides habitat for many species of plants as well as resident and migratory fish and wildlife species (> 300 wildlife species, >100 fish and shellfish species) and is vital to local, regional, and global biodiversity (Calhoun 1999, US Geological Survey 2001, National Audubon Society 2011a, b, Alford and Walker 2011). It also provides a multitude of ecosystem services (flood control, navigation, oil and gas resources, forest, fish and wildlife resources) that have been used extensively by humans as well as ecosystem services for which there are few developed

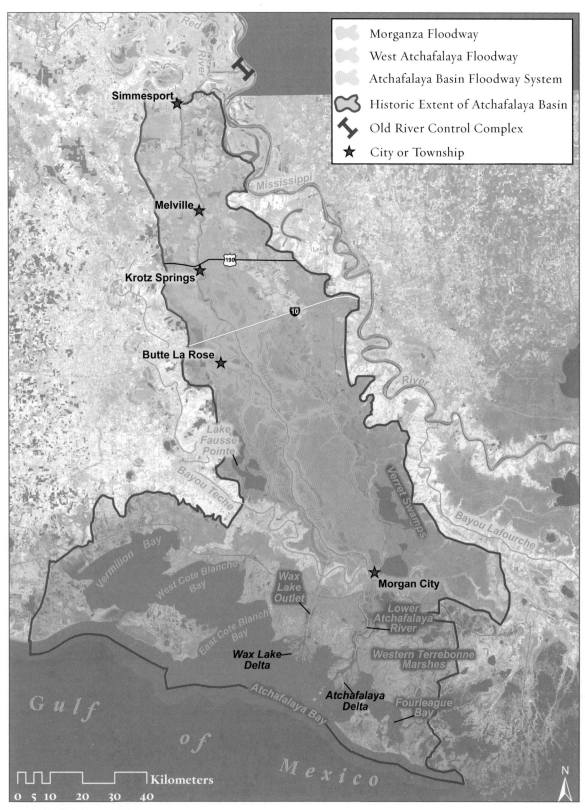

FIG. 1.2. Map showing the historic extent of the Atchafalaya River Basin as well as the system of floodways that, along with the Atchafalaya and Wax Lake Deltas, define the modern basin. Labeled areas are referenced throughout this book.

FIG. 1.3. The great Mississippi River flood of 1927 resulted in large-scale loss of property and human life and a national effort to control the Mississippi River to ensure that this type of catastrophe never happens again. Following this event, the boundaries of the Atchafalaya River Basin would be forever changed, as the Atchafalaya River Basin Floodway was constructed and became the principal distributary floodway in the Mississippi River and Tributaries Project. This photograph was taken in Melville, Louisiana (in the ARB), and shows a boat full of people and their barber chair during the great flood of 1927. Written on photo: "Portable Barber Shop, Melville, LA." Photo provided by the State Library of Louisiana.

markets (e.g., carbon sequestration, nutrient reduction; Fig. 1.6). The value of these market and non-market goods and services reaches into the billions of dollars annually (Cardoch and Day 2001, Louisiana Department of Natural Resources 2010). Therefore, the modern ARB is an expression of a unique combination of natural and anthropogenic factors which together have created a highly engineered system that still supports vast areas of remote wilderness and natural riverine processes (Fig. 1.7; McPhee 1989, Reuss 2004, Ford and Nyman 2011).

This book documents the current state of the ARB from a scientific perspective and is designed as an objective baseline of understanding from which to draw knowledge for planning the future desired condition of the ARB. Environmental conditions and ecosystem services in the Atchafalaya system have been documented in numerous scientific papers, and reviews have summarized studies performed in a particular discipline (e.g., Bryan et al. 1998, Roberts et al. 2003, Reuss 2004). However, no comprehensive synthesis has been written that provides a holistic examination of the historic and current state of the ARB, and, despite the ecological and economic importance of the Atchafalaya, there is no common scientific understanding from which to

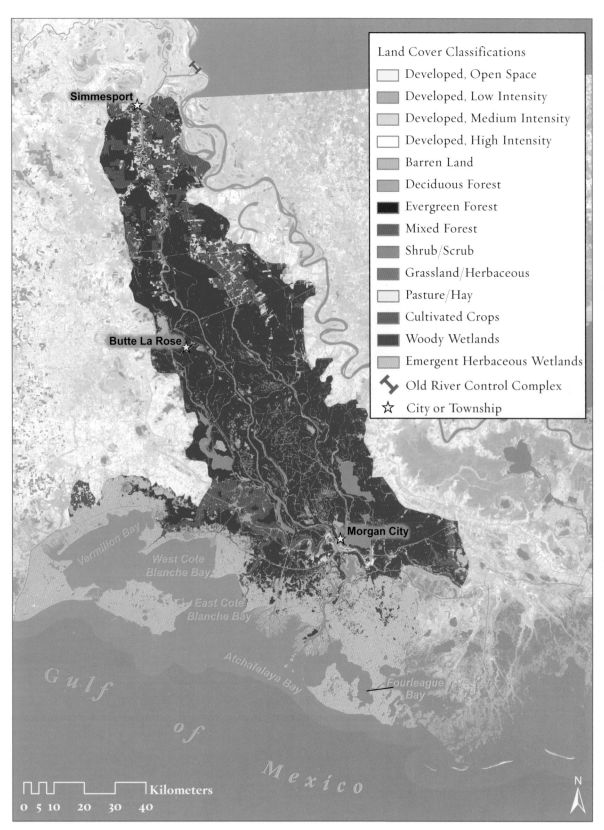

FIG. 1.4. Map showing the land cover types (2006 National Land Cover Data) located throughout the entire historical extent of the Atchafalaya River Basin in south-central Louisiana.

FIG. 1.5. The ARB is known worldwide for its expansive cypress-tupelo swamps, such as this mature forest stand in the ARB floodway. Photo © C. C. Lockwood.

view threats and assess possible solutions. The objective of this book is to provide that understanding.

Most of what is known about the habitat characteristics and biota in the historic ARB, prior to its conversion to a floodway, is anecdotal (Viosca 1927, 1928). Scientists began to appreciate the Atchafalaya as early as the 1940s, when Fisk (1944) began to study the geology of the lower Mississippi River, including the Atchafalaya. Although much of this pioneering geological research was done to assess the probability of the Mississippi River changing course into the Atchafalaya River channel and how to control it (e.g., Fisk 1952), what resulted was a deeper understanding of the geological processes that built Louisiana's deltaic coast, including the Atchafalaya River Basin. Limnological and ecological scientific study commenced in the 1960s and 1970s, resulting in a number of studies that investigated the forests and waterways of the ARB floodway (e.g., Lambou 1963, Bryan and Sabins 1979, Hern et al. 1980, Holland et al. 1983, Sabo et al. 1999a, b, and many others).

At present, a robust literature exists that describes the geology and ecology of the modern Atchafalaya River Basin from its inception at Three Rivers to the Atchafalaya and Wax Lake deltas at the coast. A much smaller, but equally compelling literature

FIG. 1.6. Oil and gas infrastructure in the Atchafalaya River Basin Floodway. The ARB is important for production (over 300 active wells in early 2000s) and transport (pipelines, shipping, and ports) of oil and gas resources in the Gulf of Mexico and the United States (US Geological Survey 2001). Currently, it produces about 1 million bbl of oil and 45,300 m³ (1.6 million ft³) of natural gas, annually (Carlson et al. 2012). This photograph shows a collection station where crude oil is piped, water and natural gas are removed, and the resulting products are stored for barge transport out of the floodway to refining facilities located on the Mississippi River or are shipped abroad. This facility, and others like it, is the hub of an oilfield and receive their supply from a system of wells located inside a maze of canals in the immediate vicinity. Photo by Richard Martin, © The Nature Conservancy.

exists for areas of the ARB outside the floodway boundaries. This literature not only provides a historical and ecological context for the ARB but also provides documentation of ecosystem services provided by these habitats and threats to this grand landscape. An integrated review of this literature, therefore, should provide a baseline state of the science and point out critical scientific questions yet unasked. It will also provide direction for future conservation and restoration action.

The modern ARB has been a place of great natural wonder and great controversy. It has been a place too inhospitable for some yet revered by others. It provides essential flood protection for millions of citizens along the Mississippi River, is one of the world's few relatively intact major river and delta systems, and contains the only accreting delta system in the Gulf of Mexico. However, the comprehensive body of scientific research, summarized in this book, clearly shows that the human need for control of this system and the water management model that has resulted—to provide flood control and navigation, as well as to acquire its natural resources—is unsustainable with regard to maintenance of the current plant community composition and structure, especially cypress-tupelo swamp forest, healthy populations of terrestrial and aquatic animal species, and production of its current suite of natural resources.

FIG. 1.7. A barge tow on the Gulf Intracoastal Waterway (GIWW) in the ARB floodway. The floodway contains four locks operated by the US Army Corps of Engineers (New Orleans District) to keep the Atchafalaya River and its associated navigation channels open for commercial barges. The ARB floodway links ports and oil and gas refining facilities on the Mississippi River (i.e., Baton Rouge, Louisiana) with important oil and gas ports in Morgan City, Louisiana, and the Gulf of Mexico. Goods transported through the ARB floodway include coal, oil and gas, fuel, chemicals, and grain. This photograph shows the GIWW located just inside the East Atchafalaya Flood Protection Levee, shown on the left. On the right side of the photo vast areas of waterways and floodplain forest span west toward the Atchafalaya River. Photo © Alison M. Jones for www.nowater-nolife.org.

This fact has spurred a movement for conservation and restoration of the ARB and an opportunity to provide comprehensive science-based solutions and progressive watershed management strategies. The science is clear that central to these solutions is a new water management model that restores natural flooding and drying cycles within the ARB floodway—both by addressing large-scale inputs at the Old River Control Structure and local barriers to flow. However, because the ARB is so dynamic, any restoration and management plans for the ARB, especially within the ARB floodway, need to recognize the impossibility and undesirability of a steady state and instead need to find new ways to manage flooding and sediment accretion to maximize the ecological value of the accretionary processes, which are now unique in the rapidly eroding Mississippi River delta plain.

The prevailing water management model introduced environmental tradeoffs to ensure flood protection, navigation, and energy resources. These tradeoffs were

FIG. 1.8. The ARB floodway protected people and property from flooding during the Mississippi River Flood of 2011. During the height of the flood, the floodway was nearly filled to capacity, as floodwaters were diverted from the Mississippi River through the Old River Control Structure Complex and the Morganza Floodway to protect people and property from flooding. Seen here is US Highway 190 on the west side of the Atchafalaya River near the town of Krotz Springs, Louisiana. Photo by David Y. Lee © The Nature Conservancy.

assumed by the public, and, especially, ARB stakeholders. Future conservation and restoration in the ARB will again involve tradeoffs. This fact makes it imperative that future conservation and restoration include a science-based decision making process that includes stakeholder values and vision for a future ARB that can continue to provide services for nature and people. This book was written to assist future conservation and restoration in the ARB by creating an objective scientific framework for that process. This synthesis is also intended to augment scientific and technical information already available through the Atchafalaya Basin Program Natural Resource Inventory and Assessment Tool (http://abp.cr.usgs.gov/), an online mapping tool, designed to (1) provide understanding of the current hydrological condition (2) provide understanding of current and proposed restoration efforts, and (3) aid in the assessment and design of hydrological restoration projects in the ARB floodway.

The Mississippi River Flood of 2011 occurred while I was writing this book. During that event, discharge through the Old River Control Structure Complex peaked around 20,000 m^3 s^{-1} (706,000 cfs), and ARB floodway successfully served its flood protection function, lowering river level and protecting people and property from flooding (Fig. 1.8; Falcini et al. 2012). The 2011 flood was large enough to warrant opening of the Morganza Floodway to protect larger downstream metropolitan areas, for only the second time in history. The flood event also prompted a large amount of new scientific research. This research, combined with the results of long-term scientific monitoring that has been done in the system, will allow effects from this flood to be determined and compared against several other smaller floods through time. However, these new studies have just begun, and comprehensive results are not yet available. Therefore, the Mississississippi River Flood of 2011 is not discussed in detail in this document. The flood is, however, referenced when appropriate, and the results of the first primary publications and some preliminary, unpublished material from the event are summarized and discussed throughout the book. Scientific information from the flood will build on what is already known about the ARB. These data will provide information on whether the system will be able to meet the future needs of nature and people, as a productive and sustainable river-floodplain system that also provides flood protection when needed.

Organization and Data

This book compiles the body of scientific knowledge for the ARB and focuses on four major questions.

1. What is the Atchafalaya River Basin, and how was it formed? This section reviews the geology and large-scale hydrologic process that created and maintain the habitats of both the historic and modern ARB.

2. Why is the Atchafalaya River Basin important? This section focuses on some of the resources and services that the system provides, including fish and wildlife resources and critical nutrient and carbon sequestration.

3. How is the modern Atchafalaya River Basin currently managed? This section details anthropogenic engineering and alteration of the basin's hydrology and natural habitats.

4. What is the current state of the Atchafalaya River Basin? This section documents the current conditions of the basin's resources.

Following this large-scale assessment is an evaluation of common patterns that exist across disciplines with respect to the current state of the ARB, as well as some important scientific questions that are yet unexplored. Answers to these questions will add to the large amount of information that is already known about the system and will assist in the future conservation and restoration of the Atchafalaya River Basin.

The ARB is a system that is extremely dynamic, serves multiple functions, and has been physically segmented by man in its modern history. Consequently the official delineation of the spatial extent encompassed by the ARB has also been variable and has decreased in size through time, affecting public understanding, investment, and involvement in decision making in the ARB (e.g., see US Army Corps of Engineers 1982, 2000, Louisiana Department of Natural Resources 1998, van Maasakkers 2009). Despite this segmentation, this book considers the ARB in the context of both its historical and modern boundaries, not only to assess the science and gain further insight into future resource trajectory but also to identify opportunities for restoration throughout the entire historical basin. The ARB floodway is of extreme importance to the modern configuration of the ARB. Therefore, the ARB floodway is given special attention throughout this book, not only because of its prominence, but also because much of the research has been done inside of its boundaries.

Throughout this book, I discuss the ARB at three main levels: 1) the historical basin, defined as the entire historic basin that exists inside the natural ecological basin boundaries (Bayou Teche Ridge, Mississippi River, Gulf of Mexico); 2) the modern ARB, defined as the combination of the ARB floodway and the Atchafalaya and Wax Lake deltas; and 3) the ARB floodway, defined as the area inside the East and West Atchafalaya Protection Levees from the Old River Control Structure complex to Morgan City. Within the ARB floodway, the Atchafalaya Basin Floodway System is also discussed, as it contains the majority of swamp habitat and has been the site of a great deal of the research done in the ARB. Lastly, because the Atchafalaya River affects both the central and western Louisiana coasts, the Atchafalaya River is discussed in terms of its relationship to Fourleague Bay and the western Terrebonne marshes, as well as Vermilion Bay, East and West Cote Blanche Bays, and the eastern end of the Louisiana Chenier Plain (Fig. 1.2).

DATA SOURCES. This review relies most heavily on primary (peer-reviewed) scientific literature but also incorporates some gray (not peer-reviewed) literature, unpublished data and information, and sources of local expert personal communication when deemed appropriate. Gray literature sources include government reports (i.e., Environmental Impact Statements, Endangered Species Act Recovery Plans, project fact sheets), databases (i.e., Louisiana Department of Wildlife and Fisheries), and natural resource atlases (i.e., breeding bird atlases). Theses and dissertations were used as references when these works did not result in peer-reviewed publications or when these works contained more detailed information that was not included in the published manuscripts.

Threat assessments and current and future trajectories not gleaned from the literature sources above were developed through analyses currently under way by US Geological Survey, US Fish and Wildlife Service, Engineer Research and Development Center of the US Army Corps of Engineers, nongovernmental organizations, and academic researchers. The Atchafalaya Basin Program Natural Resource Inventory and Assessment Tool was used to perform or assist in a number of scientific analyses, and a number of figures in this document were produced from information housed within it.

Three important sources of unpublished information include the Black Bear Conservation Coalition, the National Audubon Society, and the Louisiana Natural Heritage Program. These sources were especially relied upon for their abundance of scientific information regarding wildlife resources in the ARB, because, despite its importance as wildlife habitat and the abundance of scientific information available, there is not a great deal of primary published literature on wildlife resources (with the exception of bears) in the ARB.

The Black Bear Conservation Coalition has summarized a great deal of scientific studies and information on the ecology and status of the Louisiana black bear (*Ursus americanus luteolus*) throughout its range into fact sheets posted on its website (www .bbcc.org). This information was used to supplement peer-reviewed and other gray literature sources of information available for this subspecies.

The National Audubon Society, through its Important Bird Area program, provided a great deal of scientific information that is used in the process of identifying, defining, and prioritizing Important Bird Areas. These areas are defined as supporting either species of conservation concern (e.g., threatened and endangered species), range-restricted species (species vulnerable because they are not widely distributed), habitat-restricted species (species that are vulnerable because their populations are concentrated in one general habitat type or biome), or high-density species (species, or groups of species, such as waterfowl or shorebirds, that are vulnerable because they occur at high densities due to their flocking and foraging behavior). The process for identifying and defining these areas is science-based and data-driven. Data come from a variety of sources, including universities, private researchers, and citizen scientists. Areas selected represent the most important areas for birds throughout the country and the world. Once selected, Important Bird Areas are prioritized as continentally or globally significant by a panel of nationally recognized bird experts. The ARB contains two of the 23 recognized Important Bird Areas in Louisiana—the ARB floodway and Atchafalaya Delta/Bay (National Audubon Society 2011a, b).

The Louisiana Natural Heritage Program was founded in 1984 through a partnership between the state of Louisiana and The Nature Conservancy. Currently housed within the Louisiana Department of Wildlife and Fisheries, this program is part of the Natural Heritage Network, which is coordinated by the nonprofit NatureServe, with the goal of gathering, assessing, and distributing standardized information on the biological diversity across all 50 US states, Canada, Mexico, and also parts of Latin America. The Natural Heritage Program's mission is to maintain a database on rare, threatened and endangered species of plants and animals and natural communities for Louisiana. This database represents the most current and comprehensive resource for rare species and habitat occurrences in the state.

PART ONE

WHAT IS THE ATCHAFALAYA RIVER BASIN
AND HOW WAS IT FORMED?

GEOLOGY OF THE ATCHAFALAYA RIVER BASIN 2

THE ATCHAFALAYA RIVER. The Atchafalaya River Basin includes both lacustrine and coastal delta systems. To understand the ARB, one must understand its relationship to both the Atchafalaya (Choctaw for "long river") and Mississippi Rivers. The Atchafalaya River, located in south central Louisiana, is the fifth largest river in North America, by discharge, and is the major distributary of the Mississippi River (Table 2.1; Fig. 2.1; Kammerer 1990, Ford and Nyman 2011). The Atchafalaya River also receives the discharge from the Red River system, which is one of the major mid-continental river systems in North America (Kammerer 1990, Allison et al. 2012). Because the Atchafalaya River includes flows from both the Mississippi River and the Red River, it has the largest drainage basin in North America and shares with the Mississippi River the distinction of having the third largest drainage basin in the world—one that encompasses an area of approximately 3.3 million km² (1.2 million mi²), or about 70% of the contiguous United States and portions of two Canadian provinces (Table 2.2; Fig. 1.1; Coleman 1988, Roberts 1997, Coleman and Roberts 1991, Coleman et al. 1998, 2005, Ford and Nyman 2011). Combined, the Mississippi and Atchafalaya Rivers account for about 90% of the freshwater discharge into the Gulf of Mexico (Morey et al. 2003, Orem et al. 2007).

From its origin at Lake Itasca, Minnesota, the Mississippi River gathers the flow and suspended sediment load of many tributaries, including the Wisconsin, Illinois, Missouri, Ohio, and Arkansas Rivers, as it flows south toward the Gulf of Mexico. In southern Louisiana, the Mississippi River leaves its alluvial valley and enters the Mississippi River delta plain, an area that covers approximately 30,000 km² (11,580 mi²; Coleman 1988, Roberts 1997, Coleman et al. 2005). The Mississippi River delta plain has built up over the last 8,000 years through six distinct depositional eras that were caused by cyclic shifts in the course of the lower Mississippi River during the current interglacial period (Holocene; Fig. 2.2; Coleman and Roberts 1991, Stanley and Warne 1994). Each of these Holocene depositional events lasted approximately 1,000–1,500 years, until a delta lobe reached a threshold where the river became hydraulically inefficient in transporting its flow and sediment load. At that point, the hydraulic inefficiency would force

FIG. 2.1. The Atchafalaya River receives its flow from the Mississippi and Red Rivers and is the fifth largest river in North America according to discharge. As the major distributary of the Mississippi River, it shares the third largest drainage basin in the world and provides vital flood protection for millions of Americans. It also created the vast wetland habitats of the Atchafalaya River Basin and serves as the cultural heart of Louisiana's Cajun culture. Photo by Richard Martin © The Nature Conservancy.

a shift in river position, and the river channel would gradually be captured by a new path of greater hydraulic efficiency in a process that can take up to 100 years to complete (Fisk 1944, 1952, Frazier 1967, Roberts 1997, Coleman et al. 1998, Aslan et al. 2005). The oldest depositional lobe in the Mississippi River delta plain is the Maringouin/Sale Cypremort, dating back to almost 7500 years ago the youngest of these is the current Atchafalaya–Wax Lake lobe, which was initiated approximately 400 years ago and is the result of ongoing Mississippi River stream capture by the Atchafalaya River (Fisk 1944, 1952, Gagliano and van Beek 1975, Tye and Coleman 1989a, b, Roberts 1997, 1998, Coleman et al. 1998, Aslan et al. 2005).

Although the Atchafalaya River has been a distributary of the Mississippi River since the mid-1500s, discharges from the Mississippi into the Atchafalaya were small and sporadic in the early years (Fisk 1952). Prior to that time, the Red River and Mississippi River were separate and flowed almost parallel to each other, with the Red River flowing through the channel that would much later become the Teche segment of the

TABLE 2.1. The five largest rivers in North America, according to mean annual discharge

River	Location of Mouth	Mean Discharge at Mouth		River Length		References
		m³ s⁻¹	cfs	km	mi	
Mississippi	Louisiana, USA	16,792	593,000	3,766	2,340	Kammerer (1990)
St. Lawrence (Great Lakes)	Canada	9,854	348,000	3,058	1,900	Kammerer (1990)
Ohio	Illinois-Kentucky, USA	7,957	281,000	2,108	1,310	Kammerer (1990)
Columbia	Oregon-Washington, USA	7,504	265,000	1,996	1,240	Kammerer (1990)
Atchafalaya	Louisiana, USA	6,484	229,000	275	171	Allison, et al., (2000) Demas, et al., (2001) Alford and Walker (2011)

TABLE 2.2. The three largest drainage basins in the world

River	Continent	Drainage Basin	
		km²	mi²
Amazon	South America	7,050,000	2,720,000
Congo	Africa	4,014,500	1,550,000
Mississippi	North America	3,220,000	1,245,000

Because the Atchafalaya River is its major distributary, it shares the third-largest drainage basin in the world with the Mississippi River. Data are from Kammerer (1990).

Mississippi River and eventually modern-day Bayou Teche before emptying into the Gulf of Mexico near present-day Franklin, Louisiana (Fisk 1944; Fig. 2.3). Then, during the 15th century, a bend developed on the Mississippi River in the area now called Turnbull's Bend. This bend intercepted the Red River, which then became a tributary to the Mississippi River, and created the Atchafalaya River as a distributary (Reuss 2004, Loyola University Center for Environmental Communication 2011). Sporadic and often reversed (eastward) flows in the Atchafalaya were reinforced by an extensive logjam that developed by the early 1800s in the Red River and eventually expanding into its confluence with the Atchafalaya, obstructing flow (e.g., see Fisk 1952, Gagliano and van Beek 1975, Reuss 2004). However, a series of anthropogenic alterations that took place in the 1800s in this region now known as the "Three Rivers" (where the Red, Mississippi, and Atchafalaya Rivers converge) accelerated the diversion process (Reuss 1982, 2004, Hale 1995).

In the mid-1800s, Captain Henry Shreve, a river engineer, dug a canal that effectively cut off Turnbull's Bend to create a reliable and shortened navigation route in the

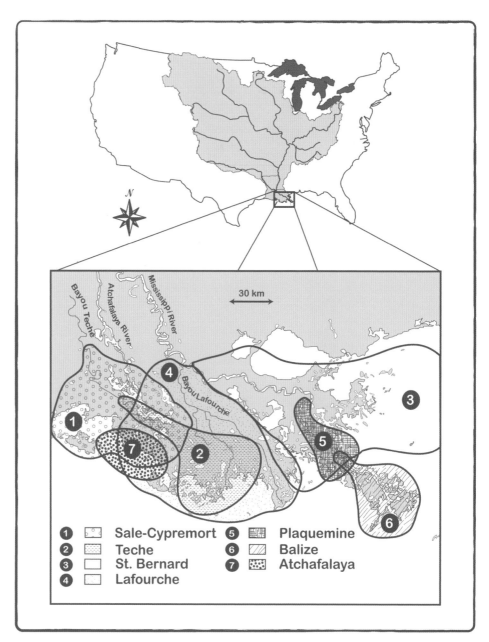

FIG. 2.2. Map showing deltaic progression on the Mississippi River delta plain during the last 8,000 years (modified from Kolb & Van Lopik 1958 and Frazier 1967). Active deltaic formation presently occurs at the mouths of the Mississippi (Balize delta lobe) and Atchafalaya Rivers. Figure taken from Sasser et al. (2009) and provided by Guerry O. Holm, Louisiana State University.

Mississippi River (Reuss 2004). This canal caused the upper half of Turnbull's Bend to silt in and accentuated the flow capture out of the Mississippi and into the Atchafalaya River through the lower half of the bend, now called Old River. At the same time, local citizens, Captain Shreve, and the federal government were gradually clearing the extensive logjam that had formed in the Red and Atchafalaya Rivers, so that by the late 1800s, the obstruction broke free (Fig. 2.4). Now flow in the Atchafalaya River seldom

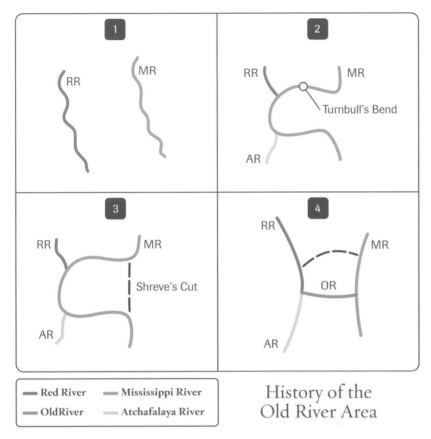

FIG. 2.3. Diagram showing the formation of the Atchafalaya River. Prior to the 1500s, the Mississippi River and Red River did not comingle, flowing instead parallel to each other (1). The Atchafalaya River formed as a distributary when the Mississippi River turned west at Turnbull's Bend (2) and intercepted the Red River, which now became a tributary. Captain Henry M. Shreve dug a canal through Turnbull's Bend in 1831 to shorten travel time down the Mississippi River (3). This river engineering caused the northern section of Turnbull's Bend to fill in with sediment. However, the southern half remained open and became known as Old River (4). Diagram adapted from Loyola University Center for Environmental Communication (2011).

reversed, and the process of stream capture began in earnest. The shorter distance to the Gulf of Mexico—220 km vs. 520 km (137 mi vs. 323 mi)—and corresponding elevation difference at the diversion site—2.4 to 5.2 m (7.9 to 17.1 ft)—presented a hydraulic advantage, and the Mississippi River began migrating into the Atchafalaya River (Martinez and Haag 1987, Allison et al. 2000, Roberts et al. 2003, Aslan et al. 2005).

The stream capture process creates a gradient advantage and usually proceeds unabated until the majority of discharge is captured—a process that typically takes about 100 years to complete (Gagliano and van Beek 1975). The distributary, which at first is too small to capture significant flow, gradually fills the receiving basin with sediment. During this time, flow capture slowly increases. At the point when the infilling process is complete, the distributary channel widens and deepens and steadily begins to capture more flow.

FIG. 2.4. "Captain Henry M. Shreve Clearing the Great Raft from Red River, 1833–38," by Lloyd Hawthorne. Image courtesy of the R. W. Norton Art Gallery, Shreveport, Louisiana.

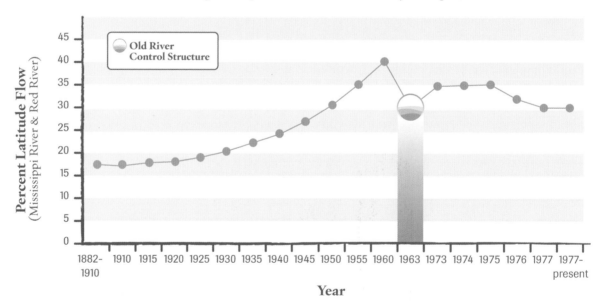

FIG. 2.5. Graph showing the percent of the daily latitudinal flow (Mississippi River plus Red River) captured by the Atchafalaya River from 1882–present. Data are from Gagliano and van Beek (1975) and Loyola University Center for Environmental Communications (2011). This stream capture process transformed the Atchafalaya River Basin from one dominated by lakes and swamps and their corresponding in situ ecological processes to one dominated by fluvial processes.

FIG. 2.6. The Old River Control Structure Complex. The Mississippi River is located on the top and left. The Atchafalaya River meets the Mississippi River at three locations and runs off to the right. Shown are the main and auxiliary structures at the top, the overbank structure in the middle, and the Sidney A. Murray, Jr. hydroelectric station in the foreground. The navigation lock is not shown. Photo © CC Lockwood.

The Atchafalaya River reached this point in 1950 after the ARB was largely filled with sediment (Fig. 2.5). It was at this time that the Atchafalaya channel began to accelerate its capture of the Mississippi's flow. Stream capture was also supported by the fact that the Atchafalaya River cuts across old, abandoned Mississippi River channels, and these abandoned channels provided ready-made conduits for accepting Mississippi River flow (Aslan et al. 2005). However, in 1963, this trajectory toward total capture of Mississippi River flow was artificially terminated by the construction of the Old River Control Structure Complex at the point of diversion (Fig. 2.6). In spite of the structures, stream capture almost became a reality in 1973–1975, which was a three-year period of record flooding in the lower Mississippi River. Without the structures, complete capture would have certainly occurred during this period (Gagliano and van Beek 1975, Roberts et al. 2003). Since the structures became operational, federal law has mandated that the daily discharge into the Atchafalaya be held static at 30% of the combined daily latitude flow of the Mississippi and Red Rivers (Fig. 2.7; Gagliano and van Beek 1975, van Heerden and Roberts 1988, Mossa and Roberts 1990, Roberts et al. 2003, Allison et al. 2012).

The Old River Control Structure Complex was originally a combination of two structures (Fig. 2.8). The first was the main low sill structure (173 m [568 ft] long with 11 gate bays, each 13 m [43 ft] wide), and the second was an overbank structure (1,023 m [3,520 ft] long with 73 bays, each 13 m [43 ft] wide). The low sill structure was designed to control daily flow, and the overbank structure was designed for use during flood events when Mississippi River discharge exceeded the capacity of the low sill structure and additional flow relief was needed (Fig. 2.9). A navigation lock was also constructed at the lower arm of Turnbull's Bend to provide stable passage between the Atchafalaya and Mississippi Rivers, while at the same time gaining flow control over this part of Old River (Fig. 2.10). The Flood of 1973 almost caused failure of the low sill structure, prompting the construction of an additional structure (auxiliary control structure) in 1986, consisting of six tainter gates, each measuring 19 m (62 ft) across (Fig. 2.11). The last addition to the complex was the Sidney A. Murray, Jr. hydroelectric station, completed in 1990 (Fig. 2.12). This structure was Louisiana's first hydroelectric plant, and is a "run-of-the-river" structure that takes advantage of the almost 5 m (16 ft) head differential between the two rivers (Bryan et al. 1998, Reuss 2004).

Since construction of the Old River structures, the Atchafalaya River carries all of the Red River flow and approximately 25% of the annual Mississippi River flow (Hupp et al. 2008, Ford and Nyman 2011). Annual mean discharge in the main channel is 6,484 m^3 s^{-1} (229,000 cfs; Demas et al. 2001, Xu 2010). The annual hydrograph parallels that of the Mississippi River and is relatively regular, with discharges in a range of 600–19,800 m^{3}s^{-1} (Fig. 2.13; 21,189–699,230 cfs; Allison et al. 2000, Xu 2010). Maximum discharge occurs in winter/spring (January–June; maximum April; mean 9,500 m^3 s^{-1} [335,489 cfs]). Minimum discharge occurs during September and October (mean 2,500 m^{3}s^{-1} [88,287 cfs]; Allison et al. 2000).

South of Morgan City, Atchafalaya River flow is discharged into an estuarine complex consisting of Atchafalaya, Fourleague, Vermilion, and Cote Blanche Bays (Fig. 2.14). The majority of the flow (about 90%) passes through the Lower Atchafalaya River and the Wax Lake Outlet into the Atchafalaya and Wax Lake deltas and onto the nearshore areas of Atchafalaya Bay (Fig. 2.15; Denes and Caffrey 1988, Lane et al. 2002). The remainder of the flow (about 10%) is distributed both east and west of Atchafalaya Bay, largely through the Gulf Intracoastal Waterway, where stage differences of more than 1 m (3 ft, NAVD88) between the lower river and the Gulf of Mexico transport Atchafalaya fresh water and sediment over 50 km (30 mi) east and west of Morgan City into wetlands in the central Louisiana coast. This Atchafalaya flow moderates salinity and provides a potential sediment resource for restoration and maintenance of coastal wetlands (Swarzenski 2003, Ford and Nyman 2011, State of Louisiana 2012).

Atchafalaya River water moving east flows into Fourleague Bay and the western Terrebonne marshes, and water flowing west enters Vermilion and Cote Blanche Bays. Water flowing eastward averages 345 m^3 s^{-1} (12,200 cfs), and either enters the Gulf Intracoastal Waterway through the Avoca Island Cutoff or is distributed east to Bayou Penchant and into Fourleague Bay and the western Terrebonne marshes before being discharged to the Gulf of Mexico through Oyster Bayou, a 180-m (590-ft) wide tidal pass (Denes and Caffrey 1988, Lane et al. 2002, Swarzenski 2003). In fact, instantaneous

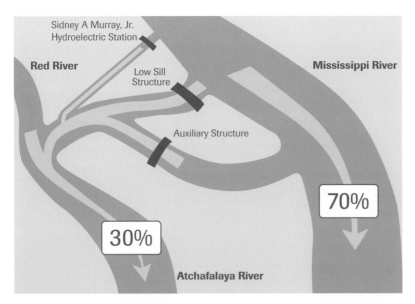

FIG. 2.7. Diagram showing the latitude flow split between the Mississippi, Red, and Atchafalaya Rivers. This flow split is federally mandated and maintained on a daily basis through a combination of flow allocations through the low sill and auxiliary structures and the Sidney A. Murray, Jr. Hydroelectric Station. During an extreme flood event, other structures and floodways are used to accomplish the necessary flow allocations to pass the flood. Diagram is adapted from Loyola University Center for Environmental Communication (2011).

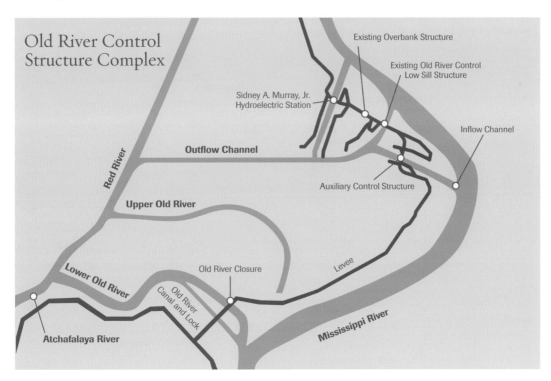

FIG. 2.8. Plan diagram of the Old River Control Structure Complex. Diagram is modified from Loyola University Center for Environmental Communication (2011).

FIG. 2.9. The Old River low sill structure. This structure contains 11 gates, operated by a Gantry crane that moves along a set of tracks (top of photo). Photo © Alison M. Jones for www .nowater-nolife.org.

FIG. 2.10. The Old River navigation lock constructed to allow reliable passage between the Mississippi and Atchafalaya Rivers. Photo © Alison M. Jones for www .nowater-nolife.org.

FIG. 2.11. The Old River auxiliary control structure that was built after the Mississippi River flood of 1973 almost destroyed the original low sill structure. This structure contains six steel tainter gates, each 19 m (62 ft) wide. Photo © Alison M. Jones for www.nowater-nolife.org.

FIG. 2.12. The Sidney A. Murray, Jr. hydroelectric station. This structure was prefabricated of steel and concrete at Avondale Shipyards in New Orleans and floated 335 km (208 mi) up the Mississippi River and installed. It consists of eight 24-megawatt turbines. It is operated as a run-of-the-river power plant and uses a portion (up to 4,814 m³ s⁻¹; 170,000 cfs) of the total flow diverted through the Old River complex. Photo © Alison M. Jones for www .nowater-nolife.org.

discharge measurements have shown more than 481 m³ s⁻¹ (17,000 cfs) of flow moving eastward through the Avoca Island Cutoff, south of the Bayou Boeuf Lock (Swarzenski 2003). An average of 268 m³ s⁻¹ (9,460 cfs) of water flows westward in the Gulf Intracoastal Waterway. Almost half of that flow enters West Cote Blanche Bay through The Jaws of Little Bay (an area in the northern portion of West Cote Blanche Bay; Denes and Caffrey 1988, Lane et al. 2002, Swarzenski 2003, Cobb et al. 2008a). Average discharge through the Jaws from July 1997 to January 1998 was 115 m³ s⁻¹ (4,061 cfs) and ranged from 16 to 173 m³ s⁻¹ (565–6,109 cfs; Cobb et al. 2008a).

The Atchafalaya River carries the entire sediment load of the Red River (suspended and bed load) and historically carried ~35% of the total sediment load of the Mississippi River (Hupp et al. 2008). Annual water and sediment budgets calculated for the Atchafalaya River during flood years 2008–2010 show that, despite contributing less than half of the water discharge, the Red River contributed an almost equal amount

Mean Monthly Atchafalaya Discharge
1963-2010

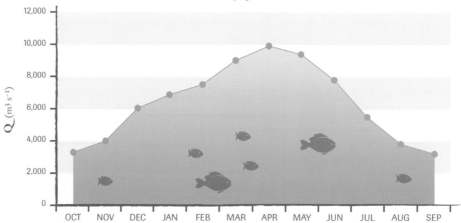

FIG. 2.13. Mean monthly discharge hydrograph for the Atchafalaya River at Simmesport, Louisiana (USGS 07381490), after construction of the Old River Control Structure complex.

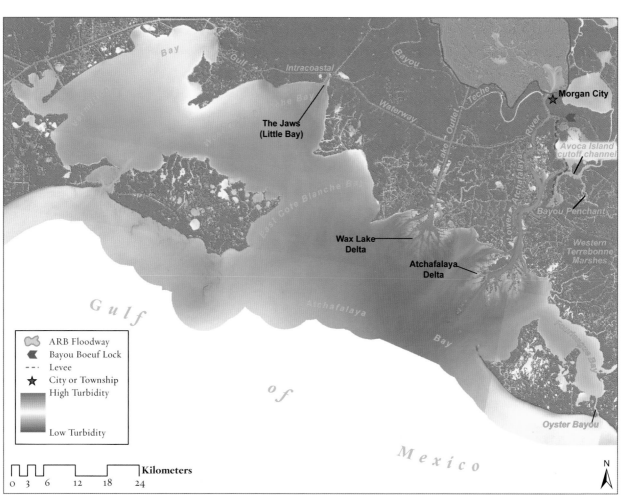

FIG. 2.14. Map showing the estuarine complex of the central Louisiana coast that is affected by flows from the Atchafalaya River. The relative degree of riverine influence is indicated by the level of turbidity. Map created from a coastal turbid water assessment (unpublished data) done by Yvonne C. Allen, US Army Engineer Research and Development Center, Baton Rouge, Louisiana.

FIG. 2.15. The Atchafalaya River delta complex, composed of the Atchafalaya and Wax Lake deltas, is the only actively building delta complex in the northern Gulf of Mexico. This photo of the Atchafalaya Delta, (looking north) shows the lower Atchafalaya River (center) and distributary channels delivering sediment-laden water and the archetypal v-shaped delta splay islands that contain a variety of wetland habitats. Photo © The Nature Conservancy.

of suspended sediment as the Mississippi and up to about four times the suspended sand discharge in the Atchafalaya River (Table 2.3; Allison et al. 2012). Modifications to the Old River structures were designed to increase Mississippi River bedload capture by the Atchafalaya River up to 60% of the total. (Mossa and Roberts 1990, Hupp et al. 2008, Daniel E. Kroes, US Geological Survey, personal communication). Mean suspended sediment load ranges from 65–84 million metric tons yr^{-1} (72–92 short tons yr^{-1}) of which 17% is comprised of sand-sized material, and the balance is finer-grained silts and clays (Allison et al. 2000, Rice et al. 2009, US Army Corps of Engineers 2011). Radioisotope analysis of seabed deposits on the continental shelf show an estimated twelve-fold increase in sediment discharge from the Atchafalaya River during high water compared with low water periods (Allison et al. 2000).

The ARB contains lacustrine and coastal delta systems that have experienced rapid sedimentation ever since the Atchafalaya River began to capture flow from the Mississippi River. This pervasive sedimentation has caused rapid seaward progradation of land in what was once a large open lake (or a series of large lakes) and has had profound effects on the landscape and hydrologic function of the ARB. For example, many open water areas (i.e., Lakes Chicot, Mongoulais, and Rond, and Grand Lake) have been filled in by lacustrine delta development, and extensive backswamp areas that formerly experienced sheet flow from seasonal flooding are now isolated from the river by natural and engineered levees. Currently, because the basin is largely filled in, most of the sediment flowing down the Atchafalaya River is not deposited within the forested wetlands of the floodway, but rather is transported to Atchafalaya Bay and near-coastal

TABLE 2.3. Total annual water and suspended sediment discharge from the Atchafalaya River during flood years 2008–2010

Station	Total Water Discharge (km³ yr⁻¹)			Total Sediment Discharge (x 10⁶ metric tons yr⁻¹)			Total Sand Discharge (x 10⁶ metric tons yr⁻¹)		
	2008	2009	2010	2008	2009	2010	2008	2009	2010
Old River Control Structure	187	141	172	40.3	28.5	35.2	6.4	3.3	4.3
Red River	57	65	88	32.5	33.2	44.8	12.2	9.5	17.8
Simmesport	244	206	260	72.8	61.7	80.0	18.6	12.7	22.1
Melville	343	206	261	39.0	30.8	42.9	16.8	10.8	15.9
Morgan City	133	113	140	30.9	22.7	30.1	13.6	8.4	12.7
Wax Lake Outlet	108	95	126	20.1	16.9	24.6	7.3	3.8	5.0

Shown are results from data stations owned and operated by the US Geological Survey and US Army Corps of Engineers at locations spanning from the Old River Control Structure Complex to Morgan City and the Wax Lake Outlet. Table created from data in Allison et al. (2012). Inputs from the Red River were calculated by subtracting the inputs from the Old River Control Structure from measurements made at Simmesport, Louisiana, as was done in the original study (Allison et al. 2012).

waters (Allen 2010, Xu 2010, Ford and Nyman 2011, Allison et al. 2012). While most of the sediment bypasses the ARB floodway, some sediment is sequestered on the floodplain, and there are localized areas within the floodway that currently receive large inputs of sediment.

The flow of river water and sediment through the ARB has defined this system throughout its entire history and continues to drive the processes, habitats, and management of the modern system today. The ARB has preserved, in its sediments and habitats, a record of its long relationship to the Mississippi River. This relationship has existed throughout the Holocene epoch and has transformed the geology and ecology of the ARB through time.

GEOLOGICAL FORMATION OF THE ATCHAFALAYA RIVER BASIN. The ARB is one of the major historic inter-distributary basins in the Mississippi River delta plain (others are the Pontchartrain, Barataria and Terrebonne basins), that are bounded by higher ground of the Pleistocene terrace on the western and eastern margins of the Mississippi River valley south of Baton Rouge, Louisiana (Fisk 1944, Tye and Coleman 1989a, Roberts and Coleman 1996). Inter-distributary basins in the Mississippi River delta plain are typically associated with preexisting alluvial ridges of ancient river courses that represent past delta lobe depositional cycles (Coleman 1966). In the case of the Atchafalaya, the basin was bounded by the natural levee ridges of the abandoned Teche-Mississippi River course (Bayou Teche) on the west and the abandoned Lafourche-Mississippi River course (Bayou Lafourche) as well as natural levees along the present Mississippi River on the east (Coleman 1966).

Lacustrine Delta Environments

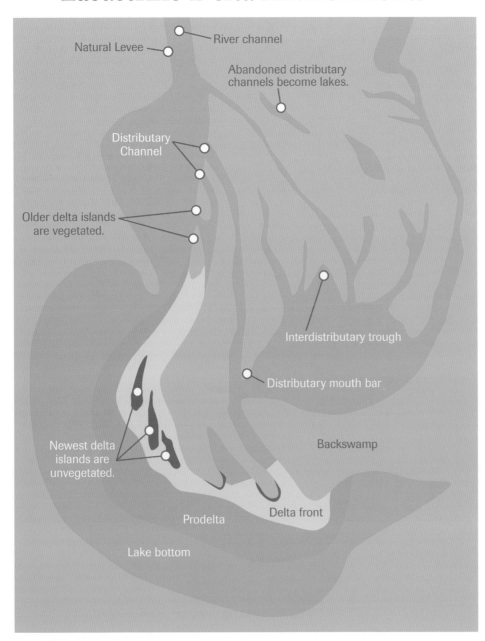

Natural Levee — River channel

Abandoned distributary channels become lakes.

Distributary Channel

Older delta islands are vegetated.

Interdistributary trough

Distributary mouth bar

Newest delta islands are unvegetated.

Backswamp

Prodelta

Delta front

Lake bottom

FIG. 2.16. Diagram showing the lacustrine delta environments. Backswamp environments are typically climax successional relicts of long periods of sediment deprivation and land subsidence. During periods of intense sedimentation, open water areas are rapidly filled with sediment and transformed into land. Figure adapted from Tye and Coleman (1989b).

Because of their position in the landscape, these interdistributary basins typically evolved through alternating cycles of sediment deprivation followed by periods of intense sedimentation (Tye and Coleman 1989b). During long periods of sediment deprivation, land subsidence processes dominated, resulting in climax successional environments that contained large bodies of open water and mature backswamp habitats. Periods of intense sedimentation that followed deprivation episodes were marked by rapid filling of open water habitats with prograding lacustrine deltas (Fig. 2.16).

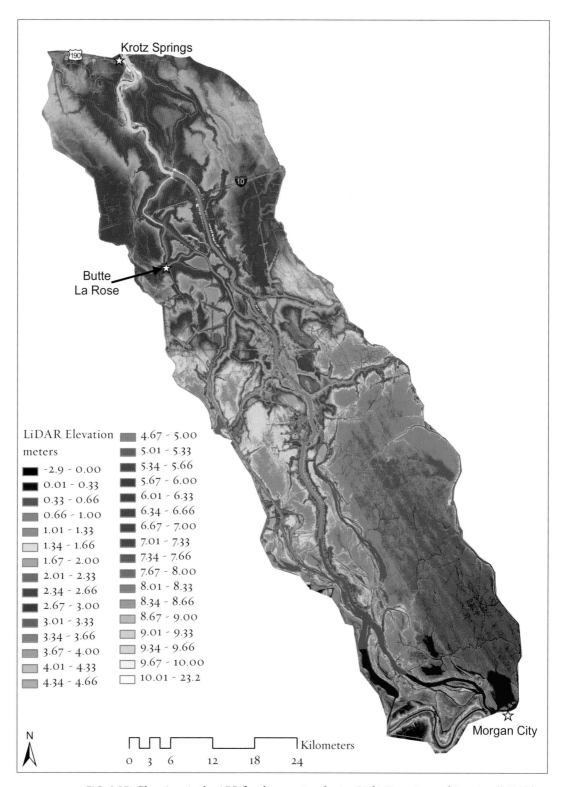

FIG. 2.17. Elevations in the ARB floodway captured using Light Detection and Ranging (LiDAR) from 2–7 December, 2010, at low water levels [stage at Butte LaRose: 1.2–1.5 m (3.9–4.9 ft)]. Note that water levels frequently interfere with capture of accurate LiDAR elevations. This dataset is the best comprehensive, detailed representation of recent elevations in the ARB floodway. Data provided by Yvonne C. Allen, US Army Engineer Research and Development Center, Baton Rouge, Louisiana.

Soil cores taken in the ARB document over 30 m (100 ft) of deltaic sedimentation that corresponds to at least four sedimentary starvation-deposition cycles (sediment starvation and compaction-induced subsidence followed by lacustrine-delta deposition and aggradation) during the construction of multiple major marine Holocene delta complexes in the Mississippi River delta plain (Fig. 2.2). The oldest deposits in the ARB are related to the Sale-Cypremort and Teche delta systems, and the most recent lacustrine delta deposits were precursors to the development of the marine Atchafalaya Delta (Fisk 1952, Coleman and Gagliano 1964, Coleman 1966, Tye and Coleman 1989a, b. Roberts and Coleman 1996). Today, elevation in the ARB ranges from 14 m (45 ft) above mean sea level in the upper basin, near Krotz Springs, Louisiana, to sea level at the Gulf of Mexico (Fig. 2.17; Coleman 1966).

GEOLOGICAL TRANSITION OF THE ATCHAFALAYA RIVER BASIN. The geological transition of the ARB represents an interaction between natural riverine and sedimentary processes and human forcing that exacerbated those processes. This interaction resulted in different physical cycles that changed the basin landscape greatly. Interestingly, many of the major landscape changes within the modern ARB happened within the lifetime of people living and working there today.

Prior to the relatively recent capture of significant flow from the Mississippi River that has resulted in the diversion of substantial amounts of Mississippi River water and sediment into the Atchafalaya, the ARB was dominated (for thousands of years) by a large lake (Lake Chetimaches—later called Grand Lake) and extensive swamp habitats that extended from the vicinity of Butte La Rose to just north of Morgan City, Louisiana (Fig. 2.18). Lake Chetimaches was actually composed of a series of large lakes—now named Lake Fausse Point, Lake Rond, Lake Chicot, Flat Lake (both), Grand Lake, Duck Lake, Six Mile Lake, and Flat Lake (Fig. 2.19; Fisk 1944, 1952, Allen 2010). South of Morgan City, the coastal portion of the basin was dominated by a series of large bays. These open water, swamp, and bay habitats had formed due to compaction-induced subsidence (about 1.4 cm yr^{-1} [0.55 in yr^{-1}]) of the old Maringouin and Teche delta lobe deposits (Fisk 1944, Gagliano and van Beek 1975, Tye and Coleman 1989a, b, Roberts 1997, Neil and Allison 2005). During this long period of time between the deltaic incursions that formed the Teche Lobe and the recent Mississippi capture, the ARB was largely isolated from significant freshwater input, except local runoff and occasional overbank flooding of the Mississippi River. Consequently, this long period of stable hydrologic conditions allowed the establishment of extensive stands of cypress-tupelo swamp that can still be seen today in the eastern portion of the floodway and areas of the historic basin outside the levees (e.g., Lake Verret; Gagliano and van Beek 1975).

After the extensive logjam was removed, the Atchafalaya began to capture more flow from the Mississippi River, and a period of intense sedimentation ensued. Large expanses of open water habitat greatly diminished flow velocities and acted as a sediment trap, and lacustrine deltas began to fill in the lakes that composed Lake Chetimaches downslope and in series (Fig. 2.19; Gagliano and van Beek 1975, Tye and Coleman 1989a, b, van Heerden 2001).

FIG. 2.18. Civil War map of the Atchafalaya River Basin showing Lake Chetimaches, now called Grand Lake. Also shown are some of the large bays (Atchafalaya Bay and Cote Blanche Bay) that dominated the coastal portion of the basin. For reference, the waterway on the left is Bayou Teche, and the waterway on the top right side of the image is the Mississippi River. Note that the Atchafalaya River is not prominent in this map, because it was still a small river, since the logjam that blocked the entrance had not yet been cleared by Captain Shreve. Original image (Atchafalaya Basin. Prepared by order of Maj. Gen. N. P. Banks. Henry L. Abbot, Capt. & Chief Top. Eng'rs., Feb. 8th, 1863 [with annotations pertaining to the campaign in April and May 1863], 02/08/1863–05/1863) created by the War Department, Office of the Chief of Engineers (1818–09/18/1947) and located at the National Archives (archives.gov; ARC Identifier 305632 / Local Identifier 77-CWMF-M99(HALF).

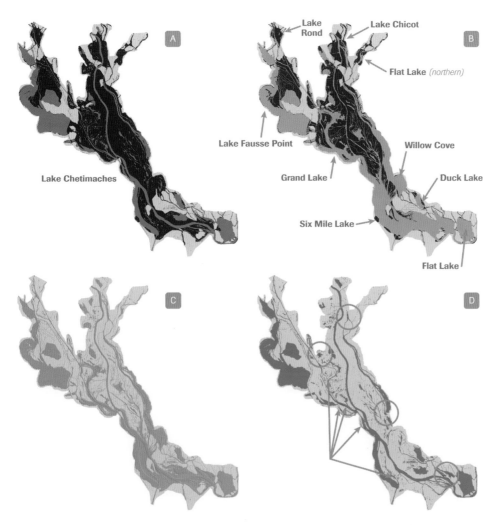

FIG. 2.19. Figure showing the geological transformation of the ARB Floodway. Panel A shows the extent of open water (Lake Chetimaches) inside the floodway in the early 1800s. Areas colored in blue have converted to land since that time, and areas in brown remain open water. Panel B shows the progression from the early 1800s to the 1950s. Areas converted from open water to land are indicated in dark blue, and areas that remained water are shown in dark yellow. Panel C shows the progression from the late 1950s to early 1980s. Areas converted from open water to land are indicated in dark yellow, and the areas that remained water are shown in blue. Panel D shows the progression from early 1980s to late 2000s. Areas that converted to open water are shown in blue, and areas that remained open water are shown in brown. Localized areas of accretion are shown with red circles and arrows. Figure modified from Allen (2010). Original provided by Yvonne C. Allen, US Army Engineer Research and Development Center, Baton Rouge, Louisiana.

Because the ARB was the site of Mississippi River distributaries in the past, sediment entered lake basins through numerous relict distributary channels that existed within prior deltas (hyperpycnal underflows), building new lacustrine deltas (Fig. 2.20). Water and sediment entered lake basins through these relict channels and delta progradation began with a long period of subaqueous (below the water surface) prodelta platform formation, created through deposition of fine-grained sediments. Rapid subaerial (above the water surface) delta building resulted from subsequent deposition of coarse-grained sediments, which eventually led to abandonment of the newly created delta (Tye and Coleman 1989a, b).

By 1917, subaerial delta features began appearing in these lakes, with historical surveys showing that Lake Chicot was already half filled with sediment at that time (Fig. 2.19; Allen 2010). By the 1930s, Lakes Chicot and Rond had almost entirely filled with sediment, and a large delta was apparent at the northern end of Grand Lake (Fig. 2.21; Fisk 1952, Allen 2010). This process, which continues today, reduced their combined surface water area nearly 80% (600–130 km²; 373–81 mi²) and decreased their mean depth from 2–3 m (6.5–9.8 ft) to less than 0.5 m (1.6 ft) in less than a century (Gagliano and van Beek 1975, Tye and Coleman 1989a, b). The delta fronts grew down-basin almost continually from that time at an average rate of about 2 km yr⁻¹ (1.2 mi yr⁻¹). The rapid sediment progradation in these lakes resulted in a lacustrine delta that covers over 240 km² (93 mi²) at an average thickness of about 3 m (9.8 ft; Tye and Coleman 1989a).

One estimate shows that the amount of sediment deposited in this system during the recent stream-capture event to be about 2,500,000,000 m³ (2,026,782 acre ft; Bryan et al. 1998), which is enough sediment to cover the entire state of Rhode Island (4,002 km² [1,545 mi²]) to a depth of 18 cm (7 in). During this period of intense sedimentation, the East and West Atchafalaya Protection Levees were completed, confining natural delta-building processes into a narrow path about one-third the original size of the basin. This confinement exacerbated sedimentation and aggradation inside the floodway, resulting in even more rapid changes to the landscape into the 1980s (about 3.8 km² yr⁻¹; Allen 2010). The protection levees also severed the connection to riverine processes outside the levees, initiating a slower phase of the delta cycle whereby landscape change would become dominated by subsidence processes.

Lake Fausse Pointe Delta is an example of a once-growing area that experienced severing of significant sediment input because of the protection levees (Fig. 2.19 and Fig. 2.21). Its growth was stopped, and its geological development was frozen in time. This delta formed between 1919 and 1932 and was fed by Grand Bayou, a small distributary channel of the Atchafalaya River. The delta grew rapidly for 13 years until the West Atchafalaya Protection Levee severed its connection to Grand Bayou, cutting off its sediment supply, and stopping its growth. However, during its growth period subaerial sediment deposition advanced 6.5 km (4.0 mi) into the lake and covered an area of 29 km² (11 mi²; Tye and Coleman 1989a). Today, that delta exists as the 2,428 ha (6,000 acre) Lake Fausse Point State Park.

By the late 1950s, Grand Lake Delta had prograded into most of Grand Lake, north of Willow Cove, confining Atchafalaya River flow through the eastern portion of Grand Lake (Fig. 2.19; Allen 2010). This flow confinement encouraged sedimentation in the southern portion of east Grand Lake and the area of Willow Cove (Allen 2010). Thus, although it had taken the lakes several hundred years to form via subsidence, within ~150 years (1970) the basin was mostly filled with sediment (Fig. 2.22; Gagliano and van Beek 1975, Tye and Coleman 1989a, b).

Additionally, by the 1970s, artificial channel engineering (a.k.a. channel training) features on the Atchafalaya River, designed to enhance navigation and flood control,

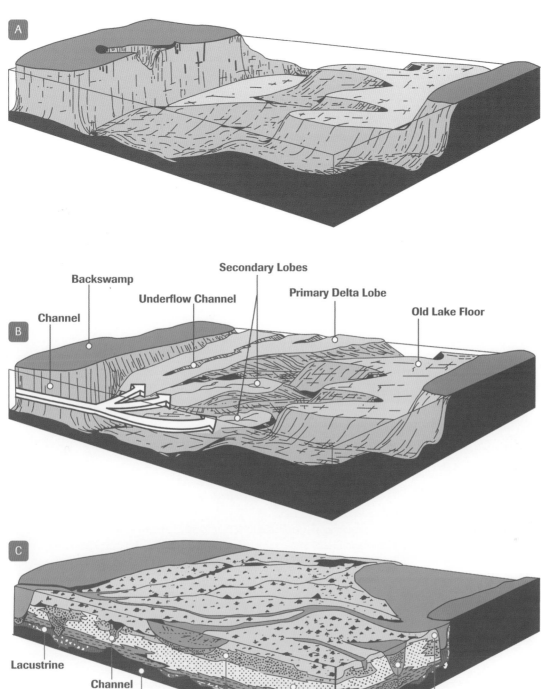

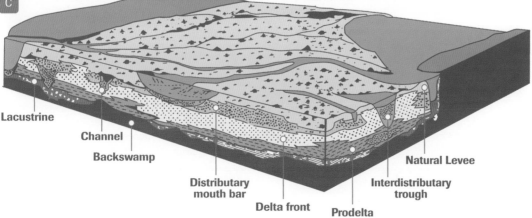

FIG. 2.20. Diagram showing the bathymetric development of lacustrine deltas in the ARB. Panel A shows the bathymetry prior to deltaic deposition with relict distributary channels present. At the intermediate stage of development (B), hyperpycnal underflows (shown with arrows) transport water and sediment into the lake basin through the relict channels and delta development proceeds until delta building is complete (C). Figure redrawn from Tye and Coleman (1989b).

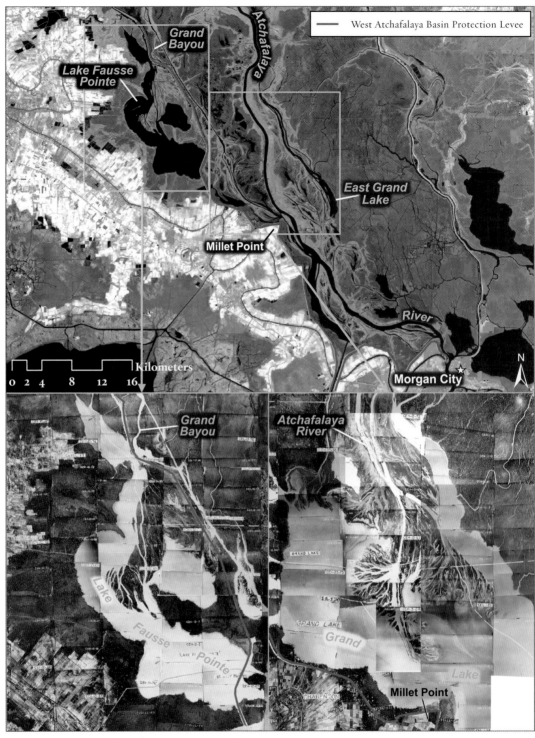

FIG. 2.21. Images showing Grand Lake and Lake Fausse Pointe delta development. On the top is the current Landsat image. On the bottom are composite aerial images showing large deltas forming in Lake Fausse Pointe (left) and the northern end of Grand Lake (right) in the early 1940s. These aerial photographs were taken after the Protection Levees were completed, so the Lake Fausse Point Delta growth was stopped and frozen in time. This same delta formation can be seen in the lake today. Composite aerial images made from original photographs by Yvonne C. Allen, US Army Engineer Research and Development Center, Baton Rouge, Louisiana.

FIG. 2.22. The lacustrine delta growing in Grand Lake (looking upstream) in 1976. The Atchafa-laya River is located in the top left, and the original lake edge is shown in the background. Note the archetypal v-shape of the splay islands and the multiple distributaries of different sizes. It is also apparent that this delta is continuing to grow, as a younger, treeless splay (foreground) expands downstream of older, mature delta deposits with trees. Photo © CC Lockwood.

resulted in construction and maintenance of a much wider and deeper river channel with higher spoil banks, which directed more water and sediment to the coast (US Army Corps of Engineers 1998). This combination of natural and anthropogenic chan-nel development also resulted in isolation of much of the river from the swamp and segmentation of hydrologic connectivity inside the swamp. For example, by the 1980s, imagery reveals that sedimentation had progressed in the lower floodway so that the eastern portion of Grand Lake had become completely isolated from flow through Six Mile Lake and the main river channel, and Willow Cove had completely filled with sediment (Fig. 2.19; Allen 2010). Sediment deposits further encroached into lower Grand and Six Mile Lakes, and Flat Lake (Allen 2010). Although there was still infill-ing in these lakes, by the 1970s, the period of rapid infilling in the Atchafalaya River Basin Floodway had ended, and most of the sediment in the Atchafalaya River was now directed to the coast. For example, it has been estimated that, between 1975 and 2004, only 9% of the total suspended sediment (170 million metric tons [187 million short tons]) in the Atchafalaya River was deposited in the floodway (Xu 2010). It was during this time that delta growth began in earnest in Atchafalaya Bay.

GEOLOGICAL TRANSFORMATION OF ATCHAFALAYA BAY. As the lacustrine deltas of the ARB were receiving course-grained (sandy) sediments, fine-grained sediments (silts and clays) remained in suspension and were transported down-basin, eventually being deposited in Atchafalaya Bay (Fig. 2.23; Fisk 1952). Thus, like the lake bottoms upstream, the bottom of Atchafalaya Bay built very slowly with fine-grained sediments from the mid- to late-1800s to the early 1950s (Fig. 2.24). Sediment deposits during this long, slow infilling period were composed of the finest silt and clay fractions and amounted to only about one-quarter of the stratigraphic column—about 100 cm (39 in)—with accumulation rates less than 1 cm yr^{-1} (0.4 in yr^{-1}; Cratsley 1975). Also, during this period of slow sediment deposition, water depths across the bay were constant at 2 m (6.6 ft; Thompson 1951, 1955). As the southern lakes in the ARB filled with sediment, the bay environment moved from the long period of slowly aggrading subaqueous deposits into a period of rapid subaqueous development (the upper prodelta phase), when the remaining three-quarters of the subaqueous sediment column was deposited in less than 30 years (Fig. 2.24; van Heerden and Roberts 1988).

The upper prodelta phase (1952–1972) was marked by rapid subaqueous growth of clay-rich sediment and deposition of upward-coarsening sediments (clays–silty clays–fine sands–medium sands) at rates averaging more than 11 cm yr^{-1} (4.3 in yr^{-1}). Throughout this time period, bathymetry in Atchafalaya Bay changed rapidly, so that by 1972 the subaqueous shoaling had reached to within 0.5 m (1.6 ft) of the surface, and deltas, although not yet subaerial, were clearly evident at both outlets (van Heerden and Roberts 1988).

The flood of 1973, and the following two years of record high discharge, marked the change from the upper prodelta phase to the bayhead delta phase (Fig. 2.24). The bayhead delta phase was a period of course-grained (bedload) sand deposition and rapid subaerial delta development (Fig. 2.25). Mean annual discharge during these three flood years exceeded long-term (1935–2001) mean values by over 4,000 m^3 s^{-1} (141,258 cfs), and sediment transport values exceeded mean values by more than 60 million metric tons (66 million short tons; van Heerden and Roberts 1988, van Heerden 2001, Roberts et al. 2003). In fact, during the 1973 flood, Atchafalaya discharge peaked at about 25,496 m^3 s^{-1} (900,000 cfs; Roberts et al. 1997), and the Old River Control Structure was almost destroyed—an event that, had it occurred, would have likely marked the completion of the stream capture process and the Atchafalaya may have become the mainstem of the Mississippi River.

During the flood years, extremely rapid accumulation rates (over 30 cm yr^{-1} [11.8 in yr^{-1}]) were documented in Atchafalaya Bay (van Heerden and Roberts 1988, Roberts et al. 2003), and estimates made in 1976 and 1977 showed about 32.5 and 40.9 km^2 (12.4 and 15.8 mi^2), respectively, of new land had emerged with an average growth rate of 6.5 km^2 yr^{-1} (2.5 mi^2 yr^{-1}; Rouse et al. 1978, Roberts et al. 1980, van Heerden and Roberts 1980). These growth rates were similar to subdelta growth rates measured by Wells et al. (1982) on the Mississippi River (Balize) delta. As the delta became more devel-

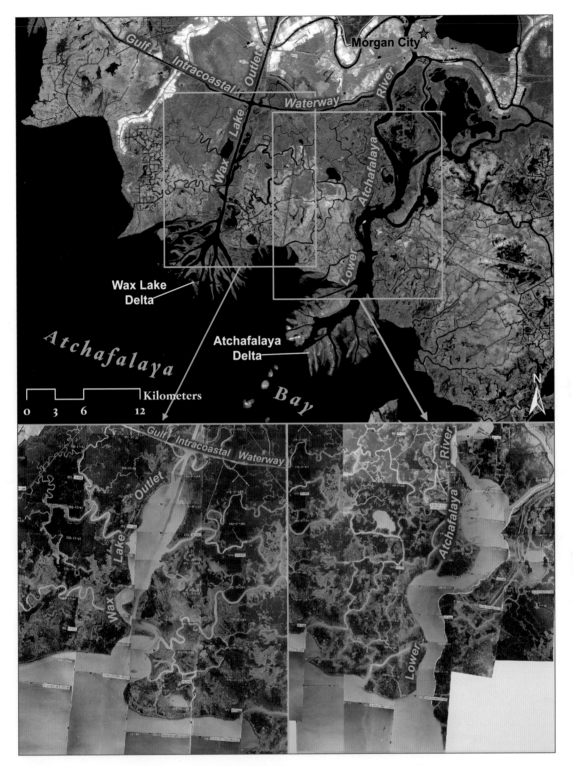

FIG. 2.23. Images showing the lower Atchafalaya River, Wax Lake Outlet, and Atchafalaya Bay. On the top is the current Landsat image. On the bottom are composite aerial images showing the Wax Lake Outlet (left image) and the lower Atchafalaya River (right image) in 1957. Note that the Atchafalaya and Wax Lake deltas are not yet emergent in Atchafalaya Bay. Composite aerial images made from original photographs by Yvonne C. Allen, US Army Engineer Research and Development Center, Baton Rouge, Louisiana.

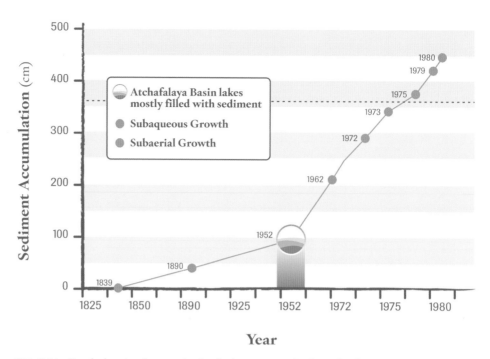

FIG. 2.24. Graph showing the magnitude of subaqueous and subaerial sediment accretion on the Atchafalaya Deltas from 1825–1980, as interpreted from stratigraphic core data. Figure modified from van Heerden and Roberts (1988).

oped, growth rates slowed a bit; from 1989–1994, growth on the Atchafalaya Delta was 3.2 km² yr⁻¹ (1.2 mi² yr⁻¹), and growth on the Wax Lake Delta was 3.0 km² yr⁻¹ (1.1 mi² yr⁻¹; Roberts et al. 2003). Growth rate estimates on the Atchafalaya (1.6 km²; 0.6 mi²) and Wax Lake (0.6 km²; 0.2 mi²) deltas from 2004–2010 suggest that rates have continued to decline (Rosen and Xu 2013). Delta facies on the Atchafalaya and Wax Lake deltas are about 2.4 m (7.9 ft) thick, and the sediment is composed mostly (70%) of sand-sized particles (van Heerden 1983, Keown et al. 1986, van Heerden and Roberts 1988, Roberts et al. 1997, Majersky-Fitzgerald 1998, Roberts et al. 2003).

With the emergence of the Atchafalaya and Wax Lake deltas, the Atchafalaya River became one of several rivers across the northern Gulf of Mexico to contain bayhead deltas (Fig. 2.26). Within a relatively short period of time the bayhead deltas of the lower Atchafalaya River and Wax Lake Outlet became large subaerial features (Fig. 2.27), and large quantities of fine-grained sediment were being transported seaward and deposited on the subaqueous clinoform (sedimentary landform on the continental shelf), considered an early shelf-phase prodelta (Fig. 2.28; van Heerden and Roberts 1988).

Some data suggest that, without major flood events, the bayhead portion of the delta building cycle has reached its peak in subaerial growth rate (van Heerden 1983, van Heerden and Roberts 1988). Still other data suggest that the bayhead deltas are rapidly expanding over the bay bottom and will eventually emerge from the bay onto the inner continental shelf, representing a progression from early lacustrine delta development, through the bayhead stage, and onto the shelf delta stage (Roberts et al. 1997, Roberts et al. 2003).

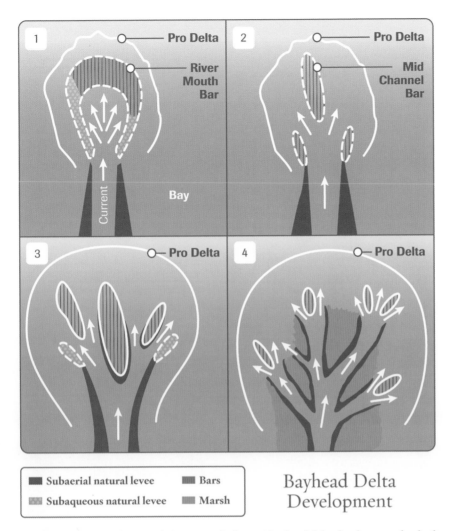

FIG. 2.25. Diagram showing the process of subaerial bayhead delta development that built the Atchafalaya and Wax Lake deltas beginning in 1973. With a prodelta platform already built, larger sand-sized sediments were transported to Atchafalaya Bay. Stream velocity slowed upon entering the unconfined bay, and sediments were dropped from the current (1). Rapid shoaling created a mid-channel bar that bifurcated the main channel into two distributaries, and the diverging current created natural levees that became subaerial (2). The same process was repeated first to form two new channels (3) and then multiple times until the bay contained a network of delta splay islands separated by branching distributaries (4). This process continues today as the deltas build seaward. Figure redrawn from van Heerden and Roberts (1980).

There is a question, however, about when this evolutionary transition will take place. Some research suggests that the shelf delta will either take a very long time (hundreds of years) to become subaerial or will never reach the subaerial stage of development, due to marine processes and sediment winnowing and export from Atchafalaya Bay during cold front passage (van Heerden and Roberts 1988, Allison et al. 2000, van Heerden 2001, Neill and Allison 2005). Other recent research demonstrates that further delta growth is possible, with large river pulses. Research during the 2008 Mississippi

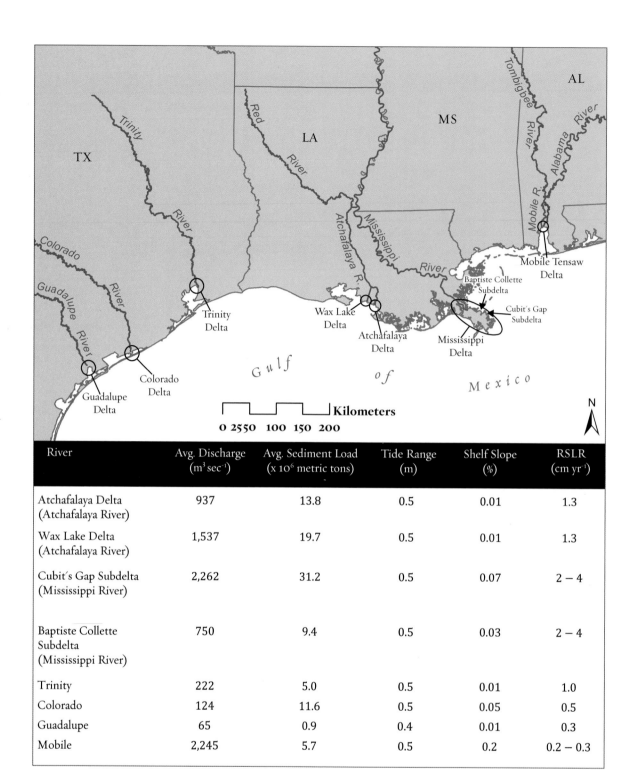

River	Avg. Discharge (m³ sec⁻¹)	Avg. Sediment Load (x 10⁶ metric tons)	Tide Range (m)	Shelf Slope (%)	RSLR (cm yr⁻¹)
Atchafalaya Delta (Atchafalaya River)	937	13.8	0.5	0.01	1.3
Wax Lake Delta (Atchafalaya River)	1,537	19.7	0.5	0.01	1.3
Cubit's Gap Subdelta (Mississippi River)	2,262	31.2	0.5	0.07	2 − 4
Baptiste Collette Subdelta (Mississippi River)	750	9.4	0.5	0.03	2 − 4
Trinity	222	5.0	0.5	0.01	1.0
Colorado	124	11.6	0.5	0.05	0.5
Guadalupe	65	0.9	0.4	0.01	0.3
Mobile	2,245	5.7	0.5	0.2	0.2 − 0.3

FIG. 2.26. Location and comparison of the Atchafalaya and Wax Lake deltas with other selected bayhead deltas along the northern Gulf of Mexico coast. Average discharge and sediment load rates represent inflow and sediment delivered to the estuary and may represent a fraction of the total river discharge and sediment load (i.e., Mississippi, Atchafalaya, and Mobile Rivers). RSLR is relative sea level rise, defined as the combination of eustatic sea level rise and land subsidence. Data from Gastaldo et al. (1987), van Heerden and Roberts (1988), Nichols (1989), Sager et al. (1992), Isphording et al. (1989, 1996), Rodriguez et al. (2010), and Donoghue (2011).

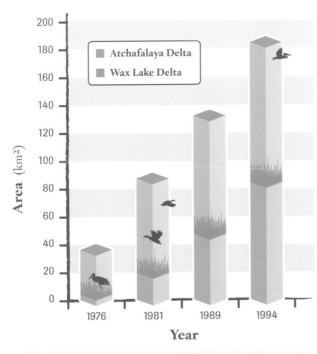

FIG. 2.27. Graph showing the subaerial extent of the Atchafalaya and Wax Lake deltas. Estimates were made with a combination of bathymetry and land elevation data collected by the US Army Corps of Engineers. Figure created with data from Roberts et al. (2003).

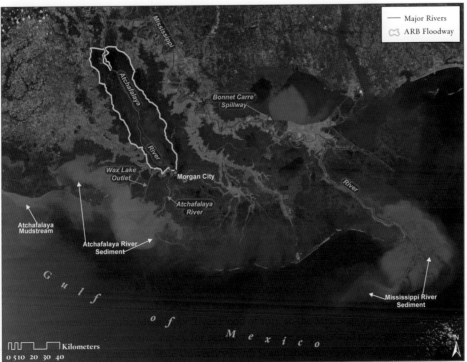

FIG. 2.28. Moderate Resolution Imaging Spectroradiometer (MODIS) true color satellite image showing the sediment plume from the Atchafalaya River and Wax Lake outlet during high discharge. Massive amounts of sediment from the river are delivered to the Atchafalaya and Wax Lake deltas, Atchafalaya Bay and onto the continental shelf in the Gulf of Mexico during high discharge. A portion of the plume can also be seen as a mud stream flowing west along the chenier plain. This image was taken during the Mississippi River Flood of 2011, so the Mississippi River plume and Bonnet Carré Floodway are also shown for perspective. Image graphics and processing were provided courtesy of the Louisiana State University Earth Scan Laboratory (www.esl.lsu.edu).

River flood shows that large river pulses produced a mineral sediment subsidy on the Wax Lake Delta that was more than 30 times greater, and vertical accretion that was more than an order of magnitude higher, than the current annual thresholds necessary for the marshes on the Mississippi delta plain to survive submergence (Azure E. Bevington, Guerry O. Holm, Charles E. Sasser, and Robert R. Twilley, Louisiana State University, unpublished data). Research after the Mississippi River Flood of 2011 showed that sediment accumulation on the Atchafalaya Delta resulting from that event was from 1.5–1.6 g cm^{-2} (0.34–0.36 ounce in^{-2}; Falcini et al. 2012, Khan et al. 2013). Another recent study following the 2011 flood showed an increase of 3.0–7.1 km^2 (1.2–2.7 mi^2) in land area on the Wax Lake Delta, depending on the datum used for reference (Carle et al., in press). While there was some elevation loss (3.5 km^2 [1.3 mi^2]), about 9 km^2 (3.5 mi^2) of already subaerial delta areas converted to a vegetation class indicative of higher elevation, and elevation on 15.2 km^2 (5.9 mi^2) remained unchanged, thus adding to the evidence of substantial maintenance and growth of the delta during this large river pulse event. While the development trajectory is not yet clear, it is uniformly accepted that delta growth is still occurring, and fine sediments have been transported off the deltas and onto a large subaqueous clinoform on the inner continental shelf, largely through storm-related resuspension of sediments from Atchafalaya Bay.

GEOLOGICAL TRANSFORMATION OF THE INNER SHELF AND CHENIER PLAIN. As with prodelta deposits in the ARB lakes and Atchafalaya Bay, subaqueous deposits of fine-grained sediment on the inner continental shelf and Chenier Plain of Louisiana were found as early as the 1950s, and growth began being monitored via satellite in the 1970s (Thompson 1951, 1955, Morgan and Larimore 1957, Rouse et al. 1978, Roberts 1998). In the 1980s it was discovered that this accretion was due to a synergistic relationship between riverine and meteorological processes and was a critical modifier not only of delta evolution in Atchafalaya Bay but also shoreline development along the Chenier Plain of Louisiana. (Morgan et al. 1953, Kemp et al. 1980, Wells and Kemp 1981, Kemp 1986, Mossa and Roberts 1990).

Remote sensing, modeling, and in situ measurements have all shown that cold front processes redistribute huge amounts of sediments from Atchafalaya Bay to bathymetric lows, sometimes up to 75 km (46 mi) offshore (Fig. 2.29). This redistribution causes an apparent land loss of subaerial delta deposits in Atchafalaya Bay, termed "cold front winnowing," especially during low discharge years (Kemp et al. 1980, Wells and Kemp 1981, van Heerden and Roberts 1988, Walker and Hammack 1999, 2000, Allison et al. 2000, Walker 2001, Roberts et al. 2003, Kineke et al. 2006, and others). Estimates of this process show substantial transport of sediment from the bay to the shelf.

One key study documented that strong northerly winds in the range of 8–15 m s^{-1} (18–34 mph) following cold front passage resulted in a process where total suspended sediment concentrations in Atchafalaya Bay were greatly increased from 150 to 800 ppm and rapid offshore currents up to 60 cm s^{-1} (1.2 knots) developed and transported the turbid waters from the bay to the inner shelf (Walker and Hammack 1999). These same researchers calculated that, during an average cold front, over 400,000 metric tons (441,000 short tons) of sediment are transported to the continental shelf opposite

CHAPTER TWO

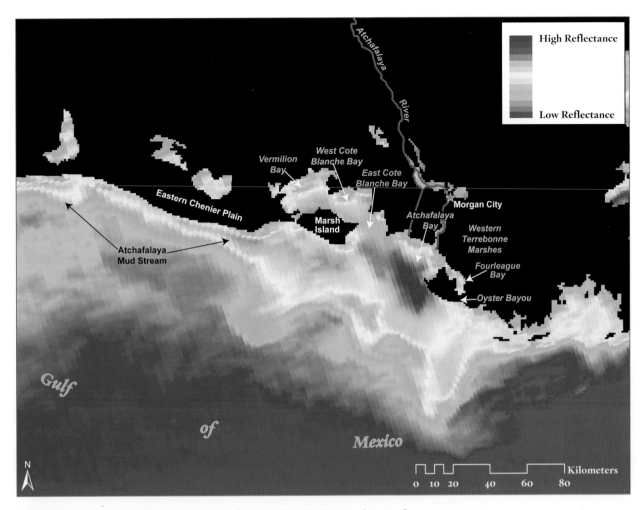

FIG. 2.29. Advanced Very High Resolution Radiometer (AVHRR) image showing the resuspension of sediment in the Atchafalaya Bay system during cold front passage. In this image, blue indicates the least amount of suspended sediments, in the water column, and red indicates the highest concentrations of sediments. Note not only the resuspension, but also the transport of sediments seaward and westward as a muddy current that runs along the chenier coast of southwestern Louisiana. Image graphics and processing were provided courtesy of the Louisiana State University Earth Scan Laboratory (www.esl.lsu.edu).

Atchafalaya Bay, and with 20–30 cold fronts per year, they estimated that over 10 million metric tons (11 million short tons) of fine-grained sediment, or 12% of the yearly sediment discharge from the Atchafalaya, are transported annually from the bay to the shelf. In another key study, a combination of X-radiographs and radioisotope analyses documented annual sediment deposition on the inner shelf that ranged from 1–3 cm yr^{-1} (0.4–1.2 in yr^{-1}) and resulted from the combined effects of high river discharge over 10,500 m^3 s^{-1} (370,000 cfs) and cold front passage (Allison et al. 2000). These same researchers also documented increased suspended sediment concentrations during a cold front event with sustained wind speeds of 15 m s^{-1} (30 mph). Suspended sediment concentrations were more than 18 times higher than baseline conditions, reaching 25,000 mg L^{-1} (25,000 ppm) in a period of 6 hours.

Tropical storms also play a large role in redistributing Atchafalaya sediment along the continental shelf (Walker 2001, Allison et al. 2005, Freeman and Roberts 2012). Depending on the track of a particular storm event, the effects on the shallow bay environments of the Atchafalaya system can either be similar to those caused by cold-front passage or cause substantial addition or redistribution of mineral sediments (Walker 2001, Turner et al. 2006, Goñi et al. 2007). For example, two remote storms in 1998 (Tropical Storm Frances—landfall Port Aransas, Texas, Hurricane Georges—landfall Biloxi, Mississippi), both of which made landfall 322–644 km (200–400 miles) on either side of Atchafalaya Bay, increased suspended sediment concentrations in the bay by an order of magnitude (75–750 mg L^{-1} [75–750 ppm]); however, destination of the suspended sediment differed (Walker 2001). In the case of TS Frances, the far westerly landfall created strong onshore flow, and resuspended sediments were driven onshore into the bays and marshes. The easterly landfall of Hurricane Georges resulted in strong northerly wind and offshore transport of resuspended sediments onto the continental shelf (Walker 2001).

Since 1985, several hurricanes (≥ Category 1, Saffir-Simpson Scale) have made landfall along the Louisiana coast, west of or at the Atchafalaya and Wax Lake deltas, presumably affecting the systems by adding and/or redistributing mineral sediments (Fig. 2.30; Allen et al. 2011). In fact, the first documented trend shift from a period of land gain (1983–2002) to a period of land loss in the Wax Lake Delta occurred with the passage of Hurricane Lili in 2002, which made landfall just west of Marsh Island (Allen et al. 2011). After the passage of Hurricane Lili, a storm-related sediment layer consisting of silty clays and fine sands was found offshore of Atchafalaya Bay on the continental shelf. The depth of that layer ranged from 0.5–20 cm, (0.2 –8 in) with most locations receiving between 3 and 10 cm (1–4 in) of deposition from the event (Goñi et al. 2006). Goñi et al. (2007) showed intense reworking (scour and deposit) of predominantly local sediments on the Louisiana shelf following Hurricanes Katrina and Rita. Deposition rates at sites offshore of the Atchafalaya River exceeded 20 cm (8 in), with the majority of the sediment deposited during Hurricane Rita. In fact, the total amounts of sediment, carbon, and nitrogen redeposited during these two extreme events far exceeded the combined annual inputs of sediments to the shelf by the Mississippi and Atchafalaya Rivers. This agrees with work done by Turner et al. (2006) that showed substantial storm-related accumulation (estimated 131 million metric tons [144 million short tons]) of inorganic resuspended sediment in Louisiana coastal marshes following Hurricane Katrina, with deposition layers reaching from 0 to 68 cm (0–27 in) deep. Tweel and Turner (2012) refined that estimate down to 68 million metric tons (75 million short tons) and also showed substantial deposition by three other hurricanes that affected Louisiana—Hurricanes Rita (48 million metric tons [53 million short tons]), Gustav (21 million metric tons [23 million short tons]), and Ike (33 million metric tons [36 million short tons]). Sediment transport from these events affected a large region, with inorganic sediment deposits observed up to 43 km (27 mi) from the coast and over 200 km (124 mi) from the storm track. Freeman and Roberts (2012) investigated storm deposits from Hurricanes Katrina and Rita and found a storm layer that was between 1 and 10 cm (0.4–4.0 in) thick in Sister Lake, about 30 miles east

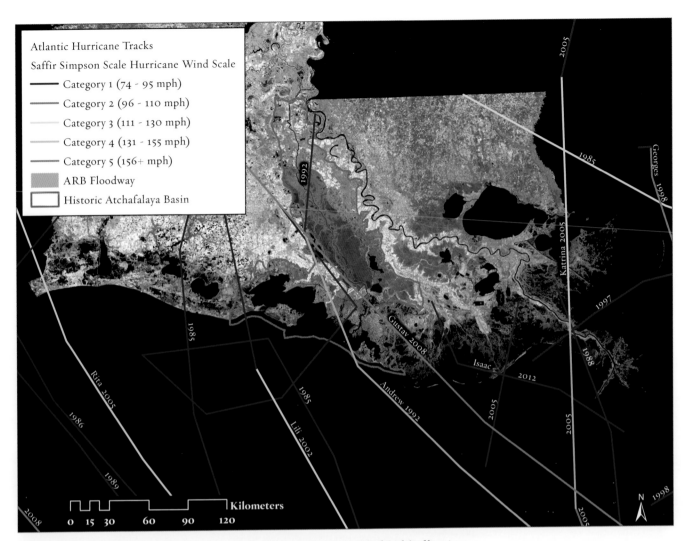

FIG. 2.30. The hurricane tracks and intensity (Saffir-Simpson Hurricane Wind Scale) of hurricanes making landfall across the Louisiana coast since 1985. Storms with named tracks are discussed further in the text. Also shown are the historical boundaries of the ARB and the ARB floodway.

of Atchafalaya Bay. This layer was an order of magnitude thicker than the long-term sedimentation rates in the lake, and the sediment that composed it originated in the Mississippi and Atchafalaya Rivers.

More recent measurements made by the National Center for Earth Surface Dynamics on the Wax Lake Delta during the 2007–2008 water-year indicate, however, that mineral sediment accretion resulting from high river discharge exceeds that resulting from a hurricane strike. In that study, 81% of the net deposition resulted from the riverine pulse, versus 19% from the hurricane surges (Azure E. Bevington, Guerry O. Holm, Charles E. Sasser, and Robert R. Twilley, Louisiana State University, unpublished data).

The final destination of resuspended sediments from Atchafalaya Bay is highly dependent upon the interaction of the river plume, tidal forcing, and wind conditions (Roberts et al. 1989, Cobb et al. 2008b). Resuspended sediment can be transported seaward onto the inner continental shelf where it flocculates into fluid mud as it meets salt water. This fluid mud bottom has been shown to dampen wave energies and may

TABLE 2.4. Comparison of the "Atchafalaya Mud Stream" with other seasonal muddy currents on three major rivers worldwide

River	Peak Discharge (x 1000 m³ s⁻¹)	Annual Sediment Discharge (x 10⁶ metric ton)	Deposition Water Depth (m)	Maximum Deposit Thickness (cm)	Deposition: Accumulation (g cm⁻² d⁻¹)	References
Atchafalaya	9.5	84	<5–15	1–3	0.38–12	Allison et al. (2000)
Eel	7–10	10–150	40–150	8	2–24	Somerfield et al. (1999); Somerfield and Nittrouer (1999)
Changjiang (Yangtze)	50–60	500	20–50	15+	10	McKee et al. (1983, 1984); DeMaster et al. (1985)
Amazon	230	1,100–1,300	10–30	150	25–1,500	Kuehl et al. (1986, 1995)

Shown in the table are flow conditions and sediment discharge, as well as offshore deposition rates on continental shelf clinoform features. Table after Allison et al. (2000).

reduce storm surge energy (Sheremet and Stone 2003, Stone et al. 2005). Some of this fluid mud is transported westward as a longshore muddy current called "The Atchafalaya Mud Stream" (Fig. 2.29). This muddy current is confined landward of the 10 m (33 ft) isobaths and moves westward at between 3 and 10 cm s⁻¹ (0.1–0.3 ft s⁻¹) along the southwest Louisiana coast. It carries a mean annual suspended sediment load of 53 million m³ (43,000 acre ft yr⁻¹)—almost half of the volume of sediment exiting Atchafalaya Bay—and sediment concentration peaks during times of peak discharge (Wells and Kemp 1981, Allison et al. 2000, Jamarillo et al. 2009). Large seasonal offshore sediment flux and longshore transport, as seen with the Atchafalaya Mud Stream, is a similar pattern seen resulting from seasonal muddy currents on a number of major rivers worldwide (Table 2.4; Allison et al. 2000, Neill and Allison 2005). The Atchafalaya Mud Stream has resulted in down-drift shoreline accretion on the eastern chenier coast, in the vicinity of Chenier au Tigre, of 60–80 m yr⁻¹ (197–262 ft yr⁻¹; Morgan et al. 1953, Morgan and Larimore 1957, Wells and Kemp 1981, Walker and Hammack 1999, 2000, Huh et al. 2001, Bentley 2002, Roberts et al. 2002, Bentley et al. 2003, Draut et al. 2005a, b, Jamarillo et al. 2009). In fact, radioisotope data have shown that seasonal deposition rates along this portion of the chenier coast, resulting from the Atchafalaya Mud Stream during high river discharges, exceeded decadal accumulation rates by over 166% (Fig. 2.28; Allison et al. 2000). While this westerly slow current dominates the net flow field and transport of fine sediments out of Atchafalaya Bay, winter storms can generate enough shear stress to resuspend sand-sized particles and transport them easterly and offshore, obscuring westerly flow for short periods of time and keeping courser sediments in the immediate vicinity of the advancing delta front (Adams et al. 1982).

Winter storms also transport resuspended sediment into marshes adjacent to Atchafalaya Bay through a synergistic relationship between riverine and winter storm-related sediment transport processes (Kemp et al. 1980, Wells and Kemp 1981, Baumann et al. 1984, Mossa and Roberts 1990, Walker and Hammack 1999, 2000, Allison et al. 2000, and many others). During late-winter and spring, as discharge and sediments are increasing in the river, cold fronts (20–30 events per year) are actively crossing the coast on a time interval of 3–7 days (mid Oct–late March/early April; Roberts et al. 1989). The combination of pre-frontal onshore winds and high sediment loads force turbid water into the marshes. The water is then forced offshore, out of the marsh, as the post-frontal winds turn northerly and the sediment is left behind. Most of the sediment deposition by this process is thought to occur in marshes on the deltas and west of the river, due to largely westerly transport of suspended sediments (Allison et al. 2000); however, a few studies have documented large seasonal transport fluxes of Atchafalaya River fresh water (1 billion m³ [811,000 acre-ft]) and sediment (172,000 metric tons [189,000 short tons]) from Fourleague Bay to the Gulf of Mexico through Oyster Bayou (east of the river) during spring high discharge events. Suspended sediment concentrations in this area during cold front passage were documented as high as 1,527 mg L⁻¹ (1,527 ppm), suggesting that sedimentation by this process may also be important east of the river, specifically in the western Terrebonne marshes (Denes and Caffery 1988, Perez et al. 2000, 2011, Lane et al. 2002).

HABITAT TYPES IN THE ATCHAFALAYA RIVER BASIN 3

I T IS IMPORTANT TO UNDERSTAND that the geological processes that filled the ARB are still active and continuously transform the distribution and composition of habitats within the system, as sediment is increasingly delivered seaward. Therefore, the current distribution of habitats in the ARB is a result of both past and present sedimentary processes, and the trajectory and maintenance of habitats is governed by seasonal water flow and sediment transport (Sabo et al. 1999a, b, Rutherford et al. 2001, Allen et al. 2008, Faulkner et al. 2009).

Three distinct landscape regions have developed within the modern ARB (Fig. 3.1; Louisiana Department of Natural Resources 2010). The upper region, located in the northern section of the basin, contains the highest land and is the only region of the basin suitable for agricultural development. Species-rich and structurally diverse bottomland hardwood forests are typical of the northern region. The middle region is composed of cypress-tupelo swamps, which reflects decreasing relative elevation and increasing flooding frequency and duration. Finally, the lower region consists of the delta plain where the Atchafalaya River meets the Gulf of Mexico. It contains delta wetland habitat and a range of emergent wetlands and bays.

The modern ARB is approximately 70% forest habitat; the remainder is open water and marsh habitats (Demas et al. 2001). It contains the largest contiguous tract of forested wetlands in the lower Mississippi River alluvial valley and is the largest block of floodplain forest in the United States (Ford and Nyman 2011). The ARB supports ~35% of the forested wetlands remaining in the lower Mississippi River floodplain, and represents about 0.2% of the remaining riverine floodplains in the world (Doyle et al. 1995, Tockner and Stanford 2002, Chambers et al. 2006). While small in comparison to the world's largest forested wetlands (i.e., Congo River Basin; approximately 190,000 km^2 [73,000 mi^2] of forests and swamps), the ARB is extremely important in North America where 90% of the river floodplains have been converted to agriculture, and it may support more forested wetlands than remain in all of Europe (Grinson and Malvarez 2002, Tockner and Stanford 2002, Campbell 2005).

Although there are individual trees that are centuries old, most forests in the ARB are relatively young (≤ 70 yr) and have regenerated since the heavy logging that occurred in the 1930s (Conner and Toliver 1990, Keim et al. 2006a, Keim and Amos 2012). These forested wetlands consist of two major types depending on hydrologic regime (depth, duration, and frequency of flooding)—bottomland hardwood forests and cypress-tupelo swamps.

BOTTOMLAND HARDWOOD FORESTS. These forests are found at higher elevations where extended flooding is rare (Fig. 3.1). This forest type covers 150,138 ha (371,000 acres) of the northern region of the ARB floodway, with about 50,586 ha (125,000 acres) in the Morganza and West Atchafalaya floodways and the rest (~ 99,147 ha [245,000 acres]) in the northern end of the Atchafalaya Basin Floodway System and on newly formed sedimentary habitats and levee ridges in the middle region of the basin (Gagliano and van Beek 1975, US Geological Survey, unpublished data). Bottomland hardwood forests in the northern region are characterized by a well-defined and relatively stable river channel bounded by large mature natural levee ridges and artificial river levees, which separate the river from the floodplain (Fig. 3.2). Although river stages are highly variable, overbank flooding is now infrequent in many areas due to the height of natural levee ridges. Therefore, outside of established river channels, water input in the upper basin is mainly from runoff and overflow events, and accretion is extremely slow. Within bottomland hardwood regions of the upper basin, substrate consists of well-oxidized soils that are dominated by clay (Gagliano and van Beek 1975).

Composition of bottomland hardwood forests is a reflection of relative degree of flooding. On the highest sites, where flooding frequency is low, forests are dominated by less flood-tolerant species such as American sweetgum (*Liquidambar styraciflua*), water oak (*Quercus nigra*), and sugarberry (*Celtis laevigata*). As elevation decreases and flooding increases (frequency, depth, duration), more flood tolerant species such as overcup oak (*Quercus lyrata*), water hickory (*Carya aquatica*), and green ash (*Fraxinus pennsylvanica*) dominate (Harms et al. 1980, Chambers et al. 2005, Faulkner et al. 2007, Faulkner et al. 2009). These habitats are intermediate in productivity (aboveground net primary productivity [NPP] about 1,000 g C m^{-2} yr^{-1}) between upland forests and swamps (Megonigal et al. 1997).

Young pioneer stands dominated by black willow (*Salix nigra*), sandbar willow (*S. interior*), box elder (*Acer negundo*), and swamp cottonwood (*Populus heterophylla*) are found on land that was most recently aggraded or disturbed and is subjected to frequent flooding such as point and channel bars and harvested tracts (Fig. 3.3; Gagliano and van Beek 1975, Doyle et al. 1995, Hupp et al. 2008). These sites are typically dominated by willow and are usually ephemeral seral stages, although they can persist for several years (up to about 50 years), depending on rates of sedimentation and degree of flooding. As forests in the middle and lower regions of the ARB age past this pioneer state, the forests become dominated by an elm-ash-maple forest type, where heavy seeded species (i.e., oaks and hickories) are found in lower numbers, even though conditions are favorable for their growth. This may be due to the relatively young age of the land and the fact that species with seeds that are broadcast by wind and birds have a

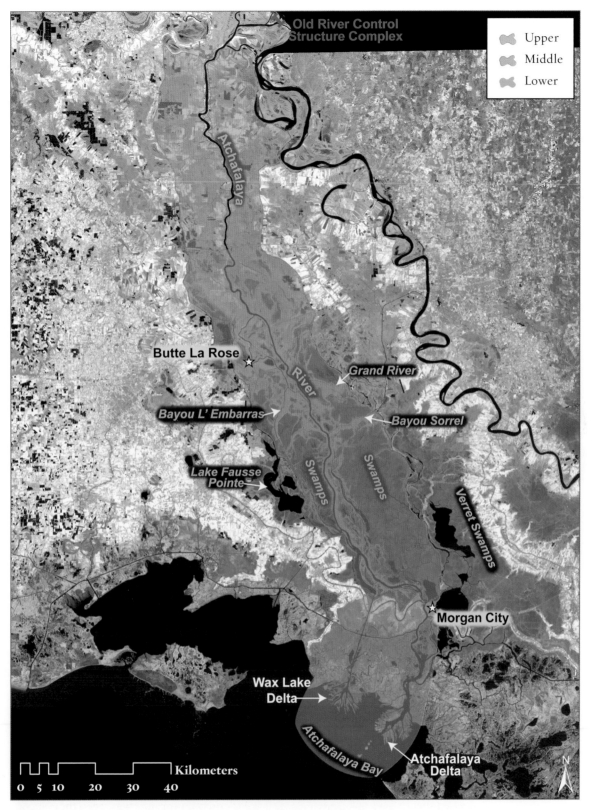

FIG. 3.1. The three landscape regions that have developed in the ARB floodway. These regions are estimated based on elevation and dominant cover type (bottomland hardwood forest, cypress-tupelo swamp, marsh), therefore the boundaries are not absolute.

FIG. 3.2. The Atchafalaya River passing through a large expanse of bottomland hardwood forest in the ARB floodway. In this region of the floodway, the river channel is well defined, and the stable river channel and mature natural levee ridges separate the river from its floodplain. Note also the abandoned distributary course, now an isolated lake, in the foreground. Photo © The Nature Conservancy.

competitive advantage (Richard F. Keim, Louisiana State University Agricultural Center, personal communication).

Bottomland hardwood forests, particularly young pioneer stands on newly accreted and largely unconsolidated sand and channel bar substrates, are very susceptible to hurricane damage (i.e., blowdown, limb loss); however long-term tree death resulting from hurricane damage appears relatively low. For example, Hurricane Andrew (see Fig. 2.30) struck the Louisiana coast in 1992 with sustained winds over 200 kph (124 mph), and its eye-path tracked inside the Atchafalaya floodway levees for about 100 km (60 mi; Doyle et al. 1995). Over 40% of the bottomland hardwood forests were severely impacted by Hurricane Andrew, with an average 21% and 25% reduction in stem density and basal area, respectively. Hardest hit were young, pioneer stands on relatively recent landforms in the Attakapas Island Wildlife Management Area, where more than 85% of the trees were blown down. Immediate tree mortality associated with the exten-

FIG. 3.3. Pioneer tree species (i.e., willow *Salix* spp.) growing on newly deposited sediment that is filling in an ARB floodway lake. Note that the trees are growing in front of a mature cypress stand. The sandbar is building from right to left, as evidenced by the shorter trees on the newest land. As this land gets progressively higher in elevation, it will be colonized by bottomland hardwood forest species. Photo by Daniel E. Kroes, US Geological Survey Louisiana Water Science Center, Baton Rouge, Louisiana.

sive damage was low (1.4%), due to large-scale resprouting of damaged and blown-down trees. For most species, survival was still high after two years, with continued high rates of resprouting and good sprout survival (Keeland and Gorham 2009). However, mortality of sandbar willow and swamp cottonwood both increased over 50% in the two years following the storm. Additionally, species composition of heavily damaged areas was affected by increased dominance of Chinese tallow (*Triadica sebifera*), an exotic tree species, which had not been observed in the plots prior to Hurricane Andrew. Chinese tallow invasion following Hurricane Andrew was also found outside the floodway levees in the Verret basin (Conner et al. 2002).

Chinese tallow is native to southeast Asia, where it is highly valued as an agricultural species grown primarily for its seed oil (Fig. 3.4; Bruce et al. 1997). It was first introduced to the southeastern United States in the late 1700s by Benjamin Franklin and to the Gulf Coast by the federal government as a possible seed crop (Jubinsky 1994, Urbatsch 2000). Chinese tallow is capable of growing across a variety of habitats, including bottomland hardwood forests; however it cannot tolerate the extended flooding conditions found in cypress-tupelo swamps (Bruce et al. 1997). This tolerance to a wide suite of conditions, as well as its high seed production, persistent seed bank, and lack of natural pests or pathogens, make it a highly invasive species that has become solidly established along the Gulf Coast. Chinese tallow is considered one of the most important nonnative invasive plants in the region and is capable of causing substantial changes to composition and structure of native plant communities (Bruce et al. 1997, Jubinsky 1994, Conner et al. 2002, Keeland and Gorham 2009). Although there are no data showing the spatial extent of Chinese tallow invasion in the ARB, it is well established in all topographically suitable regions of the ARB and is poised to aggressively colonize disturbed sites and areas where timber is harvested commercially.

Tallow not only alters plant communities, but potentially reduces habitat suitability for a wide spectrum of native wildlife. For example, Chinese tallow-dominated forest

FIG. 3.4. Chinese tallow (*Triadica sebifera*) is a highly invasive, exotic tree species that has become solidly established along the Gulf Coast. It cannot tolerate the extended flooding in cypress-tupelo swamps but is capable of colonizing disturbed areas in higher elevation bottomland hardwood areas. Although its spatial extent in the ARB has not been mapped, it is likely well established in all topographically suitable regions of the ARB. Photo by James Henson, © USDA-NRCS PLANTS Database.

stands in Louisiana support fewer wintering bird species than natural bottomland forest stands (Baldwin 2005). This may be due to the chemical composition of its leaves, which renders the trees inhospitable to most insects. Although mature Chinese tallow produce large quantities of hard kernel seeds that are rich in lipids and saturated fat, few birds are physiologically adapted to metabolize the long-chained fatty acids found in the seeds. One species that is able to metabolize the fruit pulp is the yellow-rumped warbler (*Dendroica coronata;* Baldwin 2005). While few birds can metabolize the entire seed, a number of species scrape the lipid coating off and then drop the seed in place. This suggests that birds may be a more effective dispersal agent for Chinese tallow than once thought (Conway et al. 2002a, b).

CYPRESS-TUPELO SWAMPS. Swamps exist where flooding frequency, depth, and duration are greatest. In the ARB, swamp forests are found mostly amid the meandering bayous, constructed canals, and occasional small lakes of the middle region (Fig. 3.1).

These forests are dominated by baldcypress and water tupelo, species that can persist under nearly constant flooding, although regeneration of both species requires periodic, prolonged low-water periods during the growing season.

Under relatively constant hydrological conditions, cypress-tupelo forests represent a climax vegetative community, and given exacting requirements for regeneration, cypress-tupelo forests also tend to be relatively even-age (Gagliano and van Beek 1975). Swamps in the ARB are largely dominated by second-growth, even-aged cypress-tupelo forests that mostly regenerated after large-scale clearcut logging between 1880 and the 1930s and are not subject to forest management for the most part (Keim and King 2006). Swamps are generally less productive than bottomland hardwood forests and uplands (NPP = 144 g C m^{-2} yr^{-1}–1,780 g C m^{-2} yr^{-1}; see reviews by Conner and Buford 1998, Shaffer et al. 2009) and their productivity is directly correlated to the depth and duration of flooding (Conner and Day 1976, 1992a, Conner et al. 1981, Megonigal et al. 1997, Shaffer et al. 2009, Keim and Amos 2012, Keim et al., 2012).

The ARB floodway contains large swamp areas both east and west of the Atchafalaya River (Fig. 3.1). These swamps contain just over 106,000 ha (262,000 acres) of cypress-tupelo forests, which represents almost one-third of the remaining coastal cypress-tupelo swamp forests in Louisiana, and the greatest remaining contiguous acreage in the United States (Rosson et al. 1988, Conner and Toliver 1990, Chambers et al. 2005, Lester et al. 2005, Keim and King 2006).

Swamps in the floodway receive seasonal flood pulses from the Atchafalaya River (Fig. 3.5). During flood pulses, water enters swamps over levee ridges and through natural distributary channels and constructed canals that are open to the river (Hupp et al. 2008). During high discharge, the floodplain becomes inundated at different depths depending on land elevation, river stage, and resistance to drainage. In some areas, inundation can be up to and over 3 m (10 ft; van Beek et al. 1977).

Backslope areas and swamp margins typically drain as river levels drop. However, accretion on natural levees and anthropogenic hydrologic modification have created isolated areas throughout the basin with little to no flow except during the highest discharges, are unable to drain adequately, and may remain inundated throughout the year (Gagliano and van Beek 1975, Allen et al. 2008, Hupp et al. 2008, Kroes and Kramer 2013).

Sedimentation due to water overtopping levees typically results in minimal sediment accretion in swamps; however, canals and distributary bayous that are connected to the main channel or other sediment sources (i.e., Gulf Intracoastal Waterway) may serve as conduits that more directly convey substantial mineral sediments into specific backswamp environments, resulting in rapid accretion in those areas (Fig. 3.6). In swamp areas that receive these direct sediment inputs, cypress-tupelo swamp may be transformed into higher elevation habitat (bottomland hardwoods; Gagliano and van Beek 1975, Hupp et al. 2008, Faulkner et al. 2009). Accretion in backswamps not connected to sources of mineral sediment is mainly from organic detritus (Gagliano and van Beek 1975, Bryan et al. 1998). Water temperatures in backswamps range 5–35 °C (41–95 °F; Sabo et al. 1999a, b).

Outside the floodway levees, two large isolated swamp complexes (Lake Fausse

FIG. 3.5. A large cypress-tupelo swamp area in the ARB floodway that is receiving riverine flooding. Note the brown color of the water, which contains sediment and nutrients from the Atchafalaya River. Photo by Charles Demas, US Geological Survey Louisiana Water Science Center, Baton Rouge, Louisiana.

FIG. 3.6. Canals that are connected to the Atchafalaya River (foreground) or other distributaries and cut through large forest blocks channel sediment-laden water into backswamp areas that normally would not receive large inputs of sediment. Photo by Charles Demas, US Geological Survey Louisiana Water Science Center, Baton Rouge, Louisiana.

FIG. 3.7. The East Atchafalaya Protection Levee separates large blocks of forested wetland habitat in the Lake Verret basin from the ARB floodway and fresh water, sediment, and nutrients from the Atchafalaya River. In this photo (looking south) the Lake Verret basin is on the left side, and the floodway is on the right. Photo © Alison M. Jones for www.nowater-nolife.org.

Point and Verret Swamps) occur within the natural, historic ARB boundaries (Fig. 3.1). Lake Fausse Point swamp is approximately 2,428 ha (6,000 acres) of forested wetland habitat, much of it cypress-tupelo swamp. The Lake Verret Basin extends from the agricultural fields in the northern portion of the basin south to the Gulf Intracoastal Waterway and occupies more than 162,000 ha (400,000 acres), of which about half is bottomland hardwood and cypress-tupelo swamp (Conner et al. 1993). Taken together, Lake Fausse Point and Verret swamps contain almost 90,000 ha (223,000 acres) of forested wetland habitat that once experienced the same Atchafalaya River flooding, sedimentation, and geological processes that exist inside the floodway and have been severed since the construction of the East and West Atchafalaya Protection Levees.

Construction of the protection levees altered hydrologic processes and severed riverine fresh water and sediment delivery into the Lake Fausse Point and Lake Verret basins (Fig. 3.7). Prior to construction of the protection levees, riverine flooding brought with it nutrients and mineral sediments that fertilized trees and provided accreting sediment to keep up with subsidence. After the levees were constructed, the dominant water source became runoff from adjacent lands (often from sugar cane

FIG. 3.8. A scrub-shrub community exists over large portions of the cypress swamps in the ARB floodway that does not exist in swamps outside the protection levees. This community contains only scattered cypress trees and is largely composed of extremely flood-tolerant shrub species, seen here under the cypress trees. The fact that this community shift is not seen in flooded swamps outside the levees suggests that it may be driven by flooding during the growing season; however, there are little data to indicate what processes are driving the shift. Photo by Seth Blitch © Nature Conservancy.

fields), and mineral sedimentation was replaced by the buildup of organic detritus, largely from in-situ decomposition. This combination of increased nutrient input from adjacent agricultural lands and subsidence has caused large areas of flow stagnation and poor water quality, and subsequently, reduced forest health, in both swamps areas outside the floodway levees (Gagliano and van Beek 1975, Keim and King 2006).

A particularly interesting feature of swamps in the ARB is the dominance of a scrub-shrub community that exists over large portions of the cypress swamps in the ARB floodway that does not exist outside the protection levees (i.e., Verret Swamps; Keim and King 2006, Richard F. Keim, Louisiana State University Agricultural Center, personal communication). This community, with only scattered cypress trees, is composed of water-elm (*Planera aquatica*), swamp-privet (*Forestiera acuminata*), and buttonbush (*Cephalanthus occidentalis*), all extremely flood tolerant (Fig. 3.8; Rayner 1976, Conner et al. 1981). For example, submerged water-elm seeds remain viable for at least six months and will germinate on the soil surface when floodwaters subside (Rayner 1976). Seed production in this species is apparently enhanced in trees growing in flooded, high-light conditions, especially when floodwaters cover part of the crown (Bruser 1995). Buttonbush seeds can germinate even while immersed in water (DuBarry 1963). Swamp-privet fruit floats in the water and is consumed by ducks, and consumption (and post-consumption) germination of swamp-privet fruit by channel catfish (*Ictaluras punctatus*) in a Mississippi River floodplain has been documented (Chick et al. 2003). While this pattern of an increase in flood-tolerant scrub-shrub species and a corresponding decrease in bald cypress has also been shown in permanently flooded swamps in Barataria Basin, Louisiana, what is interesting is that this pattern is not seen in the Verret Swamps where large areas of stagnant flooding also exist. This pattern seems to be driven by growing-season flooding, but there are little data to indicate the processes that are driving the community shift (Richard F. Keim, Louisiana State University Agricultural Center, personal communication).

Cypress-tupelo swamps are much more resistant to hurricane damage than bottomland hardwood forests. For example, baldcypress and water tupelo trees were largely unaffected by Hurricane Andrew, showing only about 1–2% reduction in both stem density and basal area and no immediate or delayed mortality (Doyle et al. 1995, Keeland and Gorham 2009). Location affected the amount of damage sustained by baldcypress trees; those in bottomland hardwood forests sustained more damage (25% of sampled trees) than those in cypress-tupelo swamps. However, regardless of location, baldcypress generally sustained much less damage than most other species (over 50% of sampled trees damaged). Damaged baldcypress in bottomland hardwood stands were more likely to have limb loss and snap off, indicating that adjacent falling trees had likely inflicted the damage or created more openings and reduced wind protection. However, even in the heavily damaged hardwood stands, no baldcypress trees were uprooted, thanks to their large, deep root systems, smaller canopies, rapid defoliation, and dense wood structure.

WATER BODIES. Several forms of open water exist within forested wetlands of the ARB. These include the main river channel, distributaries, bayous, lakes, and canals. While estimates are not available for areas outside the floodway, open water habitats in the ARB floodway comprise about 41,000 ha (102,000 acres; Fig. 3.9). Habitat characteristics in ARB water bodies, as well as the spatial and temporal variation in these characteristics, are driven by the quantity, quality, timing, and duration of water flow, which greatly affect dissolved oxygen concentrations and water temperatures (Bryan et al. 1998, Sabo et al. 1999a, b). The interaction between water flow, dissolved oxygen, and water temperature creates a complex mosaic of physiochemical conditions in the waterways of the ARB that can be grouped into three distinct habitat types— brown-water, green-water, and black-water (Gelwicks 1996, Rutherford et al. 2001).

Distributaries are brown-water, high-energy habitats—lotic waterways (canals and bayous) connected to the main river channel and characterized by strong currents, heavy sediment loads, relatively low water temperature, and high dissolved oxygen concentrations (Fig. 3.10). While relict distributaries are visible on the landscape out-

FIG. 3.9. A typical bayou scene in the ARB floodway. Bayous are important open water habitats that transport cool, highly oxygenated riverine water into backswamp areas and drain warm, oxygen-depleted water out. Deep, well-oxygenated bayous in the floodway are also important refuge areas for nekton during periods of hypoxia. Photo © Brittany App (www.appsphotography.com) for www.cyclingforwater.com.

FIG. 3.10. Distributary channels (foreground) carrying Atchafalaya River (background) water through large blocks of forested wetlands in the ARB floodway. Photo by Charles Demas, US Geological Survey Louisiana Water Science Center, Baton Rouge, Louisiana.

side the East and West Atchafalaya Protection Levees, active distributaries no longer exist in these forested habitats, since they have been separated from the river by both natural and artificial levees.

Green-water habitats are typically found in floodplain lakes and also are abundant in quiescent, lentic water bodies when river stages are low and water temperatures are high (Fig. 3.11). Green-water habitats are characterized by a high abundance of phytoplankton and a wide range in dissolved oxygen concentrations that is linked to the diurnal metabolic cycles of phytoplankton. During the day, these areas often become supersaturated with oxygen, creating highly productive habitats for aquatic animals (Bryan et al. 1998, Sabo et al. 1999a, b, Davidson et al. 1998, 2000, Rutherford et al. 2001).

Black-water is found in floodplain habitats and lentic water bodies (canals, bayous) in areas where flow is stagnant (Fig. 3.12). These areas are characterized by clear, slow water that is black in color (due to the lack of suspended sediment and abundance of organic tannins) with low dissolved oxygen concentrations (due to high biological oxygen demand by decomposing detritus). When the dissolved oxygen concentrations drop, the areas become hypoxic and aquatic productivity declines (Bryan et al. 1998, Sabo et al. 1999a, b, Davidson et al. 1998, 2000, Rutherford et al. 2001). Hypoxic black-water areas are found in the ARB floodway throughout the flood pulse, but become more abundant and spatially expansive during periods when water temperatures are elevated.

The distribution of brown-, green-, and black-water habitats throughout the ARB is very dynamic and is influenced by large-scale hydrologic processes, principally the distribution of variable flows throughout the basin (or severance of connectivity from riverine flooding) and local hydrologic conditions (Fig. 3.13). The interaction between large-scale and local processes determines how the physiochemical conditions in water bodies vary between high dissolved oxygen concentrations (brown, green) to low (black) dissolved oxygen concentrations.

The range of normal dissolved oxygen concentrations (normoxic conditions) within the ARB is 2.0–4.0 mg L^{-1} (2–4 ppm; 20–40% saturation). Normoxic conditions are typically found in brown- and green-water habitats. Hypoxia occurs wherever bacterial respiration (oxygen consumption) associated with decomposition of organic detritus overcomes oxygen production from photosynthesis and surface water oxygen transport and is defined by oxygen saturation values less than 2.0 mg L^{-1} (2 ppm). Hypoxia is a normal feature of southern deepwater swamps, where water stagnation often occurs in limited areas on the swamp floor (Battle and Mihouc 2000). However, in the ARB, hypoxia is not limited to internal swamp areas but can affect large areas of water. For example, lower-water column conditions in some water bodies (disconnected canals and bayous) are hypoxic (or even anoxic—less than 0.2 ppm) for much of the year (Bryan et al. 1998). Hypoxia events occur in the ARB when oxygen saturation values fall below 2.0 ppm within the remainder of the water column (upper 0.5 m [1.6 ft]), and this condition is characteristic of black-water habitats.

Hypoxia events in the ARB can be classified as acute and chronic. Acute hypoxia may occur when tropical storms and hurricanes strip leaves off trees and mix large amounts of organic matter from the swamp floor into water bodies, creating a biological oxygen demand that rapidly depletes available dissolved oxygen (Sabo et al. 1999b). Additionally, runoff resulting from massive rainfall associated with tropical storms can exacerbate hypoxia, especially if runoff is from agricultural lands. These kinds of events are sudden, typically affect a large area, and cause extensive fish kills (Fig. 3.14). After such events, dissolved oxygen concentrations return to normal within weeks, although the resilience of fish populations is somewhat variable (Sabo et al. 1999b, Perret et al. 2010, Bonvillain et al. 2011). For example, extensive fish kills resulted from acute hypoxia that occurred in the ARB floodway following passage of both Hurricane Andrew in 1992 and Hurricane Gustav in 2008 (Sabo et al. 1999b, Charlie Demas, US Geological Survey, Mike Walker, Louisiana Department of Wildlife and Fisheries, personal communication). For the most part, fish stocks rebounded rather quickly; however, largemouth bass (*Micropterus salmoides*) recovery was slow, and, although not necessary to recover the species, large restocking efforts were used to augment the process (Mike Walker, Louisiana Department of Wildlife and Fisheries, personal communication). An assessment of the effect of Hurricanes Katrina and Rita on water quality and fish abundance in the ARB floodway documented a post-storm decrease in sportfish abundance that was attributed to decreased water quality and remained for two years (Perret et al. 2010).

Chronic hypoxia is defined as a protracted period (several weeks to several months) when mild to severe hypoxia occurs over large areas of the ARB (Sabo et al. 1999b).

FIG. 3.11. Green-water habitat in the ARB floodway. This type of habitat is typically found in floodplain lakes (top) and also in quiescent bayous (bottom) during low water periods when water temperatures are high. Bodies of green water contain high abundances of phytoplankton, and, as a result, these areas often become supersaturated with oxygen, during the day, creating highly productive habitats for aquatic animals. The fact that lakes provide almost year-round green-water habitat makes them highly valuable refuges for fish when water quality conditions deteriorate and dissolved oxygen concentrations drop. Top photo by Glenn Constant, US Fish and Wildlife Service, Baton Rouge, Louisiana. Bottom photo by Seth Blitch © The Nature Conservancy.

FIG. 3.12. Aerial (top) and ground-level (bottom) views of black water in flooded baldcypress-water tupelo swamp forest in the ARB floodway. This clear, slow-moving or stagnant water looks black due to the lack of suspended sediment, which has settled out of the water column, and an abundance of organic tannins. Decomposers rapidly deplete the water of oxygen, and when dissolved oxygen concentrations decrease, the areas become hypoxic. Aquatic productivity in black-water habitats is low. Top photo by Charles Demas, US Geological Survey Louisiana Water Science Center, Baton Rouge, Louisiana. Bottom photo by Richard Martin, ©The Nature Conservancy.

FIG. 3.13. Black water draining from the swamp and mixing with brown water into Bayou Sorrel, a distributary, in the ARB floodway. The distribution of black-, brown-, and green-water habitats are spatially and temporally variable and are driven by multi-scale factors that control the input and movement of water inside the floodway. Photo by Richard Martin © The Nature Conservancy.

Inside the East and West Atchafalaya Protection Levees, the initiation, spatial extent, and persistence of hypoxia are variable and correlated to river stage (which is directly related to river discharge) and mean water temperature (Bryan et al. 1998, Sabo et al. 1999a, b). Hypoxia in the ARB floodway generally occurs when rising water associated with the spring flood extends out into the swamp. About half of the swamp floor in the floodway is inundated by rising water when the Atchafalaya River reaches 3 m (10 ft) at Butte La Rose (USGS 07381515), and by the time river stage reaches 4 m (13 ft), over 80% of the swamp floor is inundated (Fig. 3.15). As water temperatures increase (> 16.0 °C [61 °F]), increased biological oxygen demand causes the water to become critically hypoxic. This degradation of local water quality may also threaten water quality in adjoining bayous and swamps if they are in the receiving path of the hypoxic water as river levels fall through the summer and the swamp drains (Atchafalaya River stage < 1.4 m [4.6 ft] at Butte La Rose; Bryan et al. 1998, Sabo et al. 1999a, b). Spatially extensive hypoxia that can last for weeks to months occurs when flood pulses (over 3 m at Butte La Rose) arrive in late spring or early summer and persist into the late summer or early fall causing hypoxia throughout most of the lower ARB floodway (Sabo

FIG. 3.14. An extensive fish kill occurred in the ARB floodway immediately following passage of Hurricane Gustav. This fish kill was caused by acute hypoxia caused when large amounts of leaves were stripped from trees, introduced into water bodies, and then mixed into the water column along with large amounts of resuspended organic matter. The resulting biological oxygen demand rapidly depleted oxygen levels to a point where fish died in massive numbers, as evidenced by the fish floating in black water. Photo by A. Raynie Harlan, Louisiana State University.

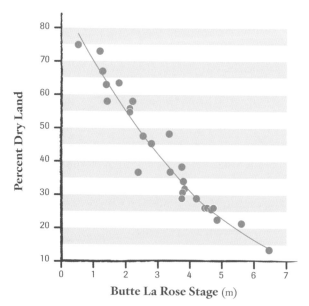

FIG. 3.15. Hydrograph showing the relationship between Atchafalaya River stage at Butte La Rose, Louisiana (USGS 07381515) and floodplain inundation in the Atchafalaya Basin Floodway System, ARB floodway. Figure redrawn from an original provided by Yvonne C. Allen, US Army Engineer Research and Development Center, Baton Rouge, Louisiana.

et al. 1999a, b). Therefore, development of hypoxic zones within the interior swamps and water bodies of the floodway appear to be related solely to the interaction between periods of elevated water stages, poor drainage conditions, and elevated water temperatures. No relationship has been shown between hypoxia and wind patterns, rainfall, or tidal conditions (Sabo et al. 1999b). Outside the East and West Atchafalaya Protection Levees, large areas of chronic hypoxia generally appear when elevated water stages resulting from precipitation and runoff cannot drain. Like the floodway, hypoxia is especially persistent in late summer and fall when water temperatures are highest, and exist until the swamp floor drains.

It is important to understand that high temperatures or high water levels alone do not cause hypoxia in this system. Rather, hypoxia results from a pattern of flood timing and duration that causes an extended interaction between high water conditions (swamps are connected to lakes, bayous, and canals) and high water temperatures. In fact, high water temperatures are also associated with the highest dissolved oxygen values in open water bodies during late summer when river levels are low and quiescent clear bayous and lakes become colonized by phytoplankton that produce large amounts of oxygen, often supersaturating the water column and creating highly productive habitats for aquatic animals (Bryan et al. 1998, Sabo et al. 1999a, b, Davidson et al. 1998, 2000, Rutherford et al. 2001).

DELTA MARSHES. The lower region of the ARB is the coastal plain. It extends from about Morgan City to the Gulf of Mexico, and the habitat in this region transitions from cypress-tupelo swamp to freshwater marsh and then on to delta marsh environments (Atchafalaya and Wax Lake deltas) and open water of Atchafalaya Bay (Fig. 3.16).

River discharge to Atchafalaya Bay occurs through the lower Atchafalaya River and the Wax Lake Outlet (Fig. 3.17). Annual maintenance dredging keeps the lower Atchafalaya River channel, below Morgan City, 6 m deep and 122 m wide (20 ft deep, 400 ft wide) to ensure that the channel remains navigable to vessels that support offshore oil and gas development (US Army Corps of Engineers 2010). This navigation channel bisects the Atchafalaya Delta, extending 30 km (~19 mi) from the head of the delta to the continental shelf (Holm and Sasser 2001). The Wax Lake Outlet was constructed between Six Mile Lake and Atchafalaya Bay in 1941 to provide a second route for Atchafalaya discharge (hydrologically connected via Six Mile Lake) to reach the Gulf of Mexico and reduce flood risk at Morgan City (Reuss 1982, 2004).

The Wax Lake Outlet was originally engineered to transport a limited amount (about 20–30%) of the total flow of the lower Atchafalaya River (Holm and Sasser 2001). This was accomplished both through channel design (dimensions were constrained) and a fixed-crest weir that was built in 1988. Flows down the Wax Lake Outlet were limited by the weir for the purpose of promoting increased flows and self-scouring within the main Atchafalaya River channel to reduce the need for dredging (Holm and Sasser 2001). A corresponding set of bank stabilization levees was also built along the main river channel from Thibodaux Chute downriver to American Pass. These levees prevented the river from overtopping due to the increased stage from the weir, which aided river-bottom scouring. The weir was removed in 1994, and this created a

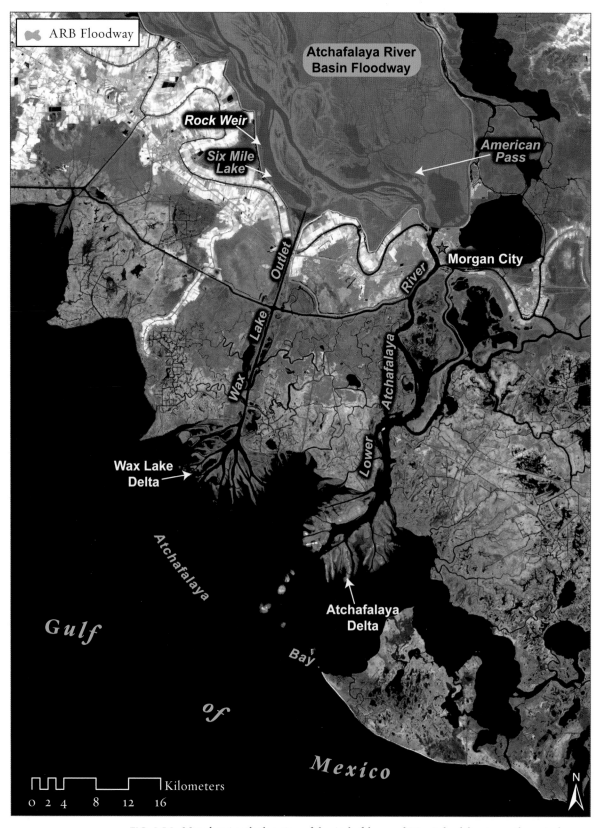

FIG. 3.16. Map showing the location of the Atchafalaya and Wax Lake deltas in south-central Louisiana.

FIG. 3.17. The lower Atchafalaya River south of Morgan City, Louisiana (in background) discharges to Atchafalaya Bay. Because this channel is important to navigation between Morgan City and the Gulf of Mexico to support offshore oil and gas development, annual maintenance dredging is used to keep the lower river channel at a depth of 6 m and 122 m wide (20 x 400 ft). Photo © Alison M. Jones for www.nowater-nolife.org.

concomitant increase in hydraulic efficiency and potential flow to approximately 45% of the total flow of the lower Atchafalaya River (Holm and Sasser 2001, Wellner et al. 2005, Allison et al. 2012). Discharge measurements from both the lower Atchafalaya River and the Wax Lake Outlet between July 1997 and January 1998 showed that the lower Atchafalaya River carried 59% (2,478 m³ s⁻¹ [87,509 cfs]) and the Wax Lake Outlet carried 41% (1,744 m³ s⁻¹ [61,589 cfs]) of the average combined flow of both channels (Cobb et al. 2008a). These measurements agree with discharge measurements made during fall 1996 when the Wax Lake Outlet was responsible for discharging 38–40% of the combined discharge (Holm and Sasser 2001). Recent measurements made during flood cycles in 2008–2010 showed that flow was divided with ~54% being discharged by the lower Atchafalaya River and the remainder (~46%) by the Wax Lake Outlet (Allison et al. 2012). While the weir was removed and flow diminished in the main channel, the bank stabilization levees, which were no longer necessary, were not removed and still remain on the landscape. The combination of reduced flow volume and the remaining stabilization levees has resulted in a significant decrease in the overbank flooding and sheet flow of river water across the adjacent swamps in this area (Alford

FIG. 3.18. The Wax Lake Delta during a spring flood pulse. Atchafalaya River water is transported onto the delta through the Wax Lake Outlet (shown at top) where it is disbursed by numerous delta channels. Natural delta islands are inundated with sediment-rich water and continue to build seaward. Photo by Azure E. Bevington, Louisiana State University.

and Walker 2011). Numerous primary and secondary distributary channels originating from the lower Atchafalaya River and Wax Lake Outlet distribute fresh water and sediment to the deltas, creating a mosaic of wetland environments including natural (delta splay) and created (dredge spoil) islands, tidally influenced backmarsh, and open bay habitat (Fig. 3.18).

Anthropogenic alteration of upstream flows from the lower Atchafalaya River and Wax Lake Outlet has affected the evolution of islands and habitats of the Atchafalaya and Wax Lake deltas, since their emergence after the 1973 flood (Roberts et al. 1997). The presence and continued maintenance dredging of the lower Atchafalaya River navigation channel has increased elevation and decreased channel complexity in the Atchafalaya Delta. The navigation channel has allowed sediment to bypass the shallow delta habitats and continue to the continental shelf. The fixed-crest weir that was constructed at the head of the Wax Lake Outlet in 1988 restricted water and sediment discharge to the Wax Lake Delta until 1994 when it was removed. Consequently, in 1994, the Atchafalaya Delta was about twice as large as the Wax Lake delta (101.5 vs. 51.1 km²). Although the Wax Lake Delta has formed at the mouth of a man-made channel, it has

FIG. 3.19. Closer view (looking seaward) of a natural delta island on the Wax Lake Delta during a spring flood pulse. Natural islands are v-shaped with highest elevations on the landward tip, along the island periphery and along smaller distributary channels where sand is deposited and creates natural ridges. These areas are colonized by black willow (*Salix nigra*) with an understory of elephant ear (*Colocasia esculenta*). Backmarsh exists behind the ridges where elevation progressively declines seaward. These areas are colonized by a number of plant species in elevational zones that correspond to each species' flood tolerance. Atchafalaya Bay can be seen in the background along the top of the photo. Photo by Azure E. Bevington, Louisiana State University.

developed a classic fan-shape with minimal anthropogenic disturbance to the delta proper. In contrast, the Atchafalaya Delta has developed as an engineered landscape with artificial channels and islands created from dredge material disposal.

Natural islands on both deltas are typically chevron-shaped, and vegetation patterns are structured by subtle changes in elevation and flooding, which affects the amount of flood stress plants endure (Fig. 3.19; Johnson et al. 1985, Shaffer et al. 1992, Piazza and Wright 2004). The upstream tip of each island is colonized by stands of black willow that reach heights of 15 m (49 ft) and contain a dense understory of the nonnative elephant ear (also known as wild taro or coco yam, *Colocasia esculenta*) that reach heights of more than 2 m (6.5 ft) during the summer, as well as rice cutgrass (*Leersia oryzoides*), climbing hempweed (*Mikania scandens*), and smartweed (*Polygonum punctatum*).

Immediately downstream of the willows, elevation decreases, and the habitat transitions to tidally influenced backmarsh habitat. Higher elevations support small willows, chairmaker's bulrush (*Schoenoplectus americanus*), softstem bulrush (*Schoenoplectus*

FIG. 3.20. An island created on the Atchafalaya Delta with dredged material excavated from the lower Atchafalaya River to maintain a channel depth and width necessary for navigation. These islands have a slightly different shape, higher elevation (at least initially), less backmarsh habitat, and less channel complexity than natural splay islands. However, these islands provide important wildlife habitat, especially for nesting seabirds and represent a beneficial use of dredged sediments. Photo by Cassidy R. Lejeune, Louisiana Department of Wildlife and Fisheries.

tabernaemontani), chickenspike (*Sphenoclea zeylandica*), broadleaf arrowhead (*Sagittaria latifolia*), purple ammannia (*Ammannia coccinea*) and looseflower water willow (*Justicia ovata*). Small, dense stands of southern cattail (*Typha domingensis*) also occur throughout this zone. The lowest elevations are dominated by broadleaf arrowhead and delta arrowhead (*S. platyphylla;* Johnson et al. 1985, Sasser and Fuller 1988, Shaffer et al. 1992, Evers et al. 1998, Holm and Sasser 2001, Castellanos and Rozas 2001).

Submerged aquatic vegetation is found in low intertidal and subtidal areas and is dominated by longleaf pondweed (*Potamogeton nodosus*) and southern waternymph (*Najas guadalupensis*). Other submerged vegetation species, including wild celery (*Vallisneria americana*) and water stargrass (*Heteranthera dubia*) are less abundant and their distributions are patchy (Castellanos and Rozas 2001). These vegetation communities also exist in island stream sides along secondary distributary channels.

Created islands on the Atchafalaya Delta are typically not chevron-shaped (Fig. 3.20). These islands are created from dredge material excavated from the navigation channel. New islands or new amendments to islands are bare sand that is quickly colonized by herbaceous vegetation that is tolerant of higher elevation such as panic grass (*Panicum* spp.). Older islands or older portions of islands contain stands of black willow with an elephant ear understory, as well as some bottomland hardwood species on the highest and oldest created islands (e.g., Big Island). At least initially, created islands

FIG. 3.21. Large stand of broadleaf arrowhead (*Sagittaria latifolia*) on an island in the Wax Lake delta. Broadleaf arrowhead is found in the tidally influenced backmarsh habitat and is an important food for wintering waterfowl. Photo by Guerry O. Holm Jr., Louisiana State University.

are typically higher elevation so they do not typically contain as much backmarsh habitat as natural splay islands. However, island stream sides are colonized by chairmaker's bulrush and submerged aquatic vegetation (Castellanos and Rozas 2001).

Although plant community composition and structure on the Atchafalaya and Wax Lake deltas tended to follow predictable successional trends governed by island elevation and flooding frequency, there have been some unexpected changes including divergence in habitat composition between the deltas. Specifically, broadleaf arrowhead, which historically comprised about 64% of the vegetated intertidal areas on delta islands since the deltas emerged (Johnson et al. 1985), experienced a substantial range contraction on the Atchafalaya Delta beginning in 1983 (Fig. 3.21). By 1986, this species only occupied 20% of the vegetated area of Atchafalaya Delta, resulting in large areas of exposed mudflats (Shaffer et al. 1992). This habitat transition has not been documented on the Wax Lake Delta and the differing conditions have been related to a combination of allogenic (river, tidal, meteorological forcing) and autogenic (herbivory, competition) factors (Shaffer et al. 1992, Sasser and Fuller 1988, Evers et al. 1998, Holm and Sasser 2001).

Broadleaf arrowhead began to decline on the Atchafalaya Delta in 1983, a year with an extremely large spring flood pulse (9 cm accretion across the delta; Shaffer et al.

FIG. 3.22. Nutria (*Myocastor coypus*), an exotic furbearing rodent, was introduced to Louisiana in the 1930s for fur farming. Native to South America (Argentina, Brazil, Bolivia, Chile, Paraguay, and Uruguay), these herbivorous semi-aquatic mammals were released into Louisiana's coastal wetlands and have caused extensive damage by eating vast amounts of native vegetation. In the Atchafalaya and Wax Lake deltas, nutria graze extensively on a number of plant species, including broadleaf arrowhead. Photo © Charles Bush (www.charlesbushphoto.com).

1992). During this large flood pulse, plants were likely subjected to protracted inundation with cold water, which retards plant growth, as well as smothering sedimentation. In 1983, the accretion rate (9 cm yr^{-1}) was nine times the accretion rate experienced between 1980–1986 (after adjusting for a delta-wide submergence rate of 0.85 cm yr^{-1}; Shaffer et al. 1992). This rapid influx of sediment likely made some areas more favorable for other higher elevation species to expand their areal coverage, decreasing the dominance of broadleaf arrowhead on the Atchafalaya Delta.

The decline in broadleaf arrowhead may be due to excessive herbivory by waterfowl and nutria (*Myocastor coypus*). Another common name for broadleaf arrowhead is duck potato, because the tubers of the species are heavily grazed by waterfowl. Together, the Atchafalaya and Wax Lake deltas are one of Louisiana's finest areas for wintering waterfowl. At any given time, from 10,000 to 20,000 ducks can be found on both deltas throughout the late fall and winter. The deltas are also important areas for wintering snow geese (*Chen caerulescens;* Larry Reynolds, Louisiana Department of Wildlife and Fisheries, personal communication). Waterfowl in both deltas numbered almost 150,000 in the winter of 1985-86, but despite having four times the land area, more than half of the birds were counted in the Wax Lake Delta where broadleaf arrowhead was still abundant (Fuller et al. 1985, Evers et al. 1998). The early 1980s was also when nutria, an exotic furbearer, began to be commonly seen on the deltas (Fig. 3.22; Shaffer et al. 1992, Evers et al. 1998). An exclosure study on both the Atchafalaya and Wax Lake deltas showed three interesting patterns: 1) grazing by waterfowl and

TABLE 3.1. Comparison of the mean aboveground biomass on the Atchafalaya and Wax Lake deltas in relation to grazing pressure

| | Mean ± SE aboveground *Sagittaria latifolia* biomass (living stems; g m⁻²) | |
Exclosure Type	Atchafalaya Delta	Wax Lake Delta
Open Grazing (no exclosure)	148 ± 39	415 ± 44
Waterfowl-grazed	436 ± 55	330 ± 37
Nutria-grazed	450 ± 84	273 ± 21
Ungrazed (complete exclosure)	909 ± 99	446 ± 37

Biomass of *Sagittaria latifolia* was studied in relation to grazing pressure by waterfowl and nutria (*Myocastor coypus*), an exotic furbearer. Data are from Evers et al. (1998).

nutria reduced aboveground biomass of broadleaf arrowhead almost nine-fold, and belowground biomass 20–40%, compared to ungrazed plots; 2) waterfowl and nutria grazing pressure was almost equal; and 3) the Wax Lake Delta did not experience the grazing loss, despite its smaller size (Table 3.1; Evers et al. 1998). Thus, although it seems that waterfowl and nutria herbivory can potentially alter the distribution of broadleaf arrowhead, herbivory alone does not appear to explain the change in arrowhead distribution in Atchafalaya Delta, since abundance of the species remained relativly stable in Wax Lake Delta (Sasser and Fuller 1988, Shaffer et al. 1992, Evers et al. 1998, Holm and Sasser 2001).

Studies of Atchafalaya Bay have demonstrated salinity pulses during low flow conditions (below 70 m³ s⁻¹ [2,500 cfs]) that are driven by onshore winds (southerly and easterly) associated with tropical storms and the setup prior to cold front passage (Holm and Sasser 2001, Li et al. 2011). For example, from 1995 to 1998, 19 salinity pulses were recorded following 7 tropical events and 12 cold fronts. These salinity pulses were tracked into the Atchafalaya Delta, but did not penetrate into the Wax Lake Delta (Fig. 3.23). During these salinity pulses, peak salinity on the Atchafalaya Delta ranged from 1.2–7.2, and peak water levels ranged from 28–80 cm (11–31 in). Salinity pulses from tropical storms lasted 2–7 days, and pulses from cold fronts lasted 2–10 days. Greenhouse experiments conducted at the same time showed that within 13 days growth of broadleaf arrowhead was impaired by salinity of 6.0, and after 36 days, above- and belowground tissues were dead (Holm and Sasser 2001).

The fact that salinity incursions occur on the Atchafalaya Delta but not the Wax Lake Delta is attributed to three major factors (Holm and Sasser 2001). First, the lower Atchafalaya River navigation channel serves as a conduit for warm, high-salinity bottom water (range 11–22) that moves inland as a salt wedge and penetrates into the shallow, secondary channels of the Atchafalaya Delta during setup events prior to cold front passage or onshore winds during tropical storm events. Next, the Wax Lake Delta may be positioned in the freshwater plume of the Atchafalaya River during setup events

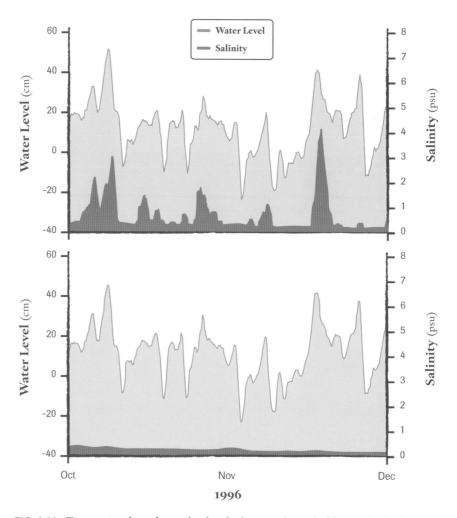

FIG. 3.23. Time-series plots of water level and salinity in the Atchafalaya Delta (top) and Wax Lake Delta (bottom) October–December 1996. Note that while water levels were identical in the two deltas, periods of increased salinity occurred in the Atchafalaya Delta but not in the Wax Lake Delta. Figure redrawn from Holm and Sasser (2001).

prior to cold front passage, inhibiting seawater incursions. Lastly, the natural channels that have developed on the Wax Lake Delta have well-developed sills at the mouth bars, and these serve as barriers to saltwater incursion.

COASTAL BAY. Atchafalaya Bay is microtidal, primarily characterized by diurnal tides that range up to 0.5 m (1.6 ft). Astronomical tides are frequently overridden, however, by both meteorological events and river discharge that force water into and out of coastal wetlands. Water temperatures in the bay range from less than 15 °C (59 °F) in the winter to over 25 °C (77 °F) during summer, and, during high discharge periods, water temperature in the bay is up to 10 °C (50 °F) cooler than surrounding bays (Hoese 1976, Castellanos and Rozas 2001). Freshwater inflow keeps salinities in the bay below 0.5 during most of the year (Orlando et al. 1993, Castellanos and Rozas 2001); however salinity can be elevated during low-flow periods (see above).

Water depths in Atchafalaya Bay reach 2 m (6.5 ft; Thompson 1951, 1955). Water is typically turbid, due to river discharge and resuspension. Because larger-grained sediments are stored on the delta islands, substrate in the bay is typically dominated by fine sediments (silts and clays; van Heerden and Roberts 1988). As late as the 1950s, the bay was ringed by three-dimensional oyster reefs (Thompson 1951, 1955, van Heerden and Roberts 1988, Stone et al. 2005). Today, due to freshwater discharge, intensive oyster harvest, and extensive shell dredging from 1914–1996, most of the historic natural reefs are gone (Louisiana Wildlife and Fisheries Commission 1968, Juneau 1984, Stone et al. 2005). However, some living reefs can still be found in the area around Point Au Fer and Marsh Island (Allison et al. 2000, Patrick D. Banks, Louisiana Department of Wildlife and Fisheries, personal communication).

PART TWO

WHY IS THE ATCHAFALAYA
RIVER BASIN IMPORTANT?

AGRICULTURE AND FOREST RESOURCES 4

WHILE MUCH OF THE ARB FLOODWAY is unsuitable for farming, limited agricultural production occurs at the extreme northern end of the floodway, in the West Atchafalaya (23,000 ha [58,000 acres]) and Morganza (7,200 ha [18,000 acres]) floodways (van Beek et al. 1979). South of the West Atchafalaya and Morganza floodways, there is about 5,260 ha (13,000 acres) of cultivated land and pasture in the floodway, but most of this occurs on the base of levees (US Geological Survey, unpublished data). Agricultural crops are also grown on the outside of the ARB levees within the historical boundaries of the ARB. Major crops in this region include sugarcane, rice, soybeans, and cotton, as well as livestock and crawfish (Fig. 4.1). The US Department of Agriculture reports the value of these agricultural products sold in the ARB parishes at almost $900 million annually, or about 45% of the state total (Louisiana Department of Natural Resources 2010).

The ARB contains extensive commercially valuable timber resources, although current timber harvest is low by historical standards. Forests in the ARB were extensively clearcut during heavy logging in the 1880s–1930s (Fig. 4.2). In fact, beginning in 1876, thousands of acres of virgin cypress forest in the ARB was purchased from the federal government for $0.25 per acre (Burns 1980). In these areas baldcypress was extensively harvested for timber. At the same time the trees were becoming easier to acquire and more valuable because of improved labor-saving and cost-cutting technologies (i.e., improved sawmill equipment, pullboat systems) and aggressive national marketing of baldcypress wood as the "wood eternal." By about 1932, the virgin baldcypress stands in the ARB were largely gone. Most forest present today is mostly composed of second-growth trees that naturally regenerated following that intensive period of harvest (Keim and King 2006). This is the case with much of the cypress-tupelo swamp forest in the ARB floodway.

Despite the historic and current importance and economic value of the forest resources in the ARB, there are little historical data that specifically document forest harvest inside basin boundaries. It is important to note that many bottomland forest tracts in the upper ARB are currently productive, commercial forests that are managed for

FIG. 4.1. Agriculture in the ARB, most of it outside the floodway levees, consists of a number of crops. In this photo, which was taken outside the floodway, sugarcane fields (background) are found on higher ground adjacent to rice fields, which are also managed to produce crawfish between rice crops. Photo ©Alison M. Jones for www.nowater-nolife.org.

FIG. 4.2. Baldcypress and water tupelo stumps in an ARB floodway lake. These giant stumps are all that remain of the extensive stands of virgin forest that were logged during the heavy lumbering era of the 1880s–1930s. These stands were purchased from the US government for $0.25 per acre. While some large cypress and tupelo remain, most of the current swamp forest in the ARB is composed of second-growth trees that regenerated naturally after the virgin stands were logged and the lumbering boom ended. Extensive lumbering of cypress-tupelo forest today would cause a loss of this forest type in the ARB, because the hydrologic conditions are not conducive to natural regeneration of baldcypress or water tupelo. Photo by Don McDowell © The Nature Conservancy.

Average Hardwood Sawtimber Harvest

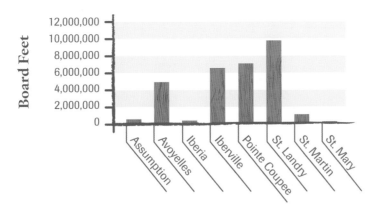

Average Hardwood Pulpwood Harvest

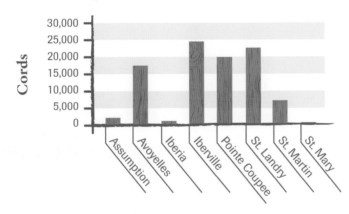

FIG. 4.3. Mean annual harvest of hardwood species (1980–2009) from Atchafalaya River Basin parishes. Shown are hardwood sawtimber (top) and pulpwood (bottom). Baldcypress is included in the hardwood tally although it is a conifer. Data are from the Louisiana Department of Agriculture and Forestry. Note that this figure is a summary of parish-wide harvest and does not show what portion of the harvest came from the ARB or the ARB floodway.

intensive wood production (Richard F. Keim, Louisiana State University Agricultural Center, personal communication). Forest severance tax data provide information on forest harvest within the eight ARB parishes (Assumption, Avoyelles, Iberia, Iberville, Pointe Coupee, St. Landry, St. Martin, St. Mary), although there is no efficient process to selectively examine data from harvests conducted in the ARB (Louisiana Department of Agriculture and Forestry 2011). From 1980–2009, an average of 3.8 million board feet (232,683–9,898,432 board feet) per year of hardwood (including baldcypress, which is actually a conifer species) sawtimber has been extracted in the eight basin parishes, with an average severance value of $651 million ($19,412–$1,583,564; Fig. 4.3; data provided in Appendix 1). Over the same time period, an annual average of 12,000 cords (469–24,751 cords) of hardwood pulpwood was harvested in the basin parishes, with an average annual severance value of $140 million ($6,011– $310,334; Fig. 4.3; data provided in Appendix 1). The parishes with the highest mean annual sawtimber harvest over the 30-year time series were St. Landry (9,898,432 board feet; $1,583,564) and Pointe Coupee (7,098,688 board feet; $1,285,407). Mean pulpwood harvest was highest in Iberville (24,751 cords; $310,334) and St. Landry (22,994 cords; $240,317).

There have been anecdotal reports of current and extensive clearcut logging of baldcypress to provide material for cypress garden mulch in the ARB, as well as many of Louisiana's coastal cypress forests. Concern over the perceived misuse of these forests caused large wholesale buyers of mulch, most notably Walmart, Lowe's, and Home Depot, to restrict their purchases of mulch sourced from Louisiana. However, due to the scale of the data, it has been impossible to determine the contribution of baldcypress harvested from the ARB to the mulch industry.

5

The ARB, like many large river systems worldwide, contains extensive fish and wildlife resources (Appendix 2). These resources are vital to local, regional, and global biodiversity and are also important cultural and economic drivers (The Nature Conservancy 2002). The Louisiana Natural Heritage Program has documented 238 unique records of 17 plant and animal species of conservation concern (7 distinct natural plant communities, 5 species of plants, 7 species of birds, 2 species of mammals, and 3 species of fish; Table 5.1).

NEKTON RESOURCES. More than 100 species of nekton (finfish and crustaceans) and mussels have been documented in the ARB (Appendix 2), and despite its current management as a floodway, the ARB floodway supports substantial fisheries resources. Biomass estimates of standing stock for fish in the floodway range from 25,500–208,000 kg km^{-2} (0.005–0.05 lbs ft^{-2}), producing an estimated harvest biomass of commercial species of around 20 million kg (44 million lbs; Lambou 1959, Bryan and Sabins 1979, Fremling et al. 1989, Lambou 1990, Bryan et al. 1998, Sabo et al. 1999b).

The tremendous fisheries production in the ARB is attributed to the interaction between the river and the floodplain that results from seasonal flooding (Fig. 5.1). As in the Mississippi and other rivers worldwide, riverine flooding in the ARB floodway provides nekton access to ephemeral habitats that are rich in food for spawning, feeding, and rapid growth of young fish and crustaceans (Pollard et al. 1983, Junk et al. 1989, Odum et al. 1995, Odum 2000, Piazza and LaPeyre 2007, 2010, Alford and Walker 2011 and others). These aggregated consumers then become available for trophic transfer throughout the food web (Fig. 5.2). In one study in Louisiana coastal wetlands, aggregated resident nekton consumers (mean 40 ± 7.2 ind m^{-2}, 1.3 ± 0.2 g DW m^{-2}) that assembled during seasonal flood pulses of the Mississippi River represented an available stock of transferrable energy of $8,164 \pm 1,490$ g m^{-2}, and over 60% of the density, biomass, and energy available was attributed to the seasonal flood pulse (Piazza and LaPeyre 2012).

Because riverine connectivity is so important to nekton production in the ARB floodway, variability in connectivity can have marked effects on species recruitment

TABLE 5.1. Louisiana Natural Heritage Program element occurrence records for the Atchafalaya River Basin, Louisiana

Common Name	Scientific Name	No. of Records	Federal Status[1]	State Status[2]	Global Ranking[3]	State Ranking[4]
Natural Communities						
Vegetated Pioneer Emerging Delta and Deltaic herbaceous vegetation	*Sagittaria latifolia, Sagittaria platyphlla, Colocasia esculenta*	1	-	-	G3,G4	S2,S3
Freshwater Marsh	-	2	-	-	G3,G4	S1,S2
Cypress-tupelo Swamp	-	7	-	-	G3,G5	S4
Cypress Swamp	-	2	-	-	G4,G5	S4
Bottomland Hardwood Forest	-	2	-	-	G4,G5	S4
Live Oak Forest	-	1	-	-	G2	S1,S2
Salt Dome Hardwood Forest	-	1	-	-	G1	S1
Waterbird Nesting Colony	-	89	-	-	GNR	SNR
Plants						
Common water-willow	*Justicia americana*	1			G5	S3
Square-stemmed monkey-flower	*Mimulus ringens*	1			G5	S2
Rooted spike-rush	*Eleocharis radicans*	1			G5	S1?
Low erythrodes	*Platythelys querceticola*	1			G3,G5	S1
Floating antler-fern	*Ceratopteris pteridoides*	1			G5?	S2
Birds			-	-		
Roseate spoonbill	*Platalea ajaja*	10	-	-	G5	S3
Osprey	*Pandion haliaetus*	6	-	-	G5	S2B,S3N
American swallow-tailed kite	*Elanoides forficatus*	11	-	-	G5	S1S2B
Bald eagle	*Haliaeetus leucocephalus*	83	Delisted	Endangered	G5	S2N,S3B
Snowy plover	*Charadrius alexandrinus*	1		-	G4	S1B,S2N
Piping plover	*Charadrius melodus*	2	LT	Threatened/ Endangered	G3	S2N
Gull-billed tern	*Gelochilidon nilotica*	8			G5	S2B,S2S3N
Mammals						
Louisiana black bear	*Ursus americanus luteolus*	1	LT	Threatened	G5,T2	S2
Manatee	*Trichechus manatus*	1	LE	Endangered	G2	SNA
Fish						
Pallid sturgeon	*Scaphirhynchus albus*	2	LE	Endangered	G1	S1
Paddlefish	*Polyodon spathula*	2		Prohibited	G4	S3
Blue sucker	*Cycleptus elongatus*	1			G3,G4	S2,S3

Examples of exemplary natural communities as well as global, federal, and state status and rankings for rare, threatened and endangered species are listed (Louisiana Natural Heritage Program 2009).

[1]LE = Listed Endangered - Species for which a final rule has been published in the Federal Register to list the species as endangered. LT = Listed Threatened - Species for which a final rule has been published in the Federal Register to list the species as threatened. Both LE and LT species are legally protected by the Endangered Species Act. (*cont.*)

and growth, and hence, overall nekton production (Fontenot et al. 2001, Engel 2003, Alford and Walker 2011). For example, annual relative abundance of largemouth bass, crappie (*Pomoxis* spp.), blue catfish (*Ictalurus furcatus*), and buffalo (*Ictiobus* spp.) are optimized when the Atchafalaya River is in flood stage for 124–157 days per year (Table 5.2; Fig. 5.3; Alford and Walker 2011). Likewise, monthly crawfish catch (sacks boat-month^{-1}) was positively related (α = 0.10) to river stage during 1993–1996 (R^2 = 0.44, $P < 0.001$) and 1999–2009 (R^2 = 0.29, $P = 0.09$) (Alford and Walker 2011).

While nekton productivity can be mostly attributed to rich detrital food sources and benthos on the floodplain, there are some species, like gizzard shad (*Dorosoma cepedianum*), that are not dependent on the connectivity between the river and floodplain (Table 5.2; Halloran 2010, Alford and Walker 2011). For these species, availability of zooplankton, rather than seasonal flooding, is critical for recruitment and growth. Open water habitats in the ARB can also support large seasonal populations of plankton that drive a planktonic food web (Hern et al. 1978, Sager and Bryan 1981, Pollard et al. 1983, Gelwicks 1996, Davidson et al. 1998, 2000, Rutherford et al. 2001).

Phytoplankton are (mostly) single-celled plants that float in the upper water column. Phytoplankton release oxygen as a product of photosynthesis and because they are often found in extremely high densities, phytoplankton are considered critical to the aquatic food web and maintenance of water quality. Studies in the ARB floodway have documented up to 107 genera of phytoplankton within five divisions (Euglenophyta–flagellates; Pyrrophyta–dinoflagellates; Chrysophyta–diatoms; Cyanophyta–blue-green algae; Chlorophyta–green algae; Hern et al. 1978, Sager and Bryan 1981). Nine genera comprised the majority of sampled phytoplankton, and, of those genera, 4 were diatoms (*Cyclotella, Melosira, Navicula, Cocconeis*), 2 were blue-green algae (*Anacystis, Anabaena*), 2 were green algae (*Scenedesmus, Crucigenia*), and 1 was a flagellate (*Euglena*). Phytoplankton density for most divisions (except diatoms) was generally greatest in lakes and swamps and least in rivers and distributaries and was positively related to calm sediment-free water conditions in warm water that was low in nutrients (Sager and Bryan 1981). It is important to note that blue-green algae are no longer

CONTINUED NOTES TO TABLE 5.1

[2]Endangered = Taking or harassment of these species is a violation of state and federal laws. Threatened = Taking or harassment of these species is a violation of state and federal laws. Threatened/Endangered = Taking or harassment of these species is a violation of state and federal laws. Prohibited = Possession of these species is prohibited. No legal harvest or possession.

[3]G1 = critically imperiled globally because of extreme rarity (5 or fewer known extant populations) or because of some factor(s) making it especially vulnerable to extinction. G2 = imperiled globally because of rarity (6 to 20 known extant populations) or because of some factor(s) making it very vulnerable to extinction throughout its range. G3 = either very rare and local throughout its range or found locally (even abundantly at some of its locations) in a restricted range (e.g., a single physiographic region) or because of other factors making it vulnerable to extinction throughout its range (21 to 100 known extant populations). G4 = apparently secure globally, though it may be quite rare in parts of its range, especially at the periphery (100 to 1000 known extant populations). G5 = demonstrably secure globally, although it may be quite rare in parts of its range, especially at the periphery (1,000+ known extant populations).

[4]S1 = critically imperiled in Louisiana because of extreme rarity (5 or fewer known extant populations) or because of some factor(s) making it especially vulnerable to extirpation. S2 = imperiled in Louisiana because of rarity (6 to 20 known extant populations) or because of some factor(s) making it very vulnerable to extirpation. S3 = rare and local throughout the state or found locally (even abundantly at some of its locations) in a restricted region of the state, or because of other factors making it vulnerable to extirpation (21 to 100 known extant populations). S4 = apparently secure in Louisiana with many occurrences (100 to 1,000 known extant populations). S5 = demonstrably secure in Louisiana (1,000+ known extant populations).

FIG. 5.1. The spring flood pulse inundating a cypress-tupelo swamp forest with river water. This newly available floodplain habitat and its rich food resources are rapidly used by nekton for spawning, feeding, and rapid growth of juvenile fish and crustaceans. Aggregated consumers will use this ephemeral habitat until the water recedes or becomes inhospitable due to declining water quality. Photo © CC Lockwood.

classified as plants, as they were at the time of these studies. Rather, they are currently classified as cyanobacteria (Order Chroococcales), bacteria that obtain energy through photosynthesis.

Phytoplankton succession in the ARB floodway is driven by the annual flooding cycle, with lowest densities during winter-spring floods as cold, nutrient-rich water inundates lakes and floodplains. After the flood pulse, as water levels recede, phytoplankton density in lakes and swamps increases with rising water temperature and decreasing nutrient levels until density peaks during the low water months between June and December (Hern et al. 1978, Sager and Bryan 1981). Phytoplankton production during these times is vital to surface water oxygenation, because bacterial respiration rates also peak in these warm swamps. Diatoms, the most abundant phytoplankton division in the ARB floodway, are the exception, because they dominate the community in all habitats and throughout the year (Hern et al. 1978, Sager and Bryan 1981). It is important to note, however, that while phytoplankton production can be locally and temporally critical to maintenance and provision of habitat in the ARB floodway dur-

FIG. 5.2. A green heron (*Butorides virescens*) eating a sunfish on the floodplain. Riverine flooding in the ARB floodway provides access to floodplain habitats that are rich in food for spawning, feeding, and rapid growth of young fish and crustaceans. These aggregated consumers then become available for trophic transfer throughout the food web. Photo © Patti Ardoin.

ing low water seasons, phytoplankton play an overall minor role in the productivity of the basin due to the high concentrations of suspended sediments in the water column (Hern et al. 1978, Sager and Bryan 1981). In areas of the ARB outside the floodway levees (i.e., Lake Verret), where riverine processes have been severed, phytoplankton densities were about an order of magnitude greater than inside the floodway (Hern et al. 1978).

Freshwater zooplankton are composed of protozoans, rotifers, and two groups of crustaceans—cladocerans and copepods. Some zooplankton, such as planktonic protozoans, have limited locomotion, but the rotifers and crustacean zooplankton are capable of some movement through the water column. Zooplankton can be herbivorous (feed on phytoplankton) or carnivorous (prey upon other zooplankton) and are extremely important prey items for planktivorous fishes (e.g., gizzard shad, skipjack herring [*Alosa chrysochloris*], white crappie [*Pomoxis annularis*]) and young-of-the-year largemouth bass in the ARB (Levine 1977, Pollard et al. 1983, Mason 2002).

Three studies have documented the spatial-temporal distribution on zooplankton in the ARB. The first of these showed that rotifers and copepods comprised the majority of the zooplankton in both permanent (canals and bayous) and ephemeral (leading edge of overflow habitat) water bodies. Permanent water bodies contained zooplankton throughout the year; however peak monthly zooplankton counts always occurred in ephemeral habitats. Abundance of cladocerans and rotifers peaked in April, and abundance of copepods and protozoans peaked in June. Rotifer populations were

TABLE 5.2. Relationship between flooding duration and annual relative abundance of finfish

Species	Flood Duration (days yr^{-1})	Fit	
		R^2	P (α = 0.10)
Largemouth bass *Micropterus salmoides*	124	0.31	0.08
Crappie *Pomoxis* spp.	127	0.60	< 0.001
Blue catfish *Ictalurus furcatus*	157	0.71	0.002
Buffalo *Ictiobus* spp.	125	0.35	0.14
Gizzard shad *Dorosoma cepedianum*	10	0.27	0.03

Study included commercially and recreationally important finfish species in the ARB floodway (from data in Alford and Walker 2011).

dominated by species in the family Brachionidae, and protozoans were dominated by rhizopods and ciliates. Cladoceran populations were dominated by different families during different seasons. Chydoridae dominated during winter, Bosminidae and Daphnidae dominated during spring and early summer, and Sididiae dominated during early fall (Pollard et al. 1983).

The second and third studies extensively documented the spatial-temporal distribution and community characteristics of crustacean zooplankton in the ARB swamp. A total of 59 crustacean zooplankton species (31 cladocerans, 28 copepods) were documented, of which 24 are considered common species. Zooplankton density ranged from 700–6,700 cladocerans and 570–4,700 copepods m^{-3}, and densities were highest in small, slow bayous in late summer. In fact, these habitats (green-water habitats) supported more than 10 times the density of zooplankton found in either distributary (high flow, turbid, brown-water habitats) waterways or in low oxygen overflow habitats (black-water habitats), and this was largely due to the seasonally high density of phytoplankton in green-water habitats (Davidson et al. 1998, 2000).

South of the floodway, 101 species of nekton have been documented in waterways and flooded tidal marsh habitats on the Atchafalaya and Wax Lake deltas (Castellanos and Rozas 2001, Thompson and Peterson 2001, 2003, 2006, Peterson et al. 2008). High rates of nekton productivity on the deltas are driven by seasonal riverine flooding that inundates large areas of marsh habitat, and the abundance of submerged aquatic vegetation. Flooded marsh and areas with submerged vegetation both provide critical areas of high-quality habitat for resident species and optimum nursery conditions for juvenile marine transient nekton (Fig. 5.4; Thompson and Deegan 1983, Castellanos and

FIG. 5.3. Black crappie (top) and black buffalo (bottom) are among a number of fish species that rely on seasonal flood pulses in the Atchafalaya River. Abundance of both species is optimized with flood duration of between 124 and 157 days. Photos © Eric Engbretson, Engbretson Underwater Photography (www.underwaterfishphotos.com).

Rozas 2001, Thompson and Peterson 2001, 2003, 2006, Kanouse et al. 2006, Piazza and La Peyre 2007, 2009, 2011, Peterson et al. 2008). Abundant nekton with high rates of growth have also been documented in submerged aquatic vegetation within marsh ponds at Marsh Island (Weaver and Holloway 1974, Kanouse 2003, Kanouse et al. 2006), and the Cote Blanche and Vermilion bay systems yield large numbers of impor-

FIG. 5.4. Stands of submerged aquatic vegetation are found in low intertidal and subtidal areas of the Atchafalaya and Wax Lake deltas. These areas include stream sides, secondary distributaries, and shallow backmarsh ponds on delta islands (shown in photo). Submerged aquatic vegetation provides critical nursery habitat for many nekton species, and the abundance of this habitat type is one factor that drives the high nekton productivity on the deltas. Photo by Gary W. Peterson, Louisiana State University.

FIG. 5.5. Photograph of Cote Blanche Bay system, showing extensive areas of vegetated wetlands, bayous, and tidal creeks. Bay systems like this yield high abundance of important estuarine and marine nekton species. The vegetated wetland habitat provides critical nursery habitat for survival and rapid growth of juvenile fish and crustaceans. Photo © Alison M. Jones for www.nowater -nolife.org.

FIG. 5.6. Energy flow through crawfish in the Atchafalaya River Basin. As polytrophic consumers and popular prey items among many of the basin's animal residents, crawfish are major transformers and movers of energy through the food web. Arrows point in the direction of energy flow. Figure modified from Bonvillain (2012).

tant estuarine forage fish (e.g, Atlantic croaker *Micropogon undulatus*) and commercial fish and invertebrate species, including brown shrimp (*Farfantepenaeus aztecus*; Fig. 5.5; Perret and Caillouet 1974, Baltz and Jones 2003).

CRAWFISH. Usage of the term crawfish vs. crayfish has a long, storied debate going back to the 1800s. Both terms are scientifically and commonly used, and usage differences are mainly a result of geography and scientific preference (Walls 2009). Because the Cajun culture and especially the Atchafalaya River Basin are synonymous with crawfish, this term will be used in this book.

Crawfishes are a diverse group of decapod crustaceans that contain more than 600 species worldwide (2 families—Astacidae and Camabaridae). More than 75% of crawfish species are found in North America and more than 300 species are endemic to the southeastern United States. In Louisiana there are 39 species and subspecies of crawfish, and 10 of those are found in the ARB (Walls 2009, Louisiana State University Agricultural Center 2011, Bonvillain 2012).

Crawfish are important ecological components of wetland systems. As highly opportunistic polytrophic consumers, crawfish are major transformers and transporters of energy through wetland food webs (Fig. 5.6). Their diet consists of a mixture of plant material, periphyton, invertebrates, and fish, and, as part-time detritivores,

FIG. 5.7. Yellow-crowned night-heron (*Nyctanassa violacea*) eating a red swamp crawfish (*Procambarus clarkii*) on the floodplain. Crawfish are eaten by many animals that inhabit the ARB and other river systems around the world. Because crawfish are intimately tied to the floodplain for survival and recruitment, seasonal flooding is critical to maintaining healthy crawfish populations in the ARB floodway. Photo © Patti Ardoin.

FIG. 5.8. The red swamp crawfish (*Procambarus clarkii*) is enjoyed as a protein food source for people and a foundation of the Cajun culture. Crawfish boils are common and enjoyed by Louisiana residents and tourists alike. Over 90% of the wild-caught crawfish in the marketplace are harvested in the ARB floodway. Photo © Erik Canning (www.escandesign.com).

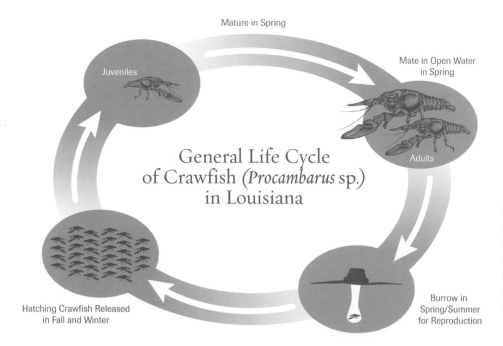

Mature in Spring

Juveniles

Mate in Open Water
in Spring

General Life Cycle
of Crawfish (*Procambarus* sp.)
in Louisiana

Adults

Hatching Crawfish Released
in Fall and Winter

Burrow in
Spring/Summer
for Reproduction

FIG. 5.9. The life cycle of crawfish in the ARB. Crawfish life history is dependent on water flow characteristics for survival and recruitment. Figure adapted from McClain et al. (2007).

crawfish shred decomposing material into finer particles that are more easily assimilated by other decomposers (i.e., bacteria, mollusks, worms, small crustaceans), which in turn can be eaten by crawfish and other consumers. Crawfish are also important prey items for fish, reptiles, amphibians, birds, and mammals, thus transporting energy into higher trophic levels, including humans (Fig. 5.7; Lambou 1961, Momot et al. 1978, Chabreck et al. 1982, Pollard et al. 1983, Dellenbarger and Luzar 1988, Bonvillain 2012).

Arguably, crawfish are the most popular crustaceans in Louisiana, and the crawfish was officially designated as the state crustacean in 1983. The red swamp crawfish (*Procambarus clarkii*) has become the cultural icon of the Cajun culture and south Louisiana, which has been designated the crawfish capital of the world. This species also serves as a major cultural foundation for tourists worldwide. Widely popular as a protein food source, especially for crawfish boils, the red swamp crawfish and white river crawfish (*Procambarus zonangulus*) are harvested extensively across southern Louisiana, especially in the ARB, thus marking a transfer of energy from the ARB (and the entire Mississippi River system) to humans across Louisiana, the United States, and the world (Fig. 5.8).

Crawfish life history is dependent on water flow characteristics (Penn 1943, 1956, Pollard et al. 1983, McClain and Romaire 2007, McClain et al. 2007, Bonvillain 2012). Studies of red swamp and white river crawfish show a generalized life cycle where mature males and females mate in open water during the spring in shallow, open water. After copulation (often with multiple males), females store sperm in a seminal receptacle on the underside of the thorax and then retreat to a burrow to spawn (Fig. 5.9).

Crawfish build burrows in late spring and summer, when water levels fall, as a safe place to spawn and to protect against drying conditions on the surface. Burrows include a tunnel down into a larger compartment that needs to be deep enough to stay moist and humid during the driest times of the season. Extracted mud is stacked in a chimney above the opening, and a mud plug at the top seals the burrow. Once inside the burrow the female's eggs mature and are expelled through the oviducts, fertilized with stored sperm, and attached to the swimmerets on the underside of the tail with an adhesive substance called glair. After about three weeks, the eggs hatch. Hatchlings stay attached to the female through two molting cycles, after which they resemble an adult crawfish, release from the female, and begin to feed themselves.

As water levels increase in the late fall and winter, females and juveniles emerge from their burrows. Juveniles grow rapidly and undergo several additional molts into and through the spring. In the ARB, water level fluctuations are variable and correspond to winter/spring pulse events on the Mississippi River. In the event of early pulses, females may emerge from burrows with eggs still attached, and spawning occurs in open water. During years of late flood pulses, mortality of juveniles in burrows may increase, due to starvation. It is important to note that, while this life cycle is generally followed, red swamp crawfish can breed throughout the year, and females are capable of producing multiple broods.

Crawfish grow in short molting episodes (11 molts to reach maturity), with long intermolt periods in between. Growth rate of juvenile crawfish is dependent on a number of factors, including water temperature, population density, dissolved oxygen concentrations, and food quantity and quality. Crawfish generally reach market size (carapace length $\geq$ 37 mm [1.4 in]) in 3–5 months, but when conditions are optimum market size can be achieved in 7–9 weeks. After reaching sexual maturity, both male and female crawfish stop growing.

Population structure of crawfish in flooded swamps is composed of hold-over adults and juveniles from the preceding season and young-of-the-year juveniles. Population density depends mostly on brood survival, which is dictated by water level fluctuations. However, harvest pressure also can affect local population densities.

Juvenile crawfish undergo substantial predation in flooded swamps in the ARB. In one study, juvenile crawfish constituted up to 86% of the food items in stomachs of many fish and were the only food items recovered in the stomachs of spotted bass (*Micropterus punctulatus*; Lambou 1961). Crawfish are also a major food source for river otters (*Lutra canadensis*) in the ARB. For example, crawfish were found in the stomachs of 34% of sampled otters, and the abundance of crawfish may be crucial to restricting dispersal of otters outside of the ARB (Chabreck et al. 1982, Latch et al. 2008). Juvenile crawfish also make up the majority of crawfish harvested by commercial fishermen in the ARB.

RIVER SHRIMP. One aquatic species deserves special attention because it is such an important prey item for a large number of fish species in the ARB. The river shrimp (*Macrobrachium ohione*) occurs in river systems throughout the central and eastern United States and has been reported throughout the Mississippi River system, from as far north as the Missouri and Ohio Rivers to the lower Mississippi and Atchafalaya Rivers (Fig. 5.10; Truesdale and Mermilliod 1979, Conaway and Hrabik 1997, Barko and Hrabik 2004, Bauer and Delahoussaye 2008, Bauer 2011, Olivier et al., 2012). In

FIG. 5.10. The river shrimp (*Macrobrachium ohione*; top photo) is an important prey item for many fishes in the Atchafalaya River, and, as such, is a vital component of the riverine food web. The species exhibits sexual dimorphism, with females (shown on the right) being much larger than the male. River shrimp also make an extraordinary seasonal migration in the Atchafalaya River (bottom photo). Adults ride the spring flood pulse downstream and release hatching larvae into estuary waters of Atchafalaya Bay. Juvenile shrimp (shown above) molt several times until they are self-feeding and then move back upstream along the river edge during late summer and early fall to reach fresher areas. Photos by Raymond T. Bauer, University of Louisiana at Lafayette.

the Atchafalaya River, this species is abundant and appears to be a major prey item of river fishes (Bauer and Delahoussaye 2008). Historically, the species was abundant enough throughout its range to support a commercial fishery for human consumption and bait, and the shrimp has long been the favorite bait for trotline fishing in the ARB (Jim Delahoussaye, Friends of the Atchafalaya, personal communication). The river shrimp population in the upper Mississippi River has declined drastically, due to a combination of factors (overfishing, habitat loss, pollution), but in the Atchafalaya, river shrimp populations are still robust (Bauer and Delahoussaye 2008). River shrimp employ an amphidromous life cycle, requiring saline water for larval development and, consequently, this species is migratory. Therefore, human activities that create barriers between freshwater and marine systems may disrupt this periodic journey and cause the species to decline.

Studies in the Atchafalaya River show that female river shrimp employ two strategies to ensure that larvae reach saline water: 1) hatching larvae are released into fresh water during the spring pulse and are then transported to estuarine habitats by the current; or 2) females migrate downstream along the river edge, where they release their larvae into brackish water of Atchafalaya Bay. Hatched larvae must reach brackish water to stimulate molting to stage-2, the first feeding stage, before their embryonic yolk supplies dwindle. Optimal molting success in this species occurs when larval drift in fresh water does not exceed three days, and those not in brackish water begin dying in significant numbers (50%) by day 7, with 100% mortality in 10 days (Bauer and Delahoussaye 2008, Rome et al. 2009). In the Atchafalaya River, hatching and female release of larvae occurs in April–June when flow velocity is 2.0–2.5 km hr^{-1} (1.1–1.3 knots), and at this velocity, stage-1 larvae drift downstream about 48–60 km day^{-1} (30–37 mi day^{-1}; Bauer and Delahoussaye 2008). Because females need to release larvae no farther than 3 days' drift time, about 140–180 km (87–111 mi), from brackish water, Butte LaRose represents about the upstream limit that stage-1 larvae should be present in the water column. However, plankton tows in the Atchafalaya River show about a tenfold increase in stage-1 larvae present on the Atchafalaya Delta as at upstream stations (Rome et al. 2009). This may be due to the fact that, during the flood pulse, the Atchafalaya Delta stays fresh, and larvae have to continue drifting some distance past the delta to find salinities high enough to stimulate molting to stage-2 (optimum salinity 6–10; Rome et al. 2009). A recent study, however, showed that a substantial number of adult females migrate downstream to release larvae into Atchafalaya Bay (Olivier and Bauer 2011, Olivier et al. 2012). Additionally, evidence of multiple brooding has been found in Atchafalaya river shrimp, so that females who have released larvae during their downstream migration ensure an optimal salinity environment for the second brood (Olivier and Bauer 2011). After molting, juveniles (3–7 mm carapace length [0.1–0.3 in]) can be found migrating in mass upstream movements along the river edge at night during the summer and early fall to reach freshwater areas where they grow to adults (Fig. 5.10).

Because this species relies so heavily on spring flooding, long stretches of river for larval transport, and the river edge for its spawning and upstream migrations, river shrimp populations may be extremely susceptible to flow barriers and channel training

FIG. 5.11. The Atchafalaya River Basin is a popular recreational fishing destination, especially for largemouth bass (*Micropterus salmoides*), sunfishes (Centrarchidae), and crappie (*Pomoxis* spp.). Photo © Matt Pardue.

structures that interrupt flow, segment river reaches, and protrude from the river edge. Other than the hydroelectric dam at Old River, which separates the Atchafalaya River shrimp population from its source population in the Mississippi River, there are no other flow obstructions in the river, and this is likely why the Atchafalaya population persists. Dams and impoundments on several rivers, including the Red River and the lower Ohio River, where the river shrimp gets its name, have been identified as factors in the decrease and even extirpation of populations of the species. Dams and control structures are thought to prevent access to naturally occurring brine springs that were likely the historic source of saline waters for larval development in upstream populations (Bauer and Delahoussaye 2008, Rome et al. 2009, Olivier et al. 2012).

RECREATIONAL FISHERIES. The ARB floodway supports the most popular freshwater recreational fishery in Louisiana (Fig. 5.11; Holloway et al. 1998). According to the Louisiana Department of Wildlife and Fisheries, recreational fishing in the floodway generates $47 million per year (Louisiana Department of Natural Resources 2010), and a survey of recreational anglers showed that 15% of respondents listed the basin as their favorite fishing destination (W.E. Kelso, Louisiana State University Agricultural Center, unpublished data). Important recreational fish species include largemouth bass, white crappie (*Pomoxis annularis*), and several other species of sunfish (Centrarchidae).

The Atchafalaya and Wax Lake deltas are popular recreational fishing destinations because anglers can catch popular freshwater species, like largemouth bass and several species of sunfish, as well as popular marine species like red drum (*Sciaenops*

FIG. 5.12. Hoop nets are commonly used to harvest finfish such as catfishes (Ictaluridae) and buffalo (*Ictiobus cyprinellus*). Photo © CC Lockwood.

ocellatus). Vermilion, Cote Blanche, and Fourleague Bays, including the western Terrebonne marshes are all extremely popular recreational fishing destinations for anglers looking for popular saltwater species, like red drum and spotted seatrout (*Cynoscion nebulosus*).

COMMERCIAL FISHERIES. The ARB floodway supports a burgeoning commercial fishery. Annual average landings of finfish and shellfish range from 5.9–11.5 million kg (13–25 million lb), and those landings support an average dockside value of $8.9–24.1 million yr[-1] (Alford and Walker 2011). The most popular commercial finfish species include catfish (Ictaluridae), buffalo, and freshwater drum (*Aplodinotus grunniens*; Fig. 5.12). However, the most important commercial fishery in terms of landings and value is crawfish.

The first recorded commercial crawfish harvest in the United States was in 1880 (10,614 kg; $2,140), and New Orleans accounted for over one-third of the catch. By the late 1930s, the annual crawfish harvest in Louisiana had increased more than 80 times to over 900,000 kg (1.9 million lb), worth about $175 million (Penn 1943, McClain et al. 2007). By 2010, the total production of crawfish in Louisiana topped 57 million kg (125 million lb). This sharp rise in production is reflected in the attitudes of Louisianans about crawfish. Louisianans regard crawfish as a household necessity rather than a luxury, and residents will not pay a lot for crawfish, easily switching to other substitute products (i.e., crabs, shrimp) if the supply of crawfish drops and price increases (income elasticity = 0.90; price elasticity = -16.94; Bell 1986, Dellenbarger and Luzar 1988).

Commercial landings in Louisiana are dominated by the red swamp crawfish (70–80%), with the other 20–30% of landings from the white river crawfish (Isaacs and Lavergne 2010, Louisiana State University Agricultural Center 2011). Prior to the 1950s,

FIG. 5.13. Crawfish support the largest fishery in the ARB, in terms of landings and revenue, and the floodway supplies over 90% of the wild-caught crawfish in the market. Because crawfish traps are set in flooded swamp forest, this fishery is dependent upon the seasonal flood cycle, making it an extremely variable fishery. Photo © CC Lockwood.

harvested crawfish were entirely wild-caught; however since that time, the commercial crawfish harvest in Louisiana comes from two sources—aquaculture and the artisanal wild fishery. Today, wild-caught crawfish make up only a small percentage (12%) of the total commercial catch, and of that percentage, almost all (~92%) is harvested in the ARB floodway (57% east of the Atchafalaya River, 30% west of the Atchafalaya River, 5% both east and west of the Atchafalaya River; Fig. 5.13; Isaacs and Lavergne 2010). While pond-raised crawfish have largely stabilized the market and protects it from the large fluctuations in supply and price inherent in artisanal fisheries, influx of crawfish from the ARB floodway has a marked effect on price. When production from the ARB enters the market, price can decline up to 25% overnight (Bell 1986, Dellenbarger and Luzar 1988).

Crawfish harvest in the ARB floodway is extremely variable. One study found that over a ten year period (1970–1980), crawfish yield fluctuated by a factor of ten, and price varied by a factor of two. Over that same time period, crawfish harvest in the floodway was positively related to both changes in water level during the spring and summer and the amount of crawfish harvested from ponds. Harvest was negatively related to unemployment rate in the Acadiana parishes, likely due to increased harvest pressure, since crawfishing provides a supplemental income source for many residents living in ARB parishes.

FIG. 5.14. A shrimp boat heads out to fish in Vermilion Bay. Photo © Alison M. Jones for www.nowater-nolife.org.

Wild crawfish harvest in the ARB floodway (1988–2010) averaged about 10 million kg (22 million lb) and ranged from ~450,000 kg to 21 million kg (1–46 million lb). This harvest supported an average dockside value of about $12 million ($1–37 million) over that same timescale (Louisiana State University Agricultural Center 2010). The crawfish fishery is the most profitable fishery and one of the most profitable industries in the ARB floodway. Over 85% of wild crawfish harvesters reported selling their crawfish to dealers inside a four-parish area (Assumption, Iberville, St. Martin and St. Mary), suggesting that the crawfish fishery in the ARB floodway is an important local source of income for communities surrounding the ARB floodway (Isaacs and Lavergne 2010).

South of the floodway, the Atchafalaya, Cote Blanche, Vermilion, and Fourleague bay systems support commercial fisheries for catfish, blue crab (*Callinectes sapidus*), and shrimp (Fig. 5.14). A limited oyster fishery is also present in the bays, mostly located around Oyster Bayou, and is regulated by seasonal freshwater inputs from the Atchafalaya River. Historically, however, the bays were extremely important for oyster production, and these bays supported some of the largest natural reefs in the state (Cary 1906). Consequently, these bays were vital resources in the development of the commercial oyster industry and were the epicenter of commercial oyster shell mining in the state (Louisiana Wildlife and Fisheries Commission 1968).

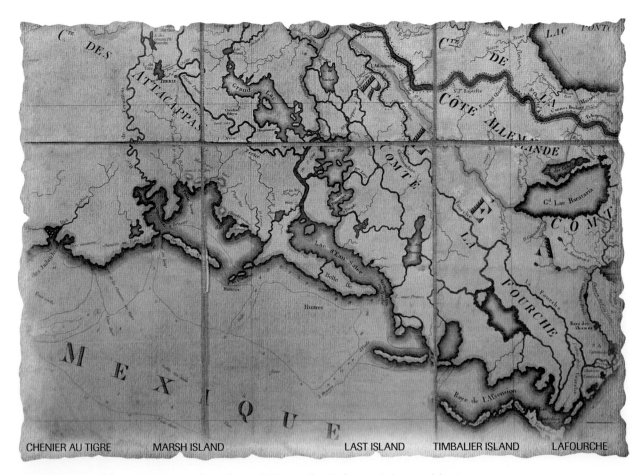

| CHENIER AU TIGRE | MARSH ISLAND | LAST ISLAND | TIMBALIER ISLAND | LAFOURCHE |

FIG. 5.15. Map of the Louisiana coast drawn by Barthélémy Lafon (Lafon 1806) showing the existence of large oyster reef shoals (huîtres) extending from the central Louisiana coast (foreground, outlined in black) into the Gulf of Mexico. Labels show the location of the present-day central coast landmarks. The modern Atchafalaya Delta is located in the area between the Marsh Island and Last Island labels. Original image source (including labels) was Condrey et al. (2010) and was provided by Richard T. Condrey, Louisiana State University.

OYSTERS. Reefs of the eastern oyster (*Crassostrea virginica*) were once plentiful on the Louisiana coast, particularly in the region of Atchafalaya Bay. An account by a Spanish surveyor in 1785 stated that the area near present-day Point au Fer was surrounded by oyster banks for a distance of 16 km (10 mi), and the areas of present-day Marsh Island and Rainey Wildlife Refuge contained oyster banks that extended out from the shore 8 km (5 miles; de Evía 1785). Barthélémy Lafon (1806) published a map that showed massive oyster shoals from Chenier au Tigre to Last Island that extended from the coastline out 4–5 leagues (13–16 mi; Fig. 5.15). These reefs likely were important for protecting the coast from storms, as de Evía (1785) remarked that even with a strong wind his vessel was protected from whitecaps by these massive oyster shoals.

One hundred years later, numerous natural oyster reefs were described in Vermilion Bay as extending out into the Gulf of Mexico (Fig 5.16; Cary 1906). Cary described a massive and productive reef that extended northeast from the entrance to Southwest

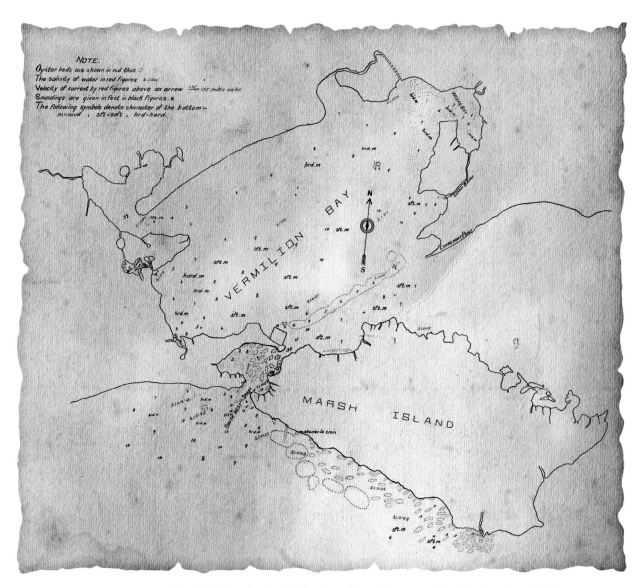

FIG. 5.16. Map showing the location of natural oyster reefs in Vermilion Bay, fringing Marsh Island, and extending out into the Gulf of Mexico. Reefs are outlined in red. Map from Cary (1906).

Pass to almost Cypremort Point. This reef was 183–549 m (600–1,801 ft) wide and was mostly subtidal, covered by 0.9–1.8 m (3–6 ft) of water. On its north end, another reef extended northwest and was 0.8 km (0.5 mi) long by 0.4 km (0.25 mi) wide. In addition to this reef, the flats on both sides of Southwest Pass were covered with long, narrow intertidal oyster reefs, and subtidal reefs extended out on both sides of Southwest Pass 1.2 km (0.75 mi) into the Gulf of Mexico. Cary also documented an almost continuous mass of oyster reefs in the open Gulf of Mexico off the southern shore of Marsh Island. These reefs started as intertidal, fringing reefs along the shore of Marsh Island and became subtidal as they extended from the shore over 1.6 km (1.0 mi) into the Gulf.

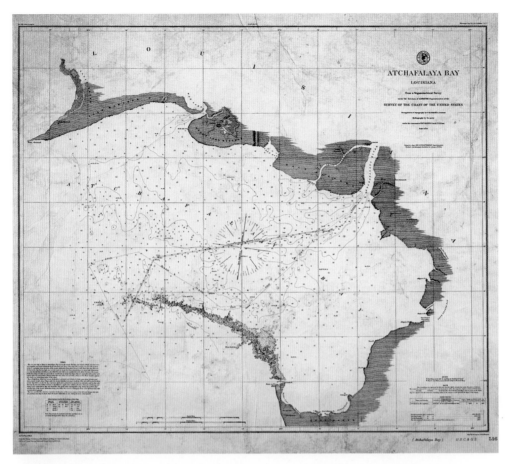

FIG. 5.17. An 1878 Coast Survey Map of Atchafalaya Bay showing the location of the Point au Fer shell reef (also called White Shell Keys Reef), a large shell reef that stretched up into the water column and extended up to 6 m (20 ft) below the bay bottom for about 32 km (20 mi). On this map, the Atchafalaya River is on the top right, and the reef is shown in the foreground extending from the southeast to the northwest.

Another large oyster reef complex called the Point au Fer Shell Reef existed off Atchafalaya Bay (Fig. 5.17). This massive reef shoal, located south of Atchafalaya Bay, extended from Point au Fer west for about 32 km (20 mi). This reef complex was very old, and in addition to extending up into the water column, it also went down to 6 m (20 ft) below the bay bottom. Interestingly, this reef was large enough to restrict freshwater discharge from the Atchafalaya River and keep the salinity in this area low, making the area unproductive for commercial oyster fishing. For this reason, a statement suggesting removal of the reef appears on an early Department of Conservation map of coastal oyster bottoms (Department of Conservation of Louisiana 1920).

The natural oyster reefs documented in Atchafalaya, Cote Blanche, and Vermilion Bays were not only extensive, but were also extremely productive, producing 107–183 spat and 38–216 young oysters (under 5 cm [2 in]) per peck of oysters (1 peck = 25 oysters; Cary 1906). As a result, these natural reefs were valuable resources to two industries. The first was the commercial oyster fishery, since the private leasing system in

Louisiana was only just beginning, and the majority of oysters were still being harvested from natural reefs (Cary 1906). The second was the commercial shell dredging industry that started in Louisiana on the Point au Fer oyster reef and spread throughout Atchafalaya, Cote Blanche and Vermilion Bays.

HISTORICAL OYSTER HARVEST. At the time of Cary's account (1906), the natural reefs in Vermilion Bay were already extremely valuable for commercial live oyster harvest. Reefs that contained saleable oysters were heavily exploited by luggers and tongers, who sold their oysters either to the large cannery at Avery Island or to markets in Franklin and Morgan City, Louisiana (Fig. 5.18). Cary documented overfishing on several of the reefs. For example, of a large reef (0.8 km [0.5 mi] wide) known at the time as "Little Hills" reef, Cary (1906) remarked: "Little Hills reef has in the past produced an abundance of oysters of large size and exceptionally good quality. At the present time the supply of oysters on this reef is very limited, due to excessive fishing, so that during the present season few of the tongers have found it profitable to work here." In another account, Cary (1906) said of a large reef called Diamond Reef: "This reef covers an area of two square miles and produces oysters of fine quality. An account of the history of this reef since its discovery is instructive in showing the rapidity with which an especially productive area may be reduced to a state of commercial barrenness by excessive fishing."

Private oyster leases were also beginning to appear during Cary's time. In Vermilion and Iberia Parishes in 1906, there were 14 documented leases covering 656 ha (1,622 acres; Cary 1906). Leasing in the area of Vermilion Bay was restricted by freshwater input and muddy bay bottoms due to the influence of the Atchafalaya River. By 1913, more than 1,700 commercial fishermen were entering the oyster trade coastwide (Cary 1906, Hart 1913). Over 6,879 ha (17,000 acres) were leased statewide by 1913, and the fishery was still considered to be in its infancy (Hart 1913). Hart (1913) remarked on the ease of entering the lucrative trade: "The cheapness of the leases and the easy conditions under which the trade can be entered make an attractive offer to anyone so inclined, and is inducing many to take up the lucrative work. The lands aggregate an annual lease rental of $1 per acre and at present there are approximately 453,888 acres of productive bottom to be chosen from."

With its combination of natural reefs and nascent leasing system, the Louisiana oyster fishery was important to the commercial oyster market, because it not only supported the entire Louisiana market but also supplied over 80% of the commercial market in Texas and much of the market in the northern and eastern portions of the country (Hart 1913). With so many entering the fishery, cultch material for growing marketable oysters on leased water bottoms was in short supply, and many natural reefs were heavily dredged for this resource. Much of this cultch material was comprised of small, unmarketable oysters that were harvested from natural reefs with densely packed and unmarketable clusters (25–35 oysters per cluster) of oysters anchored to empty shells buried in the mud (Cary 1906, Hart 1913). Shucked oyster shells were also valuable cultch resource at the time, and fishermen returned shucked shells to the water to grow more oysters. Today, very little shucked oyster shell remains in the state, because it is

FIG. 5.18. A Louisiana oyster lugger in the early 1900s. This vessel pulled an oyster dredge (port gunwale) over natural reefs or cultivated oyster beds to harvest oysters from the bay bottom. Percy Viosca, Jr. Photograph Collection, Mss. 4948, Louisiana and Lower Mississippi Valley Collections, LSU Libraries, Baton Rouge, LA.

sold by oyster processers as a value-added product and shipped out of state for a variety of uses, including chicken feed and road fill (Piazza et al. 2005).

HISTORICAL SHELL DREDGING. The extensive oyster reef shoals in Atchafalaya, Cote Blanche, and Vermilion Bays were also heavily targeted and ultimately largely removed by commercial shell dredging (Louisiana Wildlife and Fisheries Commission 1968, Juneau 1984). The shell dredging industry in Louisiana proceeded from 1914 until 1996 and focused on shells from two species—the eastern oyster and the brackish water clam (*Rangia cuneata*). Shells from these species were sought after for use as a strong but lightweight aggregate for construction base for roads, levees, parking lots, and oil drilling barge pads (US Army Corps of Engineers 1993). Because they are so light (bulk density is 1,041 kg m^{-3} [65 lbs ft^{-3}]), oyster and clam shells are particularly suited for use on soft soils in south Louisiana, and, as a result, are preferred over other popular road-bed alternatives like crushed limestone 1,602 kg m^{-3} [100 lbs ft^{-3}]), sand (1,586 kg m^{-3} [99 lbs ft^{-3}]), gravel (1,682 kg m^{-3} [105 lbs ft^{-3}]), recycled concrete (1,746 kg m^{-3} [109 lbs ft^{-3}]), and steel slag (1,842 kg m^{-3} [115 lbs ft^{-3}]) (US Army Corps of Engineers 1993). Shells were also used as a source of calcium carbonate for Portland cement, lime production, chicken feed, pharmaceuticals, chemicals, and smokestack emission control products.

Commercial shell dredging was started by the Louisiana Conservation Commission (precursor to the Louisiana Department of Wildlife and Fisheries) as a way to help fund coastal research, management, and conservation, because inadequate funding had been provided to the commission by the legislature (Louisiana Wildlife and Fisheries Commission 1968). Shell dredging leases were given to companies in exchange for a royalty fee, paid to the commission, on every cubic yard of shell harvested and a severance fee (per ton) paid to the Louisiana Department of Revenue. Leases were made for 15 years and were renewable for an additional 10 years. Lessees had the exclusive right to harvest all exposed and buried (fossilized) shell on the leased area and to sell, transfer, assign, and renegotiate leases (Louisiana Wildlife and Fisheries Commission 1968, Juneau 1984).

Oyster shells were mined with cutterhead dredges that cut trenches about 91 m (300 ft) in width and to the depth of the reef being harvested, which was up to 5–7 m (17–22 ft) below the bay bottom (Fig. 5.19; US Army Corps of Engineers 1993). Dredge slurry, containing shells, was pumped onboard and worked through a series of sprayers, shakers, and screens, and shell fragments of various sizes were collected, conveyed to a collection barge, and transported to holding yards and stockpiled. The trenches that were cut filled in with sediment over time (~10 years; US Army Corps of Engineers 1993). Each dredge was capable of working approximately 0.4 ha (1 acre) of reef per day, and this technique remained largely unchanged throughout the existence of the industry (US Army Corps of Engineers 1993).

Throughout its 80-year duration, the shell dredging industry acted under a set of laws developed in 1910 and 1912 that also remained largely unchanged (Louisiana Wildlife and Fisheries Commission 1968). In fact, until the early 1980s when a series of lawsuits were brought by environmental groups forcing the study of the environmental effects of shell dredging and the development of environmental impact statements, the industry acted largely unregulated. Consequently, the industry proceeded for about 70 years without any kind of environmental safeguards. Interestingly, conflict with the oyster industry over dredging in areas of live oysters and concern over removal of cultch material was the only substantial challenge the industry received, and, as a result, areas (much of the coast east of Point-au-Fer) of active oyster production were closed to shell dredging in 1944 (Juneau 1984). Citing concern over environmental effects, in 1996 Governor Mike Foster signed an executive order placing a statewide moratorium on shell dredging in Louisiana.

The central Louisiana coast, and particularly Atchafalaya, Cote Blanche, and Vermilion Bays were the heart of the oyster shell dredging industry and consistently produced about half of the annual statewide production of shell. The other half of Louisiana's shell production was provided by *Rangia cuneata* from an industry that centered in Lakes Pontchartrain and Maurepas and began over 15 years after oyster shell dredging.

The shell dredging industry began with sale of its first lease of 963 ha (2,379 acres) of the Point-au-Fer reef in Atchafalaya Bay (Louisiana Wildlife and Fisheries Commission 1968). This lease was sold to Alfred Mead for the removal of exposed oyster shell at a rate of $0.05 yd^{-3}. Mr. Mead soon realized that there were vast quantities of fossilized shell beneath the bay bottom and leased up huge areas of the central coast bays. From 1914–1925 the central coast was leased by only two companies, and these produced from

FIG. 5.19. A cutterhead dredge at work harvesting shell from a bay bottom. The collection barge is shown on the port side of the dredge vessel. Photo by John Messina. Image from the National Archives and Records Administration, US National Archives series: DOCUMERICA: The Environmental Protection Agency's Program to Photographically Document Subjects of Environmental Concern, compiled 1972–1977.

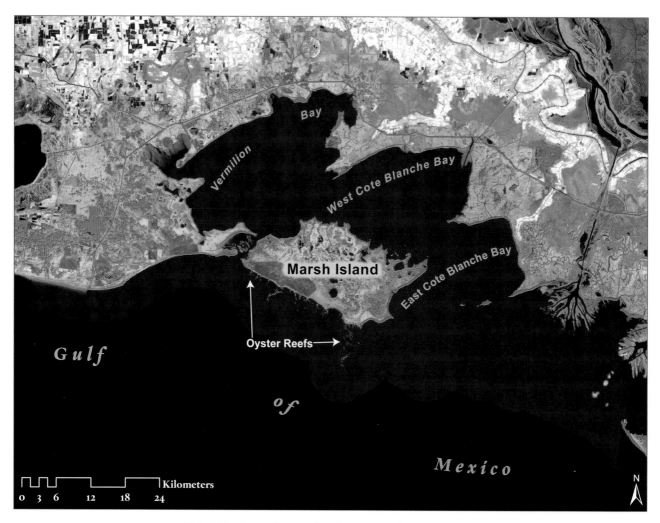

FIG. 5.20. Image showing shoreline erosion along the shore of Marsh Island and the central Louisiana coast. Orange color depicts shoreline land loss. Note that the Gulf shoreline of Marsh Island appears to be stable. This area is one of the few places along the coast with remaining relict oyster reefs (visible on image). Map created with data from Couvillion et al. (2011) and provided by Yvonne C. Allen, US Army Engineer Research and Development Center, Baton Rouge, Louisiana.

229,366 m³ yr⁻¹ (300,000 yd³ yr⁻¹) in the beginning up to 1.1 million m³ yr⁻¹ (1.5 million yd³ yr⁻¹) by 1925. After 1925, the entire western coast came under lease, and those leases were held by only 3–4 companies. Demand for oyster shells peaked in 1965 with annual production of just over 3.8 million cubic meters (5 million cubic yards). By the 1990s, with economic variability along the Gulf coast, increased regulation, and increased competition from other materials, oyster shell production decreased to 267,594 m³ yr⁻¹ (350,000 yd³ yr⁻¹; Louisiana Wildlife and Fisheries Commission 1968, Juneau 1984).

Royalty revenues received by the Louisiana Department of Wildlife and Fisheries (previously the Conservation Commission) for oyster shell ranged from $0.05 yd⁻³ (1914–1950) to $0.09 yd⁻³ (1990s), and severance tax to the Department of Revenue was fixed at $0.03 yd⁻³ (Louisiana Wildlife and Fisheries Commission 1968, Juneau 1984, US Army Corps of Engineers 1993). The top annual revenue to the state for all shell resources (oyster and clam) during its highest demand years was only $1.5 million, of which $1.2 million went to royalty revenues for fish and wildlife conservation and research. Oyster reefs provided about half of that revenue. For example, records from 1979–1983 showed that, over a five-year period, 12,288,800 m³ (16,073,144 yd³) of oyster shell was removed from the central coast reefs, producing a total royalty revenue of $2.3 million, or $460,000 yr⁻¹ (Juneau 1984).

It was not until 1987, 73 years after the industry began, that the first environmental impact statement was produced for Atchafalaya Bay and its adjacent waters (US Army Corps of Engineers 1987). Shell dredging operations on the Point-au-Fer reef proceeded, largely unregulated throughout the existence of its operations (1914–1982). Operations on the reef were closed for one year (1973), due to disputes about coastline boundaries, but it was not until the 1980s that the area was zoned and portions of the reef were within a newly created restricted zone. Shell dredging on the Point-au-Fer reef ended in 1982, and at that time, shells were being sold for $0.21 yd⁻³. By that time, the reef, like many others in the area, was largely gone.

The massive natural reef shoals that were removed from these central coast bays were important regulators of local coastal processes, and their removal had consequences for coastal erosion. After the reef complexes were removed, Atchafalaya and Vermilion Bays and their associated marshes were transformed from low-energy, protected environments to higher energy open bay environments. Numerical modeling simulations of wave height (energy) distributions show that the wave height increase in Atchafalaya Bay (and correspondingly on marshes on the Atchafalaya and Wax Lake deltas) following oyster reef removal was up to 0.7 m (2.4 ft) during fair weather conditions, and that pattern was exacerbated during storm events (Stone et al. 2005). This increase in wave energy is likely a controlling factor in erosion of marshes fringing the bay. Interestingly, the Gulf side of Marsh Island, one of the few places with remaining relict reefs, is the only place in the central coast where shoreline erosion is noticeably absent (Fig. 5.20).

PRESENT-DAY OYSTER RESOURCES. Today, Louisiana remains the leader in oyster production, producing 34% of the oyster landings in the US and over 50% of landings in the northern Gulf of Mexico (6.3 million kg [14 million lbs] of meat production in 2009; Louisiana Department of Wildlife and Fisheries 2010). However, with the loss of natural reefs, the majority of current production comes from private oyster leases, where oysters are farmed on the bay bottoms on two-dimensional "reefs" that largely lack three-dimensional relief. While this farming method produces large numbers of quality oysters for the marketplace, these two-dimensional reefs may not provide the full suite of ecosystem services (i.e., fishery habitat, shoreline protection) typically associated with three-dimensional reef structures (Piazza et al. 2005, Beck et al. 2009, 2011, Scyphers et al. 2011 and others).

At present, some natural oyster reef habitat still remains in the Atchafalaya and Vermilion bay systems; however individual reefs are much smaller than they were historically, and a much smaller proportion of the central coast bays are occupied by oyster reefs (Fig. 5.21). Most of the present-day natural reefs periodically experience significant mortality, due to high pulses of fresh water from the lower Atchafalaya River, Wax

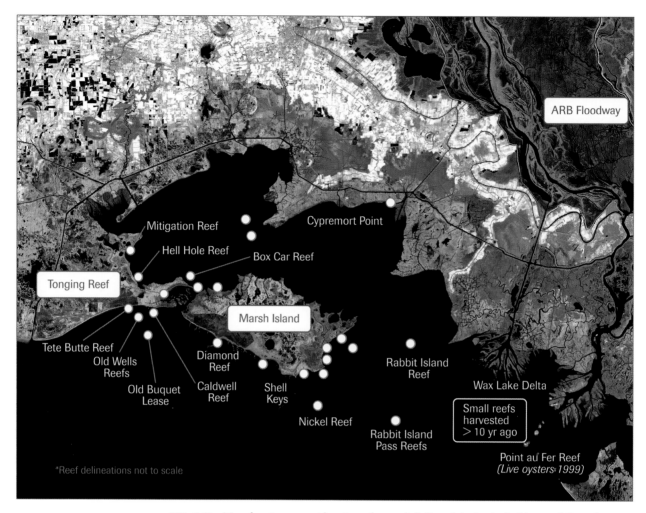

FIG. 5.21. Map showing present location of natural shell reefs in the Atchafalaya and Vermilion bay systems. Most of these reefs are still alive, although they experience significant mortality, due to fresh water from the Lower Atchafalaya River, Wax Lake Outlet, and the Gulf Intracoastal Waterway. This figure was modified from an original slide that was prepared by John Supan, Louisiana Sea Grant, and was provided by the Louisiana Department of Wildlife and Fisheries.

Lake Outlet, and the Gulf Intracoastal Waterway (Patrick D. Banks, Louisiana Department of Wildlife and Fisheries, personal communication)

The backbone of the private oyster leasing system is Louisiana's approximately 688,000 ha (1.7 million acres) of public seed grounds (Louisiana Department of Wildlife and Fisheries 2010). These public oystering areas are managed by the state and provide both seed oysters (< 8 cm [3 in]) that are transplanted to private leases for grow-out and sack oysters (> 8 cm [3 in]) for direct sale to the marketplace (Louisiana Department of Wildlife and Fisheries 2010). The ARB is part of the Vermilion/Atchafalaya Bays Public Oyster Seed Ground (Fig. 5.22). This seed ground contains 219,000 ha (541,000 acre) of water bottoms and produces generally low yields of seed (e.g., 2010 stock = 0.0–2.4 seed oysters m^{-2}) and sack (e.g., 2010 stock = 0.0–0.4 sack oysters m^{-2}) oysters, because large amounts of fresh water that are discharged from the Atchafalaya

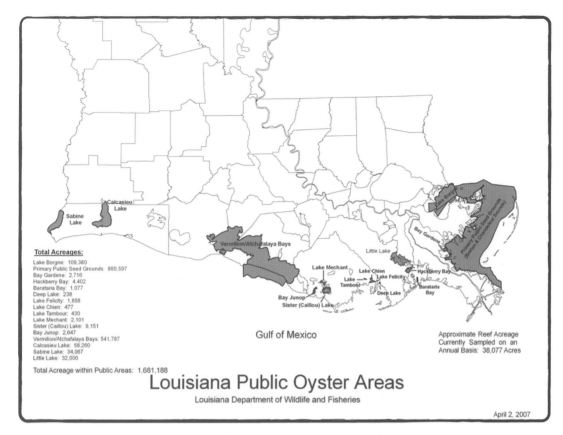

Total Acreages:
Lake Borgne: 109,380
Primary Public Seed Grounds: 880,597
Bay Gardene: 2,716
Hackberry Bay: 4,402
Barataria Bay: 1,077
Deep Lake: 238
Lake Felicity: 1,858
Lake Chien: 477
Lake Tambour: 430
Lake Mechant: 2,101
Sister (Caillou) Lake: 9,151
Bay Junop: 2,647
Vermilion/Atchafalaya Bays: 541,787
Calcasieu Lake: 58,260
Sabine Lake: 34,067
Little Lake: 32,000

Total Acreage within Public Areas: 1,681,188

Gulf of Mexico

Approximate Reef Acreage
Currently Sampled on an
Annual Basis: 38,077 Acres

Louisiana Public Oyster Areas
Louisiana Department of Wildlife and Fisheries

April 2, 2007

FIG. 5.22. Map showing the Louisiana public oyster areas. These areas total 680,352 ha (1,681,188 acres) and are located across the Louisiana coast. The ARB is part of the Vermilion/Atchafalaya Bays Public Oyster Seed Ground, a 219,000 ha (541,000 acre) complex of water bottoms that spans across the Atchafalaya, East and West Cote Blanche, and Vermilion bay systems (Louisiana Department of Wildlife and Fisheries 2010).

River causes widespread oyster mortality in the late winter and early spring (Louisiana Department of Wildlife and Fisheries 2010). During years when Atchafalaya River discharge is low, oyster survival is better, and yields of both seed and sack oysters are higher.

SPECIES OF CONSERVATION CONCERN. The ARB contains three fish species of conservation concern (Table 5.1). Two are state-listed species (paddlefish and blue sucker), and one is federally listed (pallid sturgeon).

Paddlefish (*Polyodon spathula*) are one of the most ancient freshwater fishes, having risen about 150 million years ago (Fig. 5.23). They are one of only two paddlefish species on earth (Chinese paddlefish [*Psephurus gladius*] are found in the Yangtze River) and are closely related to the sturgeons. Paddlefish are long-lived fish (maximum reported age 55 yr), quite large (maximum reported length 2.2 m [7.2 ft] eye-fork length [EFL]; maximum reported weight 90.7 kg [200 lb]) and have smooth, scaleless skin and a cartilaginous skeleton. Distinguishing characteristics include a long, flat snout (about 1/3 of the body length) and a large mouth. Both the snout, which contains electroreceptors

FIG. 5.23. The paddlefish (*Polydon spathula*). This state-listed species of conservation concern has been found in the Atchafalaya River. This photo shows both the characteristic long, flat snout and large mouth used to feed on zooplankton in the water column. Photo by Todd Stailey, Tennessee Aquarium.

to detect food, and the large mouth are used to filter feed zooplankton in the water column. Paddlefish occur in large rivers and associated lakes throughout the Mississippi Valley, including the Gulf slope drainages and their populations have declined dramatically due to a combination of waterway development, water quality deterioration, and overharvest (Reed et al. 1992, Jennings and Zigler 2000, Froese and Pauly 2011).

Paddlefish require large areas of free-flowing, riverine habitat with gravel bars or other hard substrate for spawning (Jennings and Zigler 2000). Once, uninterrupted river habitats included the preferred low-flow side channel and river-lake habitats and allowed upstream spawning migrations that were triggered by increased discharge in spring and ended at congregation areas in deep parts of the river. From there, paddlefish moved to gravel bars or other hard substrate with low current velocities to breed. Large-scale development of these river systems (dams, navigation locks, and reservoirs) for navigation and flood control has 1) flooded or severed the connection to side channels and river lakes; 2) created substantial barriers to upstream breeding migration; and 3) altered the natural hydrological regime (decreased flow velocity and increased water temperature) that is necessary to meet the exacting breeding requirements of the species. Additionally, increased sedimentation and contaminant loads can affect feeding success and also cause reduced breeding success, by reducing dissolved oxygen concentrations on gravel bars, smothering eggs, and decreasing survival of embryos and larval fish. Today, paddlefish can be found in reservoirs throughout the Mississippi River Valley; they are also found in the river channels and commonly congregate in the deep pools and tailwaters of dams.

Historically, paddlefish have been prized for their flesh and roe (Jennings and Zigler 2000). Commercial harvest of the species peaked in 1900 with an estimated 1.1 million kg (2.4 million lbs) landed from the Mississippi River Valley. By the 1920s, paddlefish stocks began to decline precipitously, after increased demand for caviar caused decimation of lake sturgeon (*Acipenser fulvescens*), thus increasing demand for paddlefish roe. The combination of harvest pressure and waterway development during the early 1900s made the paddlefish extremely susceptible to overharvest.

In 1992, the US Fish and Wildlife Service listed the paddlefish as a "species of special concern," signifying that additional study was necessary to determine if the species should be listed as endangered. However, that designation was discontinued and currently no federal protection is granted to paddlefish. In 1992, the paddlefish was protected from illegal trade by the Convention on International Trade in Endangered Species of Wild Fauna and Flora (CITES) and has varying levels of protection by the states.

Paddlefish have been extirpated from four states (New York, Pennsylvania, Maryland, North Carolina) and Canada, and currently have a 22-state range in the United States (Jennings and Zigler 2000). Its status varies within its range: 10 states (Montana, North Dakota, Oklahoma, Texas, Minnesota, Wisconsin, Ohio, West Virginia, Virginia, Alabama) list the paddlefish as endangered, threatened, or a species of special concern; 2 states (Illinois, Nebraska) report a declining population; and 10 states (South Dakota, Kansas, Iowa, Missouri, Arkansas, Indiana, Kentucky, Tennessee, Mississippi, Louisiana) report stable populations (US Geological Survey 2011). Federal and state management strategies to either recover or conserve this species include strict regulation of harvest in states where commercial or recreational harvest is allowed, stocking and reintroduction of paddlefish across its historic range, and conservation and restoration of riverine habitat, especially spawning habitats (Jennings and Zigler 2000).

In Louisiana, paddlefish can be found throughout the major river drainages and coastal lakes (Reed et al. 1992, Smith 2004). Their numbers are greatest in the Mermentau River and Lake Pontchartrain. In the ARB, paddlefish are most abundant in Henderson Lake, with fewer found in the main river channel. Paddlefish in Louisiana reservoirs and lakes are generally larger than those in rivers, likely due to the difference in zooplankton abundance (Reed et al. 1992, Smith 2004). Additionally, paddlefish in Louisiana were larger (up to 1.2 m [4.0 ft] EFL) and exhibit faster early growth rates (age-1 fish 411–455 mm [16–18 in] EFL) than other populations across its range.

Paddlefish in Louisiana also become sexually mature faster than other populations (100% females gravid by age 10), but exhibit lower fecundity (9,500 eggs kg^{-1} body weight) and higher natural mortality (26–48%; Reed et al. 1992). These trade-offs suggest that, although populations in Louisiana are stable, harvest levels that may be sustainable in other populations (i.e., 10–30%) may cause populations to decline in the state (Reed et al. 1992).

The blue sucker (*Cycleptus elongatus*) is one of 80 species in the sucker family (Catostomidae, a family of freshwater fishes endemic to North America, eastern China and eastern Siberia; Fig. 5.24). Consequently, its distinguishing characteristics are a mouth that is located on the underside of its head and thick, fleshy lips—both adaptations for feeding on bottom-dwelling organisms. Blue suckers are endemic to North America and, like most species in its family, are found in rivers. This species commonly lives more than 10 years, and is medium to large, reaching lengths (total length, TL) over 50 cm (20 in). Its maximum reported total length is 93 cm (37 in; Froese and Pauly 2011). Blue suckers are light steel-gray to black, streamlined, and extremely fast—all adaptations for surviving in swift currents. In fact, they have been described as swimming awkwardly in slow flows (Moss et al. 1983). Blue suckers are distributed throughout the Mississippi River Valley from Pennsylvania to Montana and south to Louisiana and throughout the western Gulf slope drainages to the Rio Grande (Adams et al. 2006, Froese and Pauly 2011). The species largely inhabits swiftly flowing habitats of the main channel and deep side channels (Froese and Pauly 2011). Blue sucker populations have declined dramatically, probably due to waterway development (Burr and Mayden 1999, Adams et al. 2006, Froese and Pauly 2011, NatureServe 2011).

Blue suckers require long, non-segmented river stretches with swift currents and rocky substrates, where they feed on insects in cobble areas and make extensive upstream spring runs to spawn in fast, rocky areas of the river channel (Adams et al. 2006, Burr and Mayden 1999, Froese and Pauly 2011, NatureServe 2011). While the species prefers swift-moving habitat, large numbers of larval blue suckers have been found in slackwater habitats along shorelines, in off-channel habitats, and associated with islands (Adams et al. 2006).

Like the paddlefish, in 1994, the US Fish and Wildlife Service listed the blue sucker as a "species of special concern," signifying the need for additional study to determine if it should be listed as an endangered species, and, because the designation was discontinued, there is currently no federal protection granted to blue suckers. This species is notoriously difficult to sample effectively because of its habitat use; consequently, most historical accounts of abundance are anecdotal or are from the bycatch of sampling other species (Adams et al. 2006). Based on available data, blue suckers are presumed extirpated in Ohio, critically imperiled in West Virginia, imperiled in 7 states (Montana, Illlinois, Indiana, Arkansas, Tennessee, Alabama, and Louisiana), and vulnerable in 12 states (North Dakota, South Dakota, Nebraska, Kansas, Oklahoma, Texas, Minnesota, Wisconsin, Iowa, Missouri, Kentucky, Mississippi; NatureServe 2011).

In Louisiana, blue suckers were historically distributed throughout all the major river systems. Currently, they have been reported in the Sabine, Red, and Mississippi Rivers (NatureServe 2011). Blue suckers have been documented only a few times in the Atchafalaya River, and very little is known about its distribution in the system (William E. Kelso, Louisiana State University Agricultural Center, personal communication).

The Atchafalaya River is also home to the endangered pallid sturgeon (*Scaphirhynchus albus*), a prehistoric species that has remained virtually unchanged since the Cretaceous period, 70 million years ago (Fig. 5.25). The pallid sturgeon is one of the largest freshwater fish species in North America, ranging in length (TL) from 76–150

FIG. 5.24. The blue sucker (*Cycleptus elongatus*). This state-listed species of conservation concern has been found in the Atchafalaya River, though records of its existence in this system are sparse. This photo shows the characteristics that allow this species to feed on bottom-dwelling organisms—a mouth located on the underside of its head and thick, fleshy lips. Photo © Eric Engbretson, Engbretson Underwater Photography (www.underwaterfishphotos.com).

cm (30–60 in) and weighing up to 39 kg (86 lbs; largest individual collected; US Environmental Protection Agency 2007, Froese and Pauly 2011). Pallid sturgeon are named for their pale, grayish-white coloration, which becomes whiter as fish age, and its large size and pale coloration distinguish this species from its close relative, the shovelnose sturgeon (*Scaphirhynchus platorynchus*).

Pallid sturgeon require large, turbid, free-flowing riverine habitat with rocky or sandy substrate as was historically found throughout its range—the Missouri and Mississippi Rivers, from Montana to Louisiana, and the Atchafalaya River (US Fish and Wildlife Service 1993, US Environmental Protection Agency 2007). These once large expanses of uninterrupted river habitats allowed pallid sturgeon to migrate long distances upstream to breed on extensive gravel substrates and also facilitated dispersal of larval and juvenile sturgeon on downstream currents (Kynard et al. 2002).

Large-scale development of these river systems for navigation and flood control has created substantial barriers to dispersal, fragmented the river system into discrete segments, and reduced connection between the rivers and their floodplains. Dams, navigation locks, and reservoirs have decreased flow velocity, reduced sediment load, increased water temperatures, and caused alteration of natural hydrological regimes. Virtually all 5,391 km (3,350 miles) of riverine habitat within the pallid sturgeon's range has been altered or degraded (US Fish and Wildlife Service 1993). Approximately 28% has been impounded, which has created unsuitable lake-type habitat. Another 51% has been channelized into deep, uniform channels; and the remaining 21% is downstream

FIG. 5.25. The pallid sturgeon (*Scaphirhynchus albus*). This federally listed species of conservation concern is found in the Atchafalaya River. This photo shows the characteristic gray-white color that distinguishes this species from its closest relative, the shovelnose sturgeon (*Scaphirhynchus platorynchus*). Pallid sturgeon have remained virtually unchanged since the Cretaceous period, about 70 million years ago. Photo © Joel Sartore/joelsartore.com.

of dams, which have altered the hydrography, temperature, and turbidity of the rivers. Combined with extensive loss of gravel substrate throughout its range and unregulated harvest, these factors have caused a drastic decline in pallid sturgeon populations.

In 1990, the US Fish and Wildlife Service listed the pallid sturgeon as an endangered species (US Fish and Wildlife Service 1993, US Environmental Protection Agency 2007). Recovery strategies for the species include raising fish in hatcheries, complete prohibition on harvest, restoration of connections between rivers and floodplains, incorporating fish passage into dam structures, and scientific research to better understand the species as well as the effects of recovery efforts.

In Louisiana, the pallid sturgeon can be found in the Mississippi River from New Orleans to the Old River Control Structure and in the Atchafalaya River (Constant et al. 1997, Killgore et al. 2007). These two reaches currently contain the largest populations of pallid sturgeon in the country, as well as the largest numbers of pallid-shovelnose sturgeon hybrids. There was some concern that discharge through the Old River Control Structure may allow movement of pallid sturgeon from the Mississippi River into the Atchafalaya River but prevent movement back out, thus forming a unidirectional passage barrier (US Environmental Protection Agency 2007). Marked fish from the Mississippi River have been captured in the Atchafalaya, confirming passage from the Mississippi into the Atchafalaya; however, there is no evidence for passage in the opposite direction (Constant et al. 1997, US Environmental Protection Agency 2007). Therefore, the Atchafalaya River may be a sink for adult sturgeon, where adults cannot move back upstream to spawn.

WILDLIFE RESOURCES. The extensive bottomland hardwood forests, cypress-tupelo swamps, and wetland habitats in the ARB provide excellent wildlife habitat for a number of mammals, birds, reptiles, and amphibians (Fig. 5.26; Appendix 2). Although most of the mammal species are considered non-game, the most popular game species are white-tailed deer (*Odocoileus virginianus*), eastern fox squirrel (*Sciurus niger*), eastern gray squirrel, (*Sciurus carolinensis*), swamp rabbit (*Sylvilagus aquaticus*), eastern cottontail rabbit (*Sylvilagus floridanus*), and raccoon (*Procyon lotor*; US Army Corps of Engineers 2000, Byrne and Chamberlain 2011). For example, about 450 deer are harvested on average every year from Sherburne Wildlife Management Area alone (Fig. 5.27; Louisiana Department of Wildlife and Fisheries, unpublished data). Eastern wild turkey (*Meleagris gallopavo silvestris*), American woodcock (*Scolopax minor*), and waterfowl are also hunted in the ARB, as is American alligator (*Alligator mississippiensis*).

BIRDS. The ARB provides habitat for over 250 species of birds, both residents and migrants (Fontenot 2004). It serves as valuable stopover habitat for millions of neotropical migrants, including many species of thrushes, vireos, flycatchers, warblers, buntings, and tanagers, as well as migrating waterfowl. In fact, the ARB is one of the most important areas in the United States for migrant songbirds (Barrow et al. 2005). Because Louisiana still contains large expanses of high-quality natural habitat, such as the ARB, the state supports continentally or globally significant populations of many bird species considered at risk (Fig. 5.28). For example, the ARB floodway supports continentally important populations of nesting bald eagle (*Haliaeetus leucocephalus*), little blue heron (*Egretta caerulea*), and white ibis (*Eudocimus albus*), and given the paucity of data, other breeding species, such as prothonotary warbler (*Protonotaria citrea*), are probably also found in continentally important numbers (Fig. 5.29). Additionally, the floodway also supports a continentally important post-breeding population of wood stork (*Mycteria americana*) and wintering population of American woodcock (National Audubon Society 2011a). The Atchafalaya and Wax Lake deltas and Atchafalaya Bay support continentally important wintering populations of northern pintail (*Anas acuta*), and canvasback (*Aythya valisineria*), as well as substantial populations of mottled duck (*Anas fulvigula*) and bald eagle (Fig. 5.30; Louisiana Department of Wildlife and Fisheries unpublished data, National Audubon Society 2011b). In fact, the ARB provides critically important habitat for a number of species of concern globally as well as both Louisiana Natural Heritage Program and Audubon Watch List species (Table 5.1 and 5.3).

The ARB has long been recognized as one of the few regions in the Lower Mississippi Valley capable of supporting the full suite of forest nesting birds (The Nature Conservancy 2002). The expansive forested wetland habitat in the ARB is especially important for a number of species that require large blocks of forest, such as forest dwelling raptors (Fig. 5.31; i.e., red-shouldered hawk [*Buteo lineatus*], broad-winged

FIG. 5.26. The cottonmouth (*Agkistrodon piscivorus*), also known as the water moccasin is a common semi-aquatic, venomous snake in the ARB. It is one of several species of reptiles that live in the basin's wetlands. In this photo, the snake is in its threat display, showing clearly where it gets its name. Photo © Seth Blitch.

FIG. 5.27. White-tailed deer (*Odocoileus virginianus*), found throughout the ARB, are one of the most popular game animals. They are hunted throughout the basin. Photo © CC Lockwood.

FIG. 5.28. The ARB is an important nesting area for bald eagles (*Haliaeetus leucocephalus*), because of its vast forest, large trees, and abundance of food. Photo by Matt Pardue © The Nature Conservancy.

FIG. 5.29. Forested habitats in the ARB are important areas for birds. Populations of several species that are continentally or globally at risk, such as the prothonotary warbler (*Protonotaria citrea*), an Audubon Watchlist species, are supported by the large expanses of forest . Photo by Matt Pardue © The Nature Conservancy.

hawk [*Buteo platypterus*], Cooper's hawk [*Accipiter cooperii*], Mississippi kite [*Ictinia mississippiensis*], and swallow-tailed kite [*Elanoides forficatus*]), and area-sensitive songbirds (e.g., Swainson's warbler [*Limnothlypis swainsonii*]). Bald eagle and osprey (*Pandion haliaetus*) populations in the ARB have been increasing for several years. Eastern wild turkeys use the expansive bottomland hardwood forests, and their population has been stable since the 1990s. High mean annual adult survival (male 64%; female 67%), especially during the winter, has been balanced by exceptionally low nest initiation rates (33%) and success (38%), due largely to spring flooding that destroys nests and nesting cover (Wilson et al. 2005a, b, Grisham et al. 2008, Byrne et al. 2011). However, the Mississippi River Flood of 2011 may have negatively affected turkey populations across the ARB, since widespread, deep flooding that lasted more than a month

FIG. 5.30. The ARB is an important area for wintering waterfowl. Large rafts of wintering ducks and geese can be found on the Atchafalaya and Wax Lake deltas, which support continentally important populations of northern pintails (*Anas acuta*). Also shown are green-winged teal (*Anas carolinensis*), a common winter visitor to the deltas. Photo by Matt Pardue © The Nature Conservancy.

caused turkeys to stay high in the forest canopy, causing high adult mortality (80%, mostly due to starvation) and no nesting success (Chamberlain et al. 2013).

The ARB is also important for shorebirds and wading birds. More than 30 species of rails and shorebirds have been found in the wetland habitats in the basin. Additionally, the bottomland hardwood habitats of the ARB provide ideal wintering habitat for tens of thousands of American woodcock, and their numbers in the ARB are among the highest in the nation (Fig. 5.32; Straw et al. 1984, Pace 2000).

Numerous wading bird rookeries have been documented in the ARB (Martin and Lester 1990, Green et al. 2006). Surveys conducted by the Louisiana Natural Heritage Program in May/June 2004 and 2005 documented 48 active colonial wading bird rookeries within the historical boundaries of the ARB (Fig. 5.33; Green et al. 2006). Of these, 24 occurred inside the floodway, and another 12 occurred outside the floodway levees in the area around Lake Verret. The Atchafalaya Delta contained five active rookeries, and the Wax Lake Delta contained one. In addition, 3 active rookeries were located in coastal wetlands west of the Wax Lake Delta, and 3 were found on Marsh Island (Appendix 3).

In addition to globally important numbers of white ibis (Fig. 5.34), wading bird species nesting in these rookeries include great blue heron (*Ardea herodias*), great egret

FIG. 5.31. The ARB is important for red-shouldered hawks (*Buteo lineatus*), seen here, and other forest-dwelling raptors that need large blocks of forested wetland habitat. Photo by Matt Pardue © The Nature Conservancy.

TABLE 5.3. Bird species using the Atchafalaya River Basin "Important Bird Area" during all or a portion of their life history

Common Name	Scientific name	Importance
White ibis	*Eudocimus albus*	Globally Important
Wood stork	*Mycteria americana*	Globally Important
Painted bunting	*Passerina ciris*	Globally Important / Audubon Watch List
Prothonotary warbler	*Protonotaria citrea*	Audubon Watch List
Kentucky warbler	*Oporornis formosus*	Audubon Watch List
Swainson's warbler	*Limnothlypis swainsonii*	Audubon Watch List
Wood thrush	*Hylocichla mustelina*	Audubon Watch List

Species considered as significant by the National Audubon Society's Important Bird Areas program (National Audubon Society 2011a, b).

FIG. 5.32. The ARB is an important area for wintering American woodcock (*Scolopax minor*), which can number in the tens of thousands. This species is also hunted in the bottomland hardwoods of the ARB floodway. Photo © Patti Ardoin.

(*A. alba*), roseate spoonbill (*Platalea ajaja*), snowy egret (*Egretta thula*), little blue heron (*E. caerulea*), tricolored heron (*E. tricolor*), cattle egret (*Bulbulcus ibis*), black-crowned night-heron (*Nycticorax nycticorax*) and yellow-crowned night-heron (*Nyctanassa violacea*). The rookery on Marsh Island also contained nesting colonies of Forster's terns (*Sterna forsteri*), black skimmers (*Rynchops niger*), and laughing gulls (*Leucophaeus atricilla*).

In the Atchafalaya Delta more than 19,000 individual wading bird nests were documented in 1994 and almost 12,000 in 1995 (Piazza 1997). During both years, nesting peaked in June (Fig. 5.35). The most abundant nesting birds on the delta during those two seasons were white ibis, great egrets, and the night-herons. In fact, several thousand white ibis nests were built in both 1994 and 1995 on the Atchafalaya Delta (Fig. 5.36), and the Atchafalaya Delta is one of the few places where both white-faced ibis (*Plegadis chihi*) and glossy ibis (*Plegadis falcinellus*) have been documented nesting together (Martin and Lester 1990, Piazza 1997, Piazza and Wright 2004).

Created (dredge) islands on the Atchafalaya Delta are also important nesting areas for seabirds (Martin and Lester 1990, Leberg et al. 1995, Mallach and Leberg 1999). Aerial surveys from 1990–1995 found large and growing populations of seabirds on dredge islands (580 pairs increased to 2,455 pairs). Nesting populations of gull-billed terns (*Gelochelidon nilotica*) and Forster's terns increased five-fold and black skimmer populations increased by half (Leberg et al. 1995). In 1993, about 50 breeding pairs of least terns (*Sternula antillarum*), 300 breeding pairs of gull-billed terns, 320 pairs of Forster's terns, and 1,200 pairs of black skimmers were nesting on the islands in the

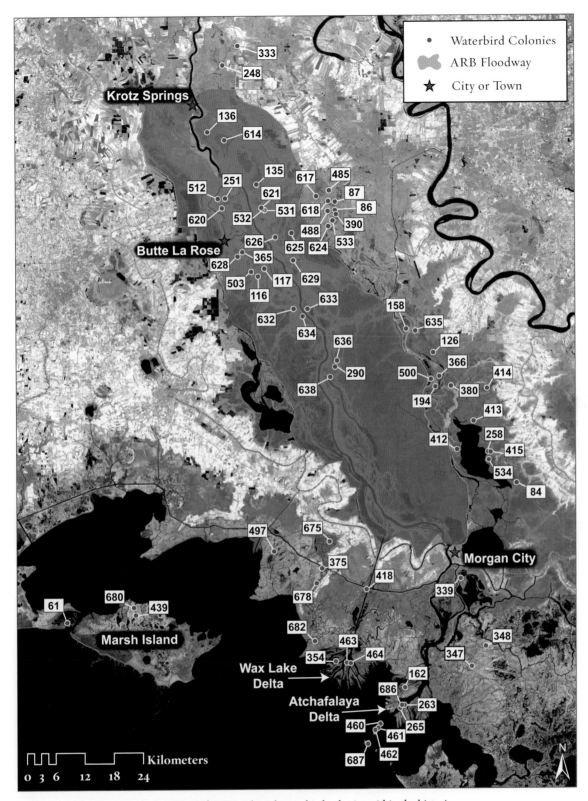

FIG. 5.33. Map showing the locations of active colonial waterbird colonies within the historic boundaries of the Atchafalaya River Basin during spring 2004–2005. Numbers refer to the colony number and correspond to the information in Appendix 3. Figure created from data in Green et al. (2006).

FIG. 5.34. Forested habitats in the ARB are important areas for breeding wading birds. Breeding bird rookeries are important for a number of species including roseate spoonbills (*Platalea ajaja*; top photo) and white ibis (*Eudocimus albus*; bottom photo). Photos © Charles Bush (www.charlesbush photo.com).

Atchafalaya Delta. Compared with other colonies in Louisiana made during that same year, abundance of nesting black skimmers and gull-billed terns was more than an order of magnitude higher on the Atchafalaya Delta (Fig. 5.37).

Nesting on dredge material islands occurred only during the first year after construction or one year after an island (or part of an island) was amended with recently dredged material (Fig. 5.38). After the first year, islands were quickly colonized by panic grass (*Panicum* spp.), paspalum (*Paspalum* spp.), goldenrod (*Solidago* spp.), purslane

Wading Bird Nests Built on the Atchafalaya Delta
1994-1995

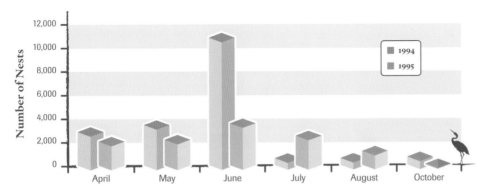

FIG. 5.35. Estimated number of wading bird nests built on six rookeries on the Atchafalaya Delta from 1994–1995. Estimates were calculated with mark-recapture techniques on a group of several thousand sample nests followed through time. Figure redrawn from Piazza (1997).

White Ibis Nests Built on the Atchafalaya Delta
1994-1995

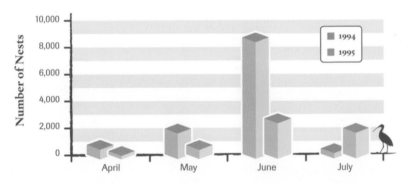

FIG. 5.36. Estimated number of white ibis (*Eudocimus albus*) nests built on six rookeries on the Atchafalaya Delta from 1994–1995. Estimates were calculated with mark-recapture techniques on a group of several hundred sample nests followed through time. Figure redrawn from Piazza (1997).

(*Portulaca oleracea*), salt bush (*Baccharis hamlifolia*), and wild coffee (*Sesbania* spp.; Leberg et al. 1995, Mallach and Leberg 1999). Average nest success (proportion of nests that hatched at least one egg) on dredge material islands was lowest for least terns (20–60%), followed by gull-billed terns (42–69%), and black skimmers (63–91%). Average hatching success (the average proportion of eggs that hatched) was also lowest for least terns (11–50%), followed by gull-billed terns (25–48%), and black skimmers (33–76%).

Louisiana's vast wetland habitats, especially its coastal marshes, provide wintering and migration habitat for two-thirds of the Mississippi Flyway waterfowl and lead the country in numbers of wintering ducks. The 30-year average number of wintering ducks in Louisiana is 3.1 million (Louisiana Department of Wildlife and Fisheries unpublished data). Like the rest of Louisiana's coastal wetlands, the Atchafalaya and Wax

FIG. 5.37. Created (dredge) islands in the Atchafalaya Delta are important nesting areas for seabirds, like these black skimmers (*Rynchops niger*). Other species that breed on the islands include least terns (*Sternula antillarum*), gull-billed terns (*Gelochelidon nilotica*), and Forster's terns (*Sterna forsteri*). Photo © Christina Evans (www. cgstudios. smugmug.com).

FIG. 5.38. A dredge material island being created in the Atchafalaya Delta. The sandy substrate does not remain unvegetated for long, so it is used extensively by breeding seabirds during the first year after construction. Because sandy beaches are not prevalent in coastal Louisiana, as they are in other Gulf of Mexico states, these islands are important wildlife habitat for seabirds, as is evident in the names of some islands (i.e., Bird Island, Avocet Island). This strategy of building islands for bird habitat is an important management tool on the Atchafalaya Delta Wildlife Management Area as a way for wildlife to benefit from the maintenance dredging program that is necessary for navigation and offshore oil and gas operational support. Photo by Cassidy R. Lejeune, Louisiana Department of Wildlife and Fisheries.

Total Number of Wintering Waterfowl Counted on the Atchafalaya and Wax Lake Deltas

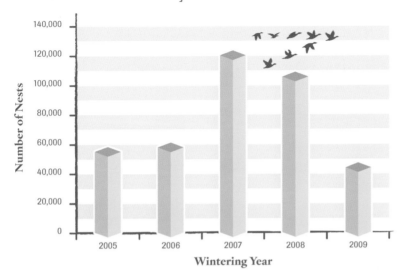

FIG. 5.39. Total number of wintering waterfowl (ducks and geese) counted on the Atchafalaya and Wax Lake deltas during wintering years 2005–2009. Wintering year was defined as September–January, and data are sums of aerial survey estimates conducted from flights taken across those months during each wintering year (i.e., November, December, January). Counts from 2005 are based on only one estimate in January. Unpublished survey data obtained from Louisiana Department of Wildlife and Fisheries.

Lake deltas are important areas for wintering waterfowl, consistently providing habitat for more than ten species of ducks and geese (Johnson and Rohwer 2000). The most common species are gadwall (*Anas strepera*), green-winged teal (*Anas carolinensis*), mallard (*Anas platyrynchos*), pintail (*Anas acuta*), canvasback (*Aythya valisineria*), and American coot (*Fulica americana;* Evers et al. 1998).

While population estimates vary from year to year, wintering waterfowl in the Atchafalaya and Wax Lake deltas consistently exceed 50,000 individuals, with numbers reaching over 100,000 during some years (Fig. 5.39; Evers et al. 1998). Aerial surveys have shown that nearly twice as many waterfowl winter in the smaller Wax Lake Delta compared to the Atchafalaya Delta (Evers et al. 1998). This may be due to a large scale loss of broadleaf arrowhead on the Atchafalaya Delta, an important food source for wintering waterfowl.

In addition to wintering waterfowl, the Atchafalaya and Wax Lake deltas may also be one of the most important nesting areas for mottled duck (*Anas fulvigula*), a non-migratory waterfowl species (Johnson et al. 1996, Holbrook et al. 2000, Johnson et al. 2002). Mottled ducks are found along the Gulf of Mexico coast, from Veracruz, Mexico, to peninsular Florida and typically nest in grassland areas adjacent to wetlands and rice fields (Moorman and Gray 1994). Recently, large numbers of nesting mottled ducks were found using islands in the Atchafalaya Delta, particularly herbaceous vegetation with scattered baccharis (*Baccharis halimifolia*) shrubs (Johnson et al. 1996, Holbrook et al. 2000). Here, the mottled duck annual nesting period spans approximately 100 days, from late March to early July (mean nest initiation mid-April; mean clutch size = 9.2 eggs). Nest density on islands (3.7–8.1 nests ha^{-1}) and across all habitats of the Delta (1.3 nests ha^{-1}) is among the highest along the Gulf Coast (Johnson et al. 1996, Holbrook et al. 2000, Johnson et al. 2002). Mean nest success (40%) on islands in the Atchafalaya Delta is also high when compared to estimates from other nesting areas along the northern Gulf coast (Holbrook et al. 2000).

Other resident waterfowl species also nest in Louisiana's coastal marshes, prairies, and forested wetland habitats including the ARB. These include black-bellied whistling duck (*Dendrocygna autumnalis*), fulvous whistling duck (*Dendrocygna bicolor*), hooded merganser (*Lophodytes cucullatus*), and wood duck (*Aix sponsa*), which nest regularly throughout the ARB in appropriate habitats. Additionally, although blue-winged teal (*Anas discors*) is typically present in the ARB as a migrant and winter resident, small numbers have also been known to nest in Louisiana's coastal marshes and prairies (Lowery 1974a, Wiedenfeld and Swan 2000), including in Atchafalaya Delta (Johnson and Rohwer 2007).

SPECIES OF CONSERVATION CONCERN. Two federally listed wildlife species occur in the ARB—the piping plover (*Charadrius melodus*; both endangered and threatened populations), and Louisiana black bear (threatened). The Florida panther (*Felis concolor coryi*), ivory-billed woodpecker (*Campephilus principalis*), and Bachman's warbler (*Vermivora bachmanii*) historically occurred in the ARB and may still exist elsewhere within their historic range. Additionally, American alligator, which is arguably the iconic wildlife species in the ARB, was formerly listed as endangered.

The piping plover has declined in abundance largely due to habitat loss on both its breeding and wintering grounds. While the importance of breeding habitat to the success of the species is obvious, this species spends two-thirds of its annual cycle on wintering grounds (Eubanks 1994). On wintering grounds, non-vegetated coastal wetland habitats, particularly tidal flats, and sandy beaches without an abundance of human disturbance and infrastructure are especially important (Withers 2002). Many of the beaches traditionally used as wintering habitat by piping plovers have been lost to commercial, residential, and recreational development. Wintering piping plovers have been documented on the Atchafalaya Delta, which contains remote and tidally available, unvegetated flats on newly emerging land, as well as newly formed and amended dredge-material islands (National Audubon Society 2011b).

The Louisiana black bear (*Ursus americanus luteolus*) is one of 16 subspecies of the American black bear (Fig. 5.40). It ranges across eastern Texas, most of Mississippi, and all of Louisiana. This subspecies was listed as threatened in 1992. The main factors in its decline were unregulated harvest and large-scale conversion of bottomland habitat to agriculture in the Lower Mississippi Valley, such that only 20% of the over 24 million acres of bottomland hardwood forest remain (Creasman et al. 1992). Black bears need large expanses of contiguous forest to meet their food and cover requirements, which includes a diversity of mast trees and fruiting plants for food and secluded den

FIG. 5.40. Louisiana black bears (*Ursus americanus luteolus*) have large home ranges and, therefore, need large expanses of contiguous forest. The ARB contains two populations of the bear. Photo by Matt Pardue © The Nature Conservancy.

sites for birthing and wintering. Black bears live 20 to 30 years and typically produce litters every other year after age 3 or 4 (US Fish and Wildlife Service 1995, Black Bear Conservation Coalition 2011). Louisiana black bears use a variety of den structures, including ground nests, woody debris piles, tree snags, and live trees (Hightower et al. 2002), and most tree dens have been documented in bald cypress trees (Fig. 5.41; Hightower et al. 2002, Benson 2005).

Because the ARB contains large expanses of swamp and bottomland hardwood forest, it continued to support bears when numbers of the species reached their lowest point. By the early 1990s, there were only 80–120 bears thought to remain in Louisiana. Currently, two breeding populations of bears exist in the ARB, and, although no reliable black bear population estimate exists for the basin, researchers have captured over 100 different individuals in the area inhabited by the two populations (US Fish and Wildlife Service 1995, Black Bear Conservation Coalition 2011)

One population is centered in the Morganza floodway. Bears in this population have a home range of approximately 33 km² (13 mi²) and with a mean litter size of 1.5 cubs (Wagner 1995, Hightower et al. 2002, Hightower 2003). Although not typically considered resident in large numbers in the lower ARB, a dense population of black bears exists just outside the West Atchafalaya Protection Levee and is centered on Bayou Teche National Wildlife Refuge. Bears in this area use the elevated forests found on salt domes and have been documented using sugarcane fields in the fall season (Nyland 1995). Home range of bears in this coastal population (mean annual occupied range

FIG. 5.41. A Louisiana black bear den tree with a sleeping bear. Large diameter trees like this honey locust (*Gleditsia triacanthos*) in bottomland hardwood forests are used as den structures during the winter. Large baldcypress and water tupelo trees are also highly used as den trees, and, therefore, are protected against harvest in the ARB floodway. Photo by Matt Pardue © The Nature Conservancy.

15 km² [6 mi²]) is smaller than the Morganza population, and mean litter size in the coastal population is higher (2.4 cubs; Wagner 1995, Hightower et al. 2002, Hightower 2003). Due to frequent flooding in the middle and lower ARB, bears are considered transient in this region and are only infrequently reported. Recently, attempts have been made to connect the Morganza population with bears found in the Mississippi Valley to the north by relocating bears in the northern reaches of the ARB, specifically in the Red River Complex (Red River and Three Rivers wildlife management areas, Lake Ophelia National Wildlife Refuge; Wagner 2003, Benson 2005, Crook 2008).

Black bears do well in naturally diverse stands of bottomland hardwood forest that include a significant hard mast component and a structurally complex forest with trees of all age classes. Forests managed in this manner maximize tree vigor and hard mast production across a variety of age classes, stand types, and vegetative composition, thus optimizing black bear foraging habitat (Black Bear Conservation Coalition 2011, Louisiana Natural Heritage Program 2011). Because large cypress and tupelo trees are important escape and denning habitats for Louisiana black bears, retention of both small isolated groups of cypress trees within bottomland hardwoods, as well as large stands of cypress and tupelo is important for bears. Therefore, under the Endangered Species Act, all large cypress and tupelo trees adjacent to water that are 91 cm dbh (36 in; diameter at breast height) and larger, with visible signs of defects (i.e., cavities, broken tops), must be protected within the range of the Louisiana black bear (US Fish and Wildlife Service 1995). Additionally, because canebrakes were historically important habitats for Louisiana black bears, as well as many wildlife species, native bamboo habitat should be favored in hardwood forest stands and restored where appropriate (i.e., abandoned agricultural fields, utility rights-of-way; Black Bear Conservation Coalition 2011).

The Florida panther was listed as endangered by the US Fish and Wildlife Service in 1967. The ARB contains suitable habitat and prey items for the panther; however only few sightings in the ARB have been considered as authentic (US Army Corps of Engineers 2000, Leberg et al. 2004).

The ivory-billed woodpecker was historically a resident of virgin bottomland hardwood and swamp forests, including the ARB, where it relied on areas with large dead trees (Fitzpatrick et al. 2005). Although uncommon (e.g., 1 pair per 16 km²; Tanner 1942), it was widespread across primary forest in the southeastern United States until the mid-19th century. Large-scale logging of old growth timber and corresponding conversion of forests into agriculture reduced this species throughout the lower Mississippi Valley. Additionally, hunting by collectors for their valuable skins added to the decline of this woodpecker in some portions of its range (Tanner 1942, Fitzpatrick et al. 2005).

The last documented population of ivory-billed woodpecker in Louisiana lived in Madison Parish, Louisiana, and following cutting of the forest where it lived, the last universally accepted sighting of this species in Louisiana was an unpaired female in 1944 (Fitzpatrick et al. 2005). The last documented ivory-billed woodpecker reports from in or near the ARB were reported by McIlhenny (1941) who last observed the species east of Avery Island in 1923. Although there were several intriguing visual encounters and sound recordings of birds identified as ivory-billed woodpeckers in the Cache River Swamp of Arkansas in 2004 and 2005 (Fitzpatrick et al. 2005), the reports were not definitively confirmed in spite of intensive searching over five years following the initial sightings, and several experts refute the original identification (e.g., Sibley et al. 2006). Unconfirmed reports of ivory-billed woodpeckers from river basins throughout Louisiana and the southeast United States continue to this day; however, most experts consider the species at least functionally extinct.

The Bachman's warbler, also considered extinct, was also likely a victim of habitat loss (Remsen 1986). This small passerine may have been dependent on insects and arthropods that lived on the large, monospecific stands of a native bamboo species (*Arundinaria gigantea*), that formed extensive "canebrakes" within major river systems across the southeastern United States (Remsen 1986, Gagnon 2006). Loss of this

FIG. 5.42. The American alligator (*Alligator mississippiensis*). This species, once considered endangered, was restored across its range and was subsequently removed from the endangered species list in 1967. The recovery and management of American alligators is considered as an American wildlife success story and serves as a model for crocodilian species worldwide. In this photo, the alligator is eating a nutria (*Myocastor coypus*), an exotic furbearing rodent, along the edge of a bayou. Photo by Matt Pardue © The Nature Conservancy.

habitat to overgrazing and agricultural conversion may have been the factor that led to the demise of this species.

The American alligator was also listed as endangered in 1967, largely due to unregulated harvest and habitat loss (Fig. 5.42; Joanen and McNeese 1989, Elsey and Woodward 2010). However, the combined efforts of the US Fish and Wildlife Service and state wildlife agencies across the South to regulate harvest and develop sustainable commercial alligator farming practices resulted in dramatic population increases and their removal from the endangered species list in 1987. The science-based approach to recovery and management of alligators in Louisiana has been an American wildlife success story and serves as a model for the sustainable management of crocodilian species worldwide (Louisiana Department of Wildlife and Fisheries 2011).

Alligator harvest and commercial trade is regulated through CITES, and while the American alligator is not endangered or threatened anywhere in the United States, it is listed on Appendix II of CITES due to its similarity of appearance to other endangered crocodilian species worldwide. Annually, CITES requires that the Louisiana Department of Wildlife and Fisheries report a finding of no detriment to the US Fish and Wildlife Service, stating that the Louisiana harvest and export of alligators are not detrimental to the survival of the species. Harvest of alligators on public and private lands in the ARB is managed under the Louisiana Department of Wildlife and Fisheries wild alligator management program (Louisiana Department of Wildlife and Fisheries 2011). Allowable harvest of alligators within cypress-tupelo swamp areas in the ARB is 1 per 500 acres, five times less than the limit allowed in preferred alligator habitat—freshwater marsh—where allowable harvest is 1 per 100 acres (Louisiana Department of Wildlife and Fisheries 2009).

NUTRIENTS, CARBON, AND POLLUTANTS 6

L IKE OTHER RIVER SYSTEMS WORLDWIDE, the ARB is an important area for nutrient and contaminant reduction and carbon sequestration. These processes are facilitated by seasonal overflow of water onto the floodplain and the immense volume of sediments that are deposited and retained in the system. About 6 million metric tons (6.6 million short tons) of sediment are retained in the ARB floodway, and more than 23 million metric tons (25 million short tons) are retained in the bay and shelf areas annually (Hupp et al. 2002, Rice et al. 2009, Xu 2010). About 12% of trapped sediment in the ARB floodway is organic material (range 7–28%; Hupp and Noe 2006). The ARB acts as a bioreactor, where nutrients (i.e., nitrogen, phosphorus) and pollutants can be stored or transformed before reaching the Gulf of Mexico. Additionally, the forests and vegetated wetlands are capable of sequestering both atmospheric- and aquatic-based carbon.

NUTRIENTS. Nitrogen indirectly controls offshore hypoxia in the Gulf of Mexico by regulating phytoplankton growth. It has been estimated that 1.5 million metric tons (1.6 million short tons) of nitrogen (~61% nitrate and ~37% organic nitrogen) are discharged annually to the Gulf of Mexico by the Mississippi and Atchafalaya Rivers, with nearly a quarter of that nitrogen discharged by the Atchafalaya River alone (Eadie et al. 1994, Xu 2006a, b, Xu and Patil 2006, Mason and Xu 2006, Lindau et al. 2008, Dale et al. 2010). The annual discharge of nitrogen by the Mississippi and Atchafalaya Rivers represents about 91% of the total annual nitrogen loading by all the rivers discharging into the northern Gulf of Mexico (Rabalais et al. 2002). This annual discharge into the Gulf of Mexico also represents about 90% of the nitrate that enters the Mississippi River, suggesting relatively low removal within the river itself and limited exposure of river water to floodplains (Alexander et al. 2000). The ARB floodway represents an area of substantial river-floodplain interaction; therefore, it is important to understand the role of these floodplains in reducing loading rates of nitrogen to the Gulf of Mexico, a major goal in the Mississippi River system (Mississippi River/Gulf of Mexico Task Force 2008).

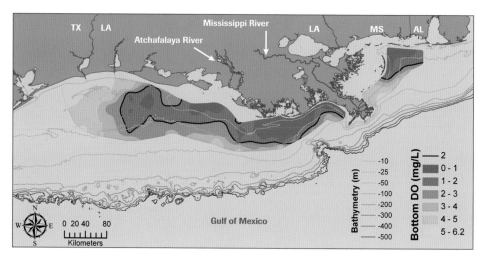

FIG. 6.1. Map showing the extent of the Gulf of Mexico Dead Zone, an area of bottom-water hypoxia that develops off the coast of Louisiana during the summer. This map depicts the distribution of bottom-water dissolved oxygen July 18–21 (east of the Mississippi River delta) and July 24–30 (west of the Mississippi River delta), 2011. Black line indicates dissolved oxygen level ≤ 2 mg L⁻¹ (2 ppm), signifying hypoxia. Figure redrawn from an original provided by Nancy N. Rabalais, Louisiana Universities Marine Consortium. Data source: Nancy N. Rabalais and R. Eugene Turner, National Oceanic and Atmospheric Administration, Center for Sponsored Coastal Ocean Research (www.gulfhypoxia.net).

Nitrogen enters the Mississippi River system mainly through non-point agricultural runoff (>70% of inputs), largely from upper basin states (Alexander et al. 2008). When discharged to the Gulf of Mexico, nitrogen causes extensive zones of hypoxia (a.k.a. dead zone) that encompass 8,000–22,000 km² (3,089–8,494 mi²) off the Louisiana-Texas coasts—an area the size of the state of New Jersey (Fig. 6.1; Turner et al. 2007). The Gulf of Mexico dead zone, second in size only to the hypoxic zone in the Baltic Sea, is a large area of bottom water with low dissolved oxygen concentrations that develops when nutrients fuel large blooms of phytoplankton which then die, sink to the bottom, and fuel bacterial decomposition that reduces oxygen concentrations (Rabalais and Turner 2001, Rabalais et al. 2002, Turner et al. 2007, and many others). Although the areal extent of the hypoxic zone is dependent on a suite of physical and biological factors, the two most important factors are nutrient inputs that fuel plankton blooms and climate, because of its effects on river discharge and water column stability (Justic et al. 2007). Hypoxia has intensified since the 1950s, concomitant with an almost threefold increase in nitrate loads in the Mississippi River system, due to increased fertilizer use and a 30% increase in Mississippi River discharge (Goolsby et al. 1999, Justic et al. 2007, Sprague et al. 2011). Hypoxia can last for several months in the summer, disrupting marine food webs, and causing massive fish kills and other ecological and economic problems (Rabalais and Turner 2001, Rabalais et al. 2002).

Total nitrogen concentration in river water consists of both organic (total Kjeldahl nitrogen; TKN) and inorganic (nitrate and nitrite; NO_x) forms of the nutrient. Several studies have shown substantial nitrogen removal by floodplain wetland systems, through a combination of sequestration by sediments and vegetation and permanent

TABLE 6.1. Estimates of annual mean nitrogen input, output and percent change in the ARB floodway

Nitrogen Species	Input	Output	Change (%)	References
Total Nitrogen				
metric tons yr^{-1}	371,044	320,500	-13.6	Xu (2006 a, b)
	368,700	330,400	-10.4	Xu (2013)
mg L^{-1}	1.86	1.75	-5.9	Mason and Xu (2006)
	1.87	1.68	-10.0	Xu (2013)
Total Kjeldahl Nitrogen (TKN)				
metric tons yr^{-1}	200,323	145,917	-27.2	Xu (2006 a, b)
	199,600	151,100	-24.2	Xu (2013)
mg L^{-1}	0.97	0.94	-3.1	Mason and Xu (2006)
	0.97	0.73	-24.7	Xu (2013)
Nitrate and Nitrite (NO$_3$ + NO$_2$)				
metric tons yr^{-1}	170,721	174,584	+2.3	Xu (2006 a, b)
	169,100	179,300	+6.0	Xu (2013)
mg L^{-1}	0.89	0.82	-7.9	Mason and Xu (2006)
	0.90	0.95	+5.5	Xu (2013)
Nitrate (NO$_3$)				
total metric tons	97,100	93,500	-4.0	BryantMason and Xu (2012)
	89,600	83,200	-7.0	BryantMason et al. (2013)

Negative rates of change signify sequestration or removal in the floodway, and positive rates signify export out of the floodway to downstream habitats and the Gulf of Mexico. Estimates from Xu (2006 a, b) were calculated using long-term (1978–2002) mass balance (inflow = Simmesport, outflow = Morgan City/Wax Lake Outlet). Xu (2013) recalculated those estimates with two additional years of data (1978–2004). Mason and Xu (2006) used long-term water quality data at upstream (Simmesport) and downstream (Morgan City) locations. Both BryantMason and Xu (2012) and BryantMason et al. (2013) calculated a mass input-output budget over the ten-week period of the Mississippi River Flood of 2011.

removal by denitrification, where nitrate is converted to nitrogen gas that returns to the atmosphere. Estimates of total nitrogen removal by these three processes range from 6–72% (Brinson et al. 1980, Watson et al. 1989, Kadlec and Knight 1996, Lane et al. 1999, 2002, Mitsch et al. 2001, Xu 2006a, b, 2013).

A number of studies have assessed the potential for the ARB floodway to remove nitrogen prior to reaching the Gulf of Mexico. Overall mass-balance estimates (inflow = Simmesport, outflow = Morgan City/Wax Lake outlet) of nitrogen retention and removal in the ARB floodway over the past 27 years (1978–2004) showed that the basin retained up to 14% of the total nitrogen inputs and 27% of organic nitrogen inputs, making the ARB floodway a major sink for organic nitrogen (Table 6.1; Xu 2006a, b,

2013). Highest retention rates of organic nitrogen corresponded to high discharge in the winter and spring. While the ARB floodway clearly acts as a sink for organic nitrogen, the dynamics and fate of inorganic forms of nitrogen in the floodway remain less clear.

Inorganic nitrogen (nitrate and nitrite) entering the ARB originates in the Mississippi and Red Rivers. Atchafalaya River water entering the floodway carries an average concentration of 1.1 mg L^{-1} (1.1 ppm) of nitrate, and this average concentration is significantly less than that of the Mississippi River (1.5 mg L^{-1} [1.5 ppm]; BryantMason and Xu 2012). Concentrations in the Atchafalaya River are likely lower due to dilution of Mississippi River water with water from the Red River, which contains average nitrate concentrations of only 0.15 mg L^{-1} (0.15 ppm; Xu and BryantMason 2011, Shen et al. 2012).

During flood events large amounts of nitrate are transported to the Gulf of Mexico through the Mississippi and Atchafalaya Rivers. For example, during the 10-week Mississippi River Flood of 2011, the Mississippi and Atchafalaya Rivers likely discharged over 292,000 metric tons (322,000 short tons) of nitrate (31% of annual nitrate input) to the northern Gulf of Mexico. The total nitrate mass input to the ARB floodway during this event represented 54% of the long-term average annual input of nitrate into the floodway (Table 6.1). However, flood-transported nitrate is typically delayed (BryantMason and Xu 2012, BryantMason et al. 2012, BryantMason et al. 2013). During large flood events in 2008 and 2011, nitrate concentrations were lowest during the period of highest discharge (0.8 mg L^{-1} [0.8 ppm]), whereas the greatest concentrations were found 1–2 months following the event, as river flow receded (2008–1.4 mg L^{-1} [1.4 ppm]; 2011–1.9 mg L^{-1} [1.9 ppm]). This delay suggests that groundwater may be an important source of nitrate in the Mississippi and Atchafalaya Rivers.

Once inside the ARB floodway, nitrate either remains in the river channel to be processed by the well-oxygenated waters or interacts with the floodplain during high flows, where it is removed by denitrification. Studies of nitrate removal have not shown clear patterns. BryantMason et al. (2012) compared isotopic signatures of nitrate in Mississippi and Atchafalaya river outflows and found that signatures did not differ by river. These results suggest minimal processing and uptake of nitrate in the Atchafalaya River, even though, unlike the Mississippi, the Atchafalaya interacts with the floodplain. Other studies that have investigated long-term records and an extreme flood event have found that the ARB floodway can be a nitrate sink (although a weak one) or a nitrate source to the Gulf of Mexico (Table 6.1; Xu 2006a, b, Mason and Xu 2006, BryantMason and Xu 2012, BryantMason et al. 2013).

Both nitrate and nitrite are exported from the ARB floodway to downstream habitats and the Gulf of Mexico. These exports typically occur from summer to early fall (June–October) when river flow was low and water temperatures are high and signals a switch to autochthonous nitrate production, because organic material is mineralized by microbes when water temperatures are high (Xu 2006b, Lane et al. 2011). Nitrate removal is also lower any time water stays in the Atchafalaya River channel and does not interact with the floodplain, because water residence time is short (36 hr from Old River Control Structure to the outlets), water is well oxygenated, and carbon is less available (BryantMason et al. 2012, 2013). Nitrate is also released from the ARB floodway during tropical storm events. This release may be due to rainfall flushing stored nitrate out of the soil and into surface waters (BryantMason et al. 2012).

Both nitrogen sequestration (plants and sediments) and removal by denitrification have also been measured in the ARB floodway (Fig. 6.2). Estimates of nitrogen accumulation in standing tree biomass show that trees accumulate less nitrogen as they age, and bottomland hardwood and cypress stands accumulate nitrogen at approximately the same rate (Table 6.2; Scaroni 2011). Potential denitrification rates (26.72–710.47 μgN g^{-1} d^{-1}), defined as the maximum rate of denitrification possible under optimal conditions (denitrifying bacteria present, ample supply of carbon, and highly reduced conditions [i.e., low dissolved oxygen concentrations]) were significantly higher than background rates (0–1.35 μgN g^{-1} d^{-1}). This suggests that denitrification can be optimal in some locations and under some flow conditions (Scaroni et al. 2010, 2011). Closer examination showed that potential denitrification in the floodway increased along a northwest–southeast vector and especially during spring flood pulse events. Potential denitrification was also significantly greater in lakes than in bottomland hardwood and baldcypress forests. When baldcypress forests were flooded, denitrification rates were similar to the lakes (200 to more than 500 g N ha^{-1} d^{-1}; Lindau et al. 2008, Scaroni et al. 2010, 2011). Denitrification rates reached maximum levels between 2 and 12 days following the onset of the flood pulse and with warmer water temperatures (Lindau et al. 2008, Scaroni et al. 2010). These results suggest that flooded environments, particularly lake habitats and flooded cypress swamps in the southeastern portion of the ARB floodway serve as denitrification hot spots during floods. Maximizing or maintaining connections between the river channel and floodplain forests and lakes can enhance the capacity for nitrogen removal by sedimentation and denitrification in the Atchafalaya River system. Conversely, natural processes and anthropogenic actions that maintain or promote further isolation of the floodplain from the river reduce the potential for nitrogen removal in the ARB.

A number of studies have shown a reduction in nitrogen concentrations in the Atchafalaya River outflow plume as it passes through coastal wetlands and bays in the Atchafalaya, Fourleague, Vermilion, and Cote Blanche bay systems (Fig. 6.3). Lane et al. (2002) calculated that these systems collectively removed 41–47% of nitrate from the Atchafalaya River plume; the majority of the removal occurred in the Atchafalaya Bay system (Atchafalaya and Wax Lake deltas), which received about 90% of the Atchafalaya River flow. Loading rates in this area ranged between 66 and 136 g m^{-2} yr^{-1} and represented 36–42% of the total nitrate carried by the river, annually. The Vermilion and Cote Blanche bay systems received only 4% of the Atchafalaya River flow and 2% of the total nitrate carried by the river (loading rate 8 g m^{-2} yr^{-1}), and nitrogen removal efficiency was 45%. The Fourleague Bay system contains the western Terrebonne marshes. This area receives only around 3% of the Atchafalaya River flow and 3% of the total nitrate carried by the river (loading rate 5 g m^{-2} yr^{-1}; Denes and Caffrey 1988, Lane et al. 2002). Removal efficiencies during high flow were calculated at 90%, likely due to the presence of large areas of vegetated marsh. Overall removal of nitrate nitrogen flowing into the western Terrebonne marshes has been estimated at 20–97% and is related to seasonal processes that affect residence time (Perez et al. 2000, 2003, 2011).

FIG. 6.2. Forested floodplains in the ARB floodway, like this baldcypress forest, sequester nutrients, carbon, and pollutants in the soil and woody tree biomass. As river water interacts with floodplain soils, nutrients are either buried or transformed and removed from the system. While these systems can be an effective nitrogen sink, they also serve as a source of nutrients during periods of the year. Plants and trees also take up nutrients and carbon from the water column and soils, as well as atmospheric carbon. Floodplains may also play an important role in the deposition, storage, and bioavailability of mercury, a neurotoxicant that is important to human health. Photo © Brittany App (www. appsphotography.com) for www.cyclingforwater.com.

TABLE 6.2. Estimates of above-ground tree biomass accumulation rates (kg ha^{-1} yr^{-1}) for nitrogen (N), phosphorus (P) and carbon (C) in the ARB floodway by stand age

	Stand Age (yr)		
	60	80	120
Bottomland Hardwoods			
N	6.98	5.93	4.05
P	0.82	0.7	0.48
C	2,305	1,960	1,337
Baldcypress Swamp			
N	6.83	5.8	3.96
P	0.77	0.65	0.45
C	2,380	2,023	1,380

Estimates are from Scaroni (2011) and are based on adjustments to models in Messina et al. (1986) and Schoch et al. (2009). It is important to note that the accumulation rate for carbon does not make a distinction between aquatic versus atmospheric sources. Table created from data in Scaroni (2011).

While these coastal systems can be effective nitrogen sinks, they also serve as a source of nitrogen during some times of the year. The highest rates of nitrogen export to the Gulf of Mexico occur during 1) winter cold-front passage, when large amounts of sediments are resuspended and then transported offshore, and 2) during summer–fall when river flow slows and warm water temperatures favor internal regeneration of nitrate in the bay (Perez et al. 2011, Lane et al. 2011).

Phosphorus is a reactive nutrient that also influences primary productivity in the Gulf of Mexico (Justic et al. 2007, Turner et al. 2007, Alexander et al. 2008). Although phosphorus is generally not considered a limiting nutrient for coastal hypoxia (Turner et al. 2006, Justic et al. 2007), phosphorus concentrations can briefly limit phytoplankton growth at salinities between 20 and 30 during periods when discharge and productivity are high (Smith and Hitchcock 1994, Lohrenz et al. 1999, Sylvan et al. 2006, Perez et al. 2011). Up to 101,000 metric tons (111,333 short tons) of total phosphorus is discharged annually to the Gulf of Mexico by the Mississippi and Atchafalaya Rivers, and the Atchafalaya River discharges more than one-third of that phosphorus (39,000 metric tons [42,990 short tons]; Goolsby et al. 1999, Turner et al. 2007). Although phosphorus concentrations in the Mississippi River vary, they exhibit little temporal trend with most studies documenting relatively stable to only slightly increased concentrations (~12%; Sylvan et al. 2006, Turner et al. 2006, 2007). Therefore, it is also

FIG. 6.3. Coastal wetlands in the ARB are important for sequestering and removing nutrients and carbon. As river water floods coastal wetlands, nutrients are either buried or transformed and removed from the system. While these coastal systems can be effective nutrient sinks, they may also serve as a source of nutrients during some times of the year. Photo by Seth Blitch © The Nature Conservancy.

important to understand the role of ARB floodplain habitats in reducing loading rates of phosphorus to the Gulf of Mexico.

Phosphorus enters the Mississippi River system mainly through non-point agricultural runoff. However, while nitrogen is largely introduced to waterways from corn and soybean agriculture, phosphorus is largely introduced from animal manure from pasture and rangelands (37–39% of inputs; Alexander et al. 2008). The main mechanism for phosphorus retention in the ARB floodway is sedimentation, and both mass-balance and sediment core data show that accumulated sediment in the ARB floodway retains an estimated 27–28% of the phosphorus loading it receives annually (Hupp and Noe 2006, Turner et al. 2007). Phosphorus is also stored in above ground tree biomass, and, like nitrogen, trees accumulate less phosphorus as they age, and bottomland hardwood and cypress stands accumulate phosphorus at about the same rate (Table 6.2; Scaroni 2011). Mass-balance estimates (inflow = Simmesport, outflow = Morgan City/Wax Lake Outlet) of phosphorus retention in the ARB floodway over the 27 years (1978–2004) showed that the basin retained, by all mechanisms, an average of 41% of total phosphorus inputs (avg. input = 0.28 mg L^{-1} [0.09 – 0.89 mg L^{-1}], avg. output = 0.19 mg L^{-1} [0.02 – 0.64 mg L^{-1}]; Xu 2013).

Coastal marshes in the ARB are both sources and sinks of phosphorus. Phosphorus concentrations decrease with distance from the Atchafalaya River plume during winter/spring high discharge events; however, concentrations increase again into summer and fall and remain high until spring (Lane et al. 2002, Perez et al. 2011). This suggests benthic remineralization of phosphate from the top few centimeters of anaerobic soils is a major source of phosphorus to the water column. This source phosphate may, if export fluxes are large enough, exacerbate hypoxic events.

CARBON. Aquatic and terrestrial carbon flow in the ARB is complex, and, from an ecosystem perspective, the basin may act as a source or sink depending on the pathways that are considered. A significant load of carbon is carried by the Mississippi and Atchafalaya Rivers, and they annually discharge about 2 million metric tons (2.2 million short tons) of carbon to the Gulf of Mexico, with the Atchafalaya River accounting for about half (Eadie et al. 1994, Xu 2006a, b, Xu and Patil 2006, Mason and Xu 2006, Lindau et al. 2008). During the Mississippi Flood of 2011, the Mississippi and Atchafalaya Rivers each discharged about 1 million metric tons (907,000 short tons) of dissolved organic carbon to the Gulf of Mexico (Bianchi et al. 2013).

Riverine organic carbon is cycled in the aquatic food web, buried in sediments, released to the atmosphere, or delivered downstream. Overall mass-balance estimates (inflow = Simmesport, outflow = Morgan City/Wax Lake Outlet) of total organic carbon retention in the ARB floodway over the past 27 years (1978–2004) showed a removal rate of up to 16% (avg. input = 1.3 million Mg yr^{-1}; avg. output = 1.1 million Mg yr^{-1}; avg. input = 0.28 ± 0.01 mg L^{-1}; avg. output 0.19 ± 0.01 mg L^{-1}; Xu and Patil 2006; Xu 2013). The amount of total organic carbon (TOC) retained in the ARB was positively related to the river discharge. Over the 27-year time series, carbon retention increased when the total monthly volume of river inflow reached 20 billion m^3 (Xu and Patil 2006, Xu 2013). The major mechanism of organic carbon retention in the ARB swamps is sedimentation of particulate organic carbon, at a rate of about 100,000 kg km^{-2} (571,000 lbs mi^{-2}) because large amounts of aquatic carbon are buried in sediments and retained in lake bottoms, backswamps, and on natural levees (Lambou and Hern 1983).

Rivers are also important components of fluxes of CO_2 to and from the atmosphere, with the river acting as a source and the continental shelf acting as a sink. Carbon dioxide enters rivers during flooding, and sources include CO_2 dissolved in soil pore water and decomposition of terrestrially derived organic material. This CO_2 is then degassed, and estimates show that streams and rivers in the United States contribute about 100 million metric tons CO_2 (110 million short tons) to the atmosphere annually (Butman and Raymond 2011). Annual estimates of CO_2 flux to the atmosphere show efflux rates of 1,454 g C m^{-2} yr^{-1} and an annual loading rate of 456,000 metric tons (503,000 short tons) from the Atchafalaya River. While efflux rates for the Atchafalaya River are similar to those in the lower Mississippi River (1,036 g C m^{-2} yr^{-1}) and across the Mississippi, Missouri, and Ohio Rivers (1,182 ± 390 g C m^{-2} yr^{-1}), annual loading rates are about half, due to its smaller size (Dubois et al. 2010). In contrast with rivers, continental shelves remove 500 million–1 billion metric tons (551 million–1.1 billion short tons) of CO_2 from the atmosphere each year through primary production, which dominates in surface waters of the coastal ocean (Bianchi et al. 2013).

During extreme flood events, efflux of CO_2 from rivers is enhanced. Not only is there increased input of stored terrestrial source material during these events, but enhanced fluxes of detritus and dissolved organic carbon to the coastal ocean change the continental shelf from a CO_2 sink to a source, because primary production is dominated by bacterial respiration (Bianchi et al. 2013). In April 2011, prior to the onset of the Mississippi River flood, large concentrations of dissolved CO_2 (up to 1,875 ppm) and corresponding fluxes of CO_2 to the atmosphere (up to 1,948 mmol m^{-2} d^{-1}) were documented in the lower Atchafalaya River (Morgan City) and into the delta. At that same time, concentration of dissolved CO_2 (about 66 ppm) in surface water on the shelf and the corresponding uptake of CO_2 (-368 mmol m^{-2} d^{-1}) from the atmosphere were very low. During the flood, dissolved CO_2 concentrations in the river were about double (up to 4,382 ppm), with degassing rates to the atmosphere up to 669 mmol m^{-2} d^{-1}, and this supersaturation and corresponding effusion extended out about 60 km (37 mi) onto the shelf. While concentrations of CO_2 declined to near normal levels post-flood, concentrations continued to be supersaturated into August. Net regional (lower Atchafalaya River to the shelf) flux of CO_2 to (positive values) and from (negative values) the atmosphere during the flood was estimated at -2,700 metric tons C d^{-1} (2,976 short tons) pre-flood, +4,400 metric tons C d^{-1} (4,850 short tons), during the flood, and -9 metric tons C d^{-1} (9.9 short tons), after the flood (Bianchi et al. 2013).

It is widely accepted that forests are also a sink for atmospheric carbon (CO_2). Sequestered CO_2 is removed from the atmosphere by plants during photosynthesis and then stored in live plant tissues or slowly decomposing soil organic matter. Due to its vast forested wetlands and vegetated marsh habitat, the ARB may also serve as a sink for atmospheric carbon. Although there are no specific data on CO_2 sequestration in the ARB, data from the lower Mississippi Valley suggest that bottomland hardwood forests sequester atmospheric CO_2 at a rate of 634 metric tons ha^{-1} (283 short tons acre^{-1}) over 70 years of forest age, with maximum sequestration occurring in forests that are between 30 and 50 years (Shoch et al. 2009). Additionally, as forests age past 70 years, they continue to accumulate carbon, albeit at a much slower rate, making them important global carbon sinks (Luyssaert et al. 2008).

The total and future amount of atmospheric carbon stored in bottomland hardwood forests in the ARB is a function of forest age, management, and potential for reforestation. If the average age of bottomland hardwood forests in the ARB floodway is 70 years, then an estimate of the currently sequestered CO_2 in the floodway would be about 95,200,000 metric tons (104,400,000 short tons). Future carbon storage potential could be increased by reforesting currently cleared agricultural and developed tracts in the floodway.

Baldcypress swamps are also potentially important global carbon sinks, because they are relatively productive and store undecomposed woody and herbaceous plant material in wet, anaerobic conditions for extended periods of time (Middleton and McKee 2004, 2005, US Geological Survey 2004). However, some debate exists on the actual value of cypress swamps (or freshwater wetlands in general) for sequestration of atmospheric CO_2. Both flooded soils and cypress knees emit methane (CH_4), a major greenhouse gas (Pulliam 1992, US Geological Survey 2004). Additionally, the growth and productivity (potential carbon storage) of cypress and tupelo is less in swamps with impaired hydrology, a condition found across many cypress swamps in the southeastern United States, including the ARB (Middleton and McKee 2004, 2005, US Geological Survey 2004, Chambers et al. 2006). Scaroni (2011) estimated carbon accumulation in standing tree biomass for both forested wetland types (bottomland hardwoods and cypress swamps) in the ARB. Whether the source of the accumulated carbon was aquatic or atmospheric cannot be determined. It was apparent, however, that carbon storage potential decreased as stands aged, and, like nitrogen and phosphorus, the accumulation rate of carbon in bottomland hardwoods and baldcypress swamps was nearly identical (Table 6.2).

When the major removal pathways for nitrogen, phosphorus, and carbon are considered (sedimentation, biomass accumulation, and denitrification), two things become apparent: 1) sedimentation is the dominant removal pathway; and 2) there are also a lot of nutrients stored in the standing trees (Table 6.3). As the ARB floodway matures and continues to fill with sediment, fewer nutrients will be stored in the sediment and biomass. However, nutrient storage should increase in the Atchafalaya and Wax Lake deltas, as those systems mature. What is clear, however, is that maximizing or maintaining connections between the river channel and floodplain forests and lakes can enhance the capacity for nutrient removal by sedimentation, biomass accumulation, and denitrification in the ARB floodway. Additionally, regeneration of young trees is critical to future nutrient removal in the ARB floodway and in large forest blocks outside the protection levees. Natural processes and anthropogenic actions that promote either further isolation of the floodplain from the river, reduced tree growth and recruitment, or conversion from forest to open water will reduce potential storage capacity.

TABLE 6.3. Nitrogen (N), phosphorus (P), and carbon (C) removal rates and total removed by area in each major habitat in the ARB floodway

	Area (ha)	Sedimentation		Biomass Accumulation		Denitrification		Total Removal
		(g ha^{-1} yr^{-1})	(metric tons yr^{-1})	(g ha^{-1} yr^{-1})	(metric tons yr^{-1})	(g ha^{-1} yr^{-1})	(metric tons yr^{-1})	(metric tons yr^{-1})
Bottomland Hardwoods	396,900							
N		110,000	43,660	4,050–6,980	1,607–2,770	1,971–2,956	782–1,173	46,049–47,603
P		50,000	19,850	480–820	190–325			20,040–20,175
C		1,630,000	646,950	1,337,000–2,305,000	530,655–914,855			1,177,605–1,561,805
Baldcypress Swamp	106,227							
N		200,000	21,250	3,960–6,830	421–726	1,424–3,650	150–388	21,821–22,364
P		20,000	2,120	450–770	4,882			2,168–2,202
C		3,270,000	347,360	1,380,000–2,380,000	146,593–252,820			493,953–600,180
Lakes	63,873							
N		80,000	5,110	–	–	511–4,380	30–280	5,140–5,390
P		40,000	2,550	–	–			2,550
C		900,000	57,490	–	–			57,490

Estimates for biomass removal are for 60–100 year-old forest stands, and there is no distinction between carbon sourced from aquatic and atmospheric sources. Estimates for denitrification are given for the range from background (no nitrate added) to potential (1 mg NO_3-N L^{-1}). Table after Scaroni (2011).

POLLUTANTS. Sedimentation is also effective at binding and removing some pollutants from the water column. One pollutant that is especially important to human health is mercury (Hg) because, in its methylated form (CH_3Hg), it is a neurotoxicant that can cause developmental deficiencies in children and significantly increased risk of heart disease in adults. Elemental mercury readily bonds to the dissolved organic matter and can be removed from aquatic systems through deposition. Methyl-mercury is formed from inorganic mercury by anaerobic bacteria and can readily enter the food web. Methyl-mercury can bioaccumulate and become a threat to humans and other higher-order predators through fish and shellfish consumption (Marvin-Dipasquale et al. 2000, Rice et al. 2009, Sunderland et al. 2009). The dominant anthropogenic inputs of inorganic mercury into the food chain are atmospheric deposition (exacerbated by burning fossil fuels, particularly coal) and transportation into streams and rivers via runoff from adjacent uplands. The majority of the mercury in aquatic systems accumulates on the dissolved organic matter in sediments, where it is bound and either buried or methylated by sulfate-reducing bacteria (Marvin-Dipasquale et al. 2000, Rice et al. 2009, Gu et al. 2011). The combination of relatively high sedimentation rates, large concentrations of organic matter, and rapid burial rates, as well as the abundance of sulfate-reducing bacteria in the top few centimeters of the sediment, makes coastal areas potential hotspots for mercury deposition and storage; however, periodic fluctuations in the redox state (aerobic vs. anaerobic conditions), may also make floodplains a source of mercury methylization and transfer to the food web (Marvin-Dipasquale et al. 2000, Rice et al. 2009, Sunderland et al. 2009, Gu et al. 2011). The redox state of both the soil and sulfur, as well as the concentration of dissolved organic matter in the soil all interact to influence the interaction between mercury and floodplain soils (Gu et al. 2011).

The Mississippi and Atchafalaya Rivers are major sources of elemental mercury to the northern Gulf of Mexico (Mississippi River = 3.5 ± 2 million metric tons Hg_{ss} yr^{-1} [3.9 ± 2.4 million short tons Hg_{ss} yr^{-1}]; Atchafalaya River = 3.3 ± 2 million metric tons Hg_{ss} yr^{-1} [3.6 ± 2.4 million short tons Hg_{ss} yr^{-1}]). However, most (~80%) of the mercury entering the system is bound to suspended sediments and buried with other sediments deposited on the Mississippi and Atchafalaya deltas, or transported into deeper water on the continental shelf (Rice et al. 2009).

There are no estimates of mercury burial in the freshwater areas of the ARB floodway, and there is some question about the properties of mercury in floodplain systems. Some studies suggest that freshwater systems are excellent mercury sinks because of low methylization due to lack of sulfates. Other studies suggest that the highly reduced nature of anaerobic floodplain soils promotes methylization of mercury. It is currently thought that the combination of the strength of anaerobic reduction in the soil and of the sulfur, as well as the amount of organic matter in the soil, all can affect the burial, reduction, and methylization rates of mercury. Additionally, the role of seasonal or tidal water fluctuations in the biogeochemistry of mercury is not understood, and understanding this role is especially important to understanding the relationship between water management and the biogeochemistry of mercury and the trade-offs implicit in this relationship. For instance, when sediments dry and organic matter oxidizes, mercury that had been bound to that organic matter in the top layer of the sediment can return to its elemental form and be mobilized. In this form, the mercury can also be volatilized by sunlight, returning it in the atmosphere, where it can be deposited elsewhere. Therefore, water management that more closely approximates seasonal flooding-drying cycles may reduce local mercury levels in the sediment.

While local mitigation of mercury is a potentially positive thing, at larger scales this may represent a tradeoff because the mobilized mercury, both in the atmosphere and the water, will eventually settle somewhere else. This remobilization of mercury by oxidation and volatilization and the relationship of these processes to flooding-drying cycles are poorly understood components of the mercury cycle in riverine systems. More research is needed to understand mercury deposition, methylization, mobilization, and biomagnification in riverine watersheds, particularly in freshwater wetlands.

Other pollutants have also been reported in the ARB, particularly outside the floodway levees, in areas of agricultural and residential development. High concentrations of atrazine, an herbicide that is commonly used for the control of broadleaf weeds in agricultural fields (grain, sugarcane) and residential lawns, was reported in three sub-watershed areas outside the East Atchafalaya Floodway Protection Levee (Tri-Parish Partnership 2009). High concentrations of atrazine have also been found in the drinking water, requiring extensive filtration to remove the pollutant. As expected, atrazine was sourced to agricultural and residential runoff, particularly in the spring when application rates are highest. Elevated concentrations of chloride have also been found in the heavily agricultural areas outside the protection levees (Tri-Parish Partnership 2009). This non-point source pollutant has been linked to runoff from both irrigated and non-irrigated crop production, and estimates of up to 53% reductions in chlorides have been required south of the Morganza Floodway to meet minimum daily load values. Fecal coliform pollution is also a problem in some areas outside the floodway levees, and the largest source is on-site wastewater treatment (i.e., septic systems). Elevated fecal coliform bacterial counts have been found in 9 of 14 sub-watershed areas outside the East Atchafalaya Flood Protection Levee, and estimates of 20–92% reductions are required to meet minimum daily load standards in these areas (Tri-Parish Partnership 2009).

STAKEHOLDERS IN THE ATCHAFALAYA RIVER BASIN

7

THE IMPORTANCE OF THE ARB as a provider of natural resources, livelihoods, recreation, and social culture transcends into a deep connection to the ARB by people who use it, live in it, and appreciate it (Fig. 7.1). This deep sociocultural connection is well documented in oral histories and a broad literature, as well as by the people that currently live, work, and play in the ARB. As a consequence, ARB stakeholders are deeply connected to the ecological processes that have and continue to shape the ARB, as well as changes to those processes, brought about by past and future decisions that will further shape the landscape.

The ARB encompasses all or part of 13 Louisiana parishes (counties), and about 10,000 Louisiana residents live in the ARB along narrow strips of land outside the protection levees (van Maasakkers 2009). Many basin-parish and ARB residents directly or indirectly use the ARB for work and recreation. While many stakeholders live and work in the ARB, many others do not live in close proximity.

STAKEHOLDER GROUPS. The Consensus Building Institute (2010a) identified clusters of stakeholder groups in the ARB based on their similar interests and priorities: land interests (individual and corporate landowners), fishing (commercial fishermen and buyers and recreational fishermen), environmental groups (with different missions and approaches), flood protection systems (US Army Corps of Engineers, Vicksburg District, Atchafalaya Basin Levee District), government agencies (habitat management and public use), government agencies (research and science—includes academic scientists), local and tribal government, navigation, oil and gas, timber, tourism and recreation (e.g., hunting, fishing, bird watching and canoeing, ecotourism), and coastal interests (i.e., coastal parish governments, coastal landowners). Three additional stakeholders did not group with other clusters because of their divergent interests or responsibilities: US Army Corps of Engineers, New Orleans District and Louisiana Department of Natural Resources (agencies responsible for management and restoration in ARB floodway) and the Sidney A. Murray, Jr. Hydroelectric Station (interest in maximizing power production for profit). Stakeholders or stakeholder groups make

FIG. 7.1. Stakeholders in the ARB. The ARB is valued by a large and growing number of people who live, work, and play inside its boundaries. Therefore stakeholders are very connected to the land and deeply vested in decisions made about its current and future management. In this photo a family enjoys their floating camp, while a friend, also an oil industry worker in the ARB, pays them a visit. In the background, recreational fishermen fish along the edge of a forest that is owned by private landowners. Photo © CC Lockwood.

different basic assumptions about the boundaries of the ARB, speak very differently about its "naturalness" and history, and feel differently (and strongly) about what is important in terms of its use (van Maasakkers 2009). These differences have, at times, fueled conflict, controversy, and litigation regarding the rights to basin resources, and these conflicts have affected the ability to find consensus and implement efforts to improve the ecology and management of the ARB (Reuss 2004, van Massakkers 2009, Consensus Building Institute 2010a). It is important to note that despite these marked differences in opinions, recent research shows broad areas of shared interest and agreement, within and among stakeholders, about the importance of the ARB's resources and its current ecological state, and concern about the future trajectory of the ARB (Louisiana Department of Natural Resources 2002, van Maasakkers 2009, Consensus Building Institute 2010a).

LAND OWNERSHIP. About half (161,874 ha [400,000 acres]) of the land in the ARB floodway is publicly owned, with the remainder (177, 252 ha [438,000 acres]) pri-

FIG. 7.2. A typical floating camp in the ARB floodway. Photo by Richard Martin © The Nature Conservancy.

vately owned. Publicly owned land inside the floodway levees includes state lands and water bottoms, state wildlife management areas (Sherburne and Attakapas Island) and federally owned lands (Indian Bayou Wildlife Management Area, Atchafalaya National Wildlife Refuge). Two additional state wildlife management areas (Elm Hall and Thistlewaite) are located outside the floodway levees, but within the historical extent of the ARB. Additional state-owned land outside the floodway levees but inside the historical basin boundaries is the Bayou Teche Scenic Bayou that includes Lake Fausse Pointe State Park (2,428 ha [6,000 acres]). Below Morgan City, Louisiana, the Atchafalaya and Wax Lake deltas are owned and managed by the State of Louisiana (Atchafalaya Delta Wildlife Management Area).

There is no permanent habitation within the ARB floodway. However, there are a number of land-based and floating camps that are maintained for recreational purposes, mostly hunting and fishing (Fig. 7.2). Camps are typically basic, since structures are periodically flooded by the river. Currently, new camp construction is not allowed inside the floodway but established camps may remain.

The Morganza Floodway is privately owned, however, under the authority of the Flood Control Act of 1928, the US Army Corps of Engineers purchased perpetual comprehensive easements on all lands within the floodway (US Army Corps of Engineers 1998). Habitation within the floodway is not permitted, but use of the land for farming, timber and mineral harvest, and other purposes not in conflict with flood control are permitted with prior approval.

Perpetual flowage easements were acquired for all lands in the West Atchafalaya Floodway (US Army Corps of Engineers 1998). These easements allow for full use of the lands for flood control; however, they are less restrictive than the comprehensive easements in the Morganza Floodway. Landowners retain their right to improve and inhabit the land, farm, and harvest timber and minerals.

PART THREE

HOW IS THE ATCHAFALAYA
RIVER BASIN MANAGED?

HUMAN ENGINEERING IN THE ATCHAFALAYA RIVER BASIN   8

FROM THE TIME OF EUROPEAN SETTLEMENT in the ARB, people have tried to control seasonal flooding and make life along the river more stable and predictable (Reuss 2004). Settlers wanted assurance that crops growing in the fertile alluvial soils would not be flooded. Steamboat operators supported construction of channel cutoffs and maintenance of artificially high water levels to ensure shorter, more dependable navigation routes within the basin, as well as assurance that transport between the Atchafalaya and Mississippi Rivers was possible throughout the year. Consequently, many small earthen levees were built by settlers to protect land parcels. River cutoffs were built in the main channel (e.g., Shreve's Cutoff) to shorten travel time, and private interests partnered with the US government to keep the river free of log rafts. However, it was not until the catastrophic flood in 1927 and the subsequent Flood Control Act of 1928 that large-scale modification of the Atchafalaya River Basin began in earnest.

The Flood Control Act of 1928 established the Mississippi River and Tributaries Project and directed the US Army Corps of Engineers to develop and implement flood control on the Mississippi River. The goal of this massive flood control project was to safely pass a flood of up to 84,950 m^3 s^{-1} (3 million cfs, "project flood") to the Gulf of Mexico, and the tools used to successfully accommodate this flow would be levees, floodways, structures, and other channel and basin improvements (Fig. 8.1; Reuss 2004). As part of this project, the ARB floodway was constructed. The floodway was designated as the principal floodway for the lower Mississippi River and was designed to carry one-half of project flood discharge—42,475 m^3 s^{-1} (1.5 million cfs)—from the lower Mississippi River to the Gulf of Mexico. To put this in perspective, consider that this volume of water would completely fill the Louisiana Superdome in 1 minute or would flood the entire city of New Orleans to a depth of 2.4 m (8 ft) in less than 12 hours (US Geological Survey 2001). The ARB floodway is actually managed as a system of three floodways: the West Atchafalaya and Morganza (East Atchafalaya) floodways on the north and the Atchafalaya Basin Floodway System (also called the Lower Atchafalaya Basin Floodway) to the south (Fig. 8.2).

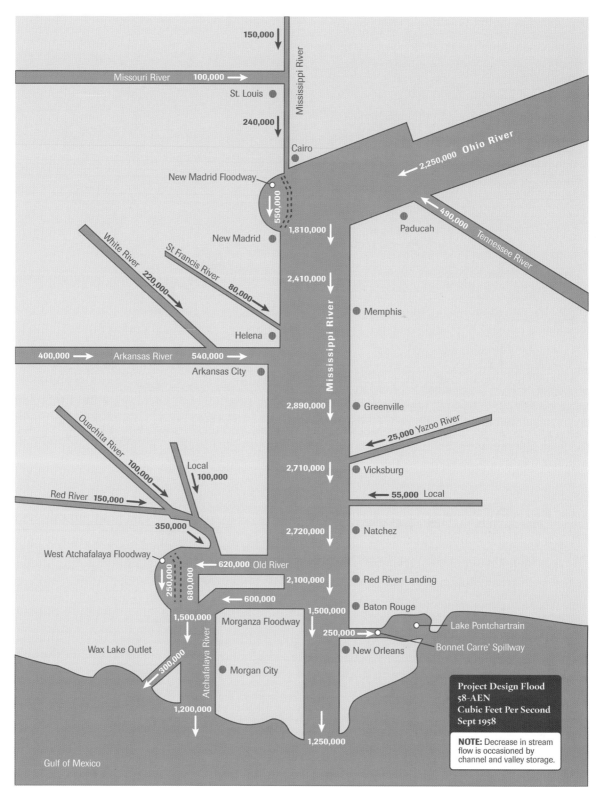

FIG. 8.1. US Army Corps of Engineers design plan diagram for discharge routing during Project Flood (84,950 m³ s⁻¹ [3 million cfs]) on the lower Mississippi River. The ARB floodway is designed to pass one-half of the total project flood discharge. Diagram reports discharge in cubic feet per second (cfs). Figure redrawn from US Army Corps of Engineers (1982).

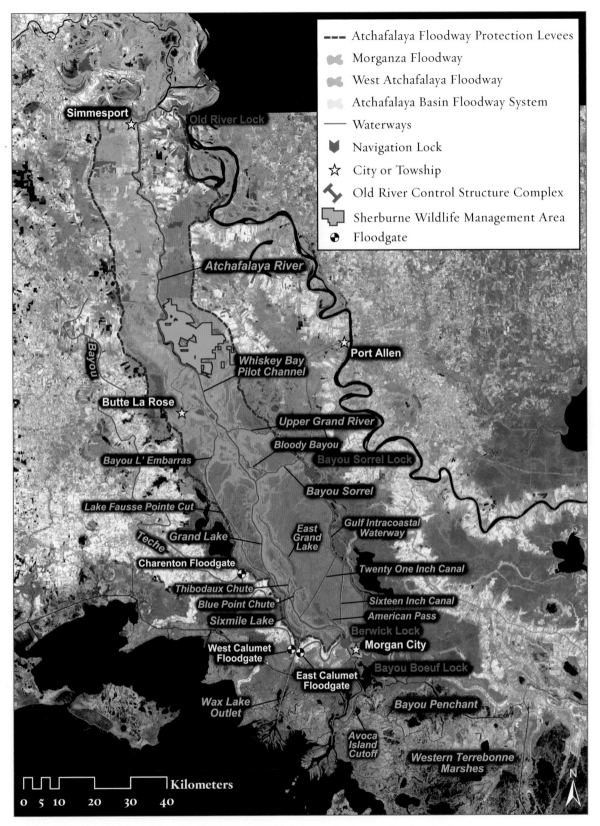

FIG. 8.2. Map showing the engineering features created for drainage, flood control, and navigation in the modern ARB.

FIG. 8.3. The Morganza Floodway structure discharging water into the floodway during the Mississippi River Flood of 2011. This flood event marked only the second time in history that this structure was opened to lower river level and protect downstream municipalities (e.g., Baton Rouge, Louisiana). At the height of the flood, the structure passed up to 5,154 m³ s⁻¹ (182,000 cfs), or about 30% of its capacity. Photo by David Y. Lee © The Nature Conservancy.

The Atchafalaya River and the Atchafalaya Basin Floodway System are always active and handle most discharge events up to ~28,000 m³ s⁻¹ (1 million cfs) combined flow from the Red River (100% of discharge directly into the Atchafalaya River) and the Mississippi River, through the Old River Control Structure complex. The Morganza Floodway to the south is used when Mississippi River flow reaches 42,475 m³ s⁻¹ (1.5 million cfs) at Red River Landing (USACE 01120), located 35 river km (22 river mi) to the north, and the floodway is accessed by opening a series of floodgates in the overbank structure.

The Morganza Floodway, which includes extensive forested habitat as well as agricultural land, is designed to discharge ~16,990 m³ s⁻¹ (600,000 cfs). At the time of this writing, this floodway has been used only twice—1973 and 2011 (Fig. 8.3). In 1973, it discharged half its capacity. During the Mississippi River Flood of 2011, 17 of its 125 gates were opened and it discharged up to 5,154 m³ s⁻¹ (182,000 cfs), or about 30% of its capacity (Fig. 8.4).

The West Atchafalaya Floodway was designed to be activated only in a worst-case scenario. It is normally protected from riverine input by a fuse-plug levee at its north end. To date, the floodway has never been opened. If this floodway is needed for flood control, the fuse-plug levee would be blown, allowing access by 7,079 m³ s⁻¹ (250,000 cfs) of discharge. Because this floodway is only activated as a last resort and has never been operated, this area is largely kept in agriculture.

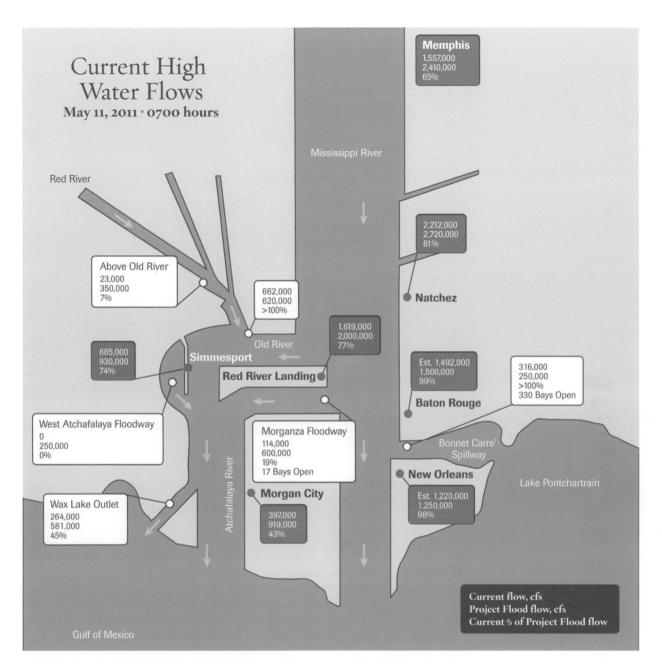

FIG. 8.4. US Army Corps of Engineers diagram showing discharge routing on the lower Mississippi River on the morning of May 21, 2011, during the Mississippi River Flood of 2011. This diagram shows full operation of the Old River Control Structure Complex and partial operation of the Morganza Floodway, with both discharging into the ARB floodway. Partial opening of the Morganza Floodway began on May 14, 2011, when flows reached 42,475 m³ s⁻¹ (1.5 million cfs) at Red River Landing, and closure began on May 24, 2011. This diagram represented conditions during the period of peak flood crest at Red River Landing. Diagram modified from the US Army Corps of Engineers, Operation Floodfight.

FIG. 8.5. A river levee adjacent to the upper portion of the Atchafalaya River. These levees exist only in the upper portion of the ARB floodway, confine flow to the river channel, and isolate the West Atchafalaya (left side of photo) and Morganza floodways from seasonal flooding. Photo by David Y. Lee © The Nature Conservancy.

The levees along the Atchafalaya River are designed to keep these three inflows (Atchafalaya River, Morganza Floodway, and West Atchafalaya Floodway) separate until they merge at the Atchafalaya Basin Floodway System (Fig. 8.5). Extensive engineering and modification has occurred in the ARB and includes the use of levees and structures for water control and drainage as well as extensive dredging and channelization for channel training.

LEVEES AND STRUCTURES. There are a total of 723 km (449 mi) of federal levees in the ARB (Fig. 8.2; US Army Corps of Engineers 2011). East and West Atchafalaya Basin Protection levees, located approximately 25 km (15 mi) apart, run the length of the ARB floodway from the structure complex at Old River to Morgan City and define the outermost boundaries of the Atchafalaya Basin Project (see Fig. 3.7). The protection levees restrict the floodway to about 26% of the historic 8,345 km^2 (3,222 mi^2) basin (Lambou 1990, Sabo et al. 1999a, b). While these levees permit relatively unrestricted sheet and channel flow inside the levees, they separate a large portion of the historic basin from the natural riverine flooding and flushing processes associated with seasonal floods on the Atchafalaya River (Gagliano and van Beek 1975).

FIG. 8.6. An internal levee adjacent to the Atchafalaya River channel at Butte La Rose, Louisiana, the southernmost town inside the ARB floodway. While inside the floodway, the town sits high atop the riverbanks and constructed levee, far above all but the highest water marks (i.e., Mississippi River Flood of 2011). Also shown is Interstate Highway 10, which is elevated as it traverses across the width of the ARB floodway. Photo © Alison M. Jones for www.nowater-nolife.org.

A second set of major internal levees has been constructed adjacent to the main river channel of the Atchafalaya River, from its inception at the Old River Control Structure to Sherburne Wildlife Management Area on the east bank and past Butte La Rose on the west bank (Fig. 8.6). These levees confine and channelize flow in the upper basin and divide the upper basin floodway areas (see above). In addition to confining flood flows to within the Atchafalaya channel, the levees cut off natural distributaries from the river (e.g., Alabama Bayou) and isolate portions of the former Atchafalaya River floodplain within the West Atchafalaya and Morganza Floodways from seasonal inundation and high-flow pulses. South of the main channel levees, flood water is able to inundate the historic Atchafalaya floodplain and associated swamp forests of the Atchafalaya Basin Floodway System. The extent of that inundation is dependent upon river stage, land elevation, and associated hydrologic factors. However, flood flow is contained within the East and West Protection Levees and, therefore, restricted to the middle one-third of the former extent of the historic ARB. In the lower basin, the Avoca Island Levee, which was constructed south of Morgan City in 1953 for the purpose of preventing backwater flooding south and east of Morgan City, may impede freshwater flow eastward into the Terrebonne marshes (Reuss 2004).

FIG. 8.7. The Bayou Sorrel navigation lock. This lock is built into the East Atchafalaya Protection Levee and permits navigation of barge traffic between Baton Rouge and Morgan City, through the ARB floodway. Note the sediment-laden water in the Gulf Intracoastal Waterway on the inside of the floodway (bottom half) compared with that of the waterway on the outside of the levee. Photo © Alison M. Jones for www.nowater-nolife.org.

Four locks and several structures have been built in the ARB floodway for drainage, flood control, and navigation. Navigation channels are found at Old River, Bayou Sorrel, Bayou Boeuf, and Berwick and facilitate passage of a combined cargo tonnage of more than 45 million metric tons yr⁻¹ (50 million short tons yr⁻¹; US Army Corps of Engineers 2001). The Old River lock is a part of the Old River Control Structure complex and permits navigation between the Mississippi and Atchafalaya Rivers. Bayou Sorrel lock is the busiest lock in the ARB and is part of the East Atchafalaya Protection Levee (Fig. 8.7). This lock permits navigation from the Mississippi River at Port Allen, Louisiana, to Morgan City, Louisiana, via the Gulf Intracoastal Waterway Alternate Route inside the East Atchafalaya Protection Levee. The Bayou Boeuf lock is a part of the East Atchafalaya Protection Levee south of Morgan City and permits east-west travel through the Gulf Intracoastal Waterway at all water stages in the ARB. Because Bayou Boeuf once served as a distributary for water eastward from the river, the lock now prevents this direct eastward flow of lower Atchafalaya River water through Bayou Boeuf. However, there is evidence that shows instantaneous discharge greater than

FIG. 8.8. The Point Coupee drainage structure and pump station. This structure, one of three in the ARB floodway, pumps water out of a large agricultural area north of the Morganza Floodway into the Atchafalaya River (top of photo) south of Simmesport, Louisiana. In this photo, the Morganza Floodway is located to the left of the structure. Photo by David Y. Lee © The Nature Conservancy.

481 m³ s⁻¹ (17,000 cfs) moving eastward through the Avoca Island Cutoff, south of the Bayou Boeuf lock (Swarzenski 2003). Some of this flow is distributed east to marshes of upper Fourleague Bay, while some is distributed northeastward to the Gulf Intracoastal Waterway and then through Bayou Penchant to the western Terrebonne marshes (Lane et al. 2002, Swarzenski 2003). The Berwick lock is a part of the West Atchafalaya Protection Levee. It serves both as a navigation link and a floodgate for the lower Bayou Teche (US Army Corps of Engineers 2011).

Drainage structures are found at Bayou Courtableau and Bayou Darbonne and just outside the Morganza Floodway levee (Point Coupee Drainage Structure; Fig. 8.2). The drainage structures at Bayou Courtableau and Bayou Darbonne are used to direct drainage intercepted by the West Atchafalaya Protection Levee into the ARB floodway. Those two structures are also operated to retain water for irrigation in the Teche-Vermilion basin during low flow periods. Three floodgates (Charenton, East and West Calumet) are used to control the flow of fresh water into Bayou Teche and allow navigation through the Atchafalaya levees. The Pointe Coupee structure is used to evacuate water from the system of canals that drain a large agricultural area (Fig. 8.8).

481 m^3 s^{-1} (17,000 cfs)

FIG. 8.9. Dredging has been widely used in the ARB floodway to ensure that the system can reliably pass floodwaters and to ensure a stable navigation route. In this photo, a dredge works on the Lower Atchafalaya River, south of Morgan City to offset the effects of sedimentation and maintain the channel for navigation. Photo by Cassidy R. Lejeune, Louisiana Department of Wildlife and Fisheries.

CHANNEL ENGINEERING. Intensive dredging (1955–1966) has accelerated the development of the Atchafalaya River and offsets the effects of sedimentation, provides a stable navigation route, and creates canals for oil and gas exploration and transport (Fig. 8.9; Hebert 1966, Sabo et al. 1999a, Reuss 2004, US Army Corps of Engineers 2001). It is important to note that channel development is a natural process in deltaic systems, and, as deltas build up, river channels stabilize by deepening within progressively taller riverbanks. As delta development proceeds, floodplains become progressively more isolated from river flooding, except at the highest river stages. In fact, well-developed river channels are characteristic of bottomland hardwood systems, where elevation and riverbanks are highest (Gagliano and van Beek 1975). The engineered deepening of the main Atchafalaya River channel (and corresponding bank stabilization levees along the lower river from Thibodaux Chute to American Pass) has accelerated this natural channel stabilization process causing unnaturally rapid and extreme separation of swamp habitats from the river so that higher discharges are necessary to reach overbank flow in the river and distributaries and flood swamp forests that normally would flood seasonally via overbank flow (Bryan et al. 1998, Alford and Walker 2011).

While one component of dredging involved deepening and channelizing the main river channel, another component involved dredging new channels to shunt water downstream as quickly as possible. The Whiskey Bay Pilot Channel is an example of a channel dredged for navigation. Originally built to serve as a shorter path for navigation, this channel now captures the majority of Atchafalaya flow and is significantly larger and deeper than the main Atchafalaya River at Butte La Rose. The Lake Fausse Pointe Cut is an example of a channel dredged to shunt water downstream. When the

FIG. 8.10. Bayou Sorrel is one of the few distributaries remaining in the ARB floodway. It is important for flood control because it directs water downstream. It also distributes large amounts of water from the Atchafalaya River to swamp forests on the east side of the floodway. In this photo, taken during a river flood, excess water is overtopping the banks and flowing across the floodplain. Photo by David Y. Lee © The Nature Conservancy.

West Atchafalaya Protection Levee was built and severed the Lake Fausse Pointe delta from the rest of the ARB, the natural distributaries (Bayou L'Embarras, Grand Bayou) that moved water down-basin were also severed. Therefore, the Lake Fausse Pointe Cut was dredged to move water south to Grand and Six Mile Lakes.

An additional component of this channel engineering was the closure of most of the natural bayous and distributaries (22 closures) from the main trunk on the lower 91 km (56 mi) of the Atchafalaya River, so that only eight natural distributaries now divert water out of the main channel and into the swamp (Fig. 8.2; Sparks 1992, Sabo et al. 1999a, Reuss 2004). Distributaries vary in size and orientation to the main channel, controlling the quantity of water input as well as the direction of water movement in swamp habitats. While other canals and bayous can serve as distributaries at various river stages, the eight major natural distributaries are Bayou L'Embarras, Grand River, Bloody Bayou/Jake's Bayou, Bayou Sorrel, Blue Point Chute, American Pass, and Crook Chene (Fig. 8.10).

FIG. 8.11. An oil field in the ARB floodway. Note the array of canals oriented in a number of directions. Some pipeline canals, (i.e., top inset photo) run in an east-west direction and impede natural flow patterns. Other canals that are connected to the river channel or a distributary (i.e., bottom inset photo) channel sediment-laden water directly into lakes and backswamps. Top photo by Richard Martin © The Nature Conservancy. Bottom photo by Anne Hayden, Southern Illinois University, Carbondale, Illinois.

Channel engineering and distributary closure has been used to evacuate water as rapidly as possible through the ARB floodway mainly down three main drain paths: 1) main Atchafalaya River channel; 2) lateral distributaries to shunt water east to the Gulf Intracoastal Waterway; and 3) lateral distributaries to shunt water west down the Fausse Point Cut, through Six Mile Lake, and to the Wax Lake Outlet (Bryan et al. 1998).

Oil and gas exploration has also resulted in a great deal of channelization in some portions of the ARB floodway (Fig. 8.11). Large oil field and mineral transport canals (e.g., Twenty-One Inch Canal, Sixteen Inch Canal) and their associated levees are often oriented in a direction that impedes natural flow patterns. Additionally, many canals are connected to the mainstem river or high-energy distributaries, which serve to channel sediment-laden and oxygenated water directly into backswamp habitat that would naturally not receive large sediment inputs (Hupp et al. 2008). The combination of rapid water inflow and impeded hydrology means that water entering swamp areas via canals often cannot exit and becomes stagnant and hypoxic (Sabo et al. 1999a, b).

The effects of channel engineering have increased the volume share of the main river channel and the few remaining distributary channels. This, combined with canalization and flow obstructions from mineral extraction, has caused unnatural inundation patterns and sediment inputs to the floodplain in the ARB floodway, resulting in exacerbated flooding and sedimentation in some swamp habitat and increased isolation of other swamp habitat except when flood stages are at their peak (Bryan et al. 1998). These factors and the existing large-scale water management model that has kept flow allocation static and unchanged since construction of the Old River Control Structure have been identified as causing an array of problems in the ARB floodway, ranging from accelerated habitat conversion to large-scale water quality problems.

WATER MANAGEMENT UNITS IN THE ARB FLOODWAY. A system of 13 water management units (originally called environmental and hydrologic units) were developed by the Environmental Protection Agency in the 1970s and identified by the US Army Corps of Engineers in 1982 (in the Atchafalaya Basin Floodway System Project) as a means to address the hydrologic problems (water quality, persistent flooding, and sedimentation) in the ARB floodway and to improve crawfish production (Fig. 8.12; US Army Corps of Engineers 2001, Reuss 2004). Although overall water distribution and quality inside the units are driven mainly by the water levels in the river and distributaries, each unit is unique and hydrologically disconnected from the others for the most part (Bryan et al. 1998, Allen et al. 2008, Kaller et al. 2011).

The four principal objectives for the water management units were: 1) to mimic, as nearly as possible, historical natural water overflow patterns; 2) to provide proper water movement through the units; 3) to restrict sediment movement and deposition within the units; and 4) to supply nutrient and organic matter to estuarine and Gulf areas, all with the goal of improving water quality (US Army Corps of Engineers 1982, Reuss 2004). Originally, the strategy for implementing these objectives was to manage each unit independently with a system of interior levees and water control structures. However, this particular strategy was never pursued in earnest, due to insufficient evidence that the structures would achieve the goals and concern about further destruction of the intrinsic beauty and nature of the ARB floodway (Bryan et al. 1998). Although the original strategy was rejected, the approach to water management in the individual units (or multiple units) has evolved based on an improved scientific understanding of hydrology, sedimentation, and ecology of the ARB floodway, such that two of the congressionally authorized pilot water management unit projects (Buffalo Cove and Henderson Lake) are currently being designed and implemented. Furthermore, the state of Louisiana, through the Atchafalaya Basin Program, is designing, planning, and implementing projects to manage water movement and improve hydrology within single and multiple management units. These projects are designed to improve water quality and reduce the loss of aquatic habitats and are being rigorously planned and assessed by a panel of scientific and technical experts with the help of the Atchafalaya Basin Program Natural Resource Inventory and Assessment Tool (http://abp.cr.usgs.gov/). More information,

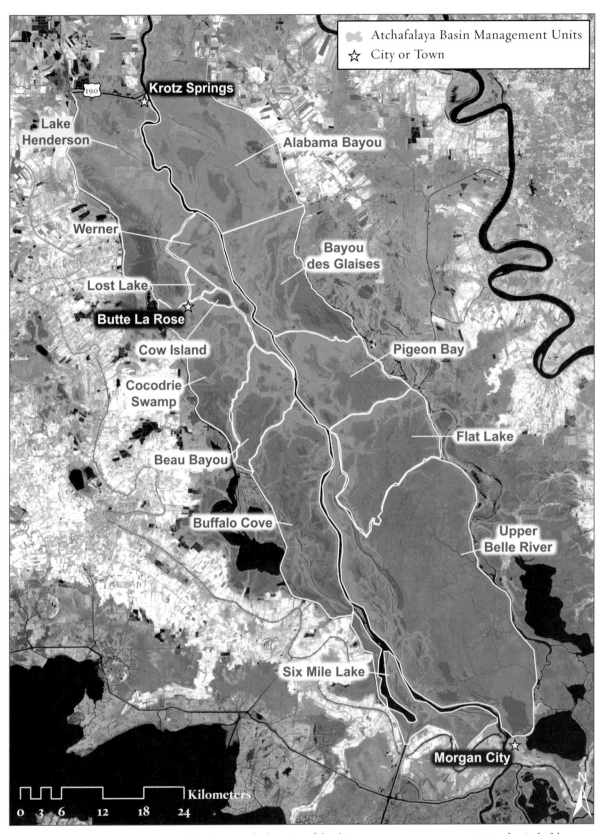

Legend:
- Atchafalaya Basin Management Units
- ☆ City or Town

Map labels:
- Krotz Springs
- Lake Henderson
- Alabama Bayou
- Werner
- Bayou des Glaises
- Lost Lake
- Butte La Rose
- Cow Island
- Pigeon Bay
- Cocodrie Swamp
- Flat Lake
- Beau Bayou
- Buffalo Cove
- Upper Belle River
- Six Mile Lake
- Morgan City

Kilometers
0 3 6 12 18 24

FIG. 8.12. Map showing the location of the thirteen water management units in the Atchafalaya Basin Floodway System in the ARB floodway.

including the current Atchafalaya Basin annual plan containing a current list of approved projects can be found at the Atchafalaya Basin Program website (http://dnr. louisiana.gov/index.cfm?md=pagebuilder&tmp=home&pid=494&pnid=0&nid=273).

WATER MANAGEMENT IN THE ARB FLOODWAY. As a result of floodway and navigation features and management, water movement through the ARB floodway is controlled by hydraulic features at three distinct scales (Sabo et al. 1999a). At the floodway scale, the quantity of water entering the system is controlled by the structures at Old River and relatively unregulated flow from the Red River system. The Old River structures are set to a daily discharge schedule and determine the overall water exchange rate among the river, swamp, and delta wetland habitats. The flow allotment is distributed between the control structure and the hydroelectric plant, and that distribution is set daily, depending on river stage, and power needs (US Army Corps of Engineers 2009). Next, the quantity and flow direction of river water delivered to specific regions (one or multiple water management units) of the ARB floodway is controlled by the degree of development of the main river channel as well as the size and orientation of the remaining distributaries. Lastly, conveyance of delivered water through the swamp is controlled by the smaller water courses, such as low-energy channels and lakes. Due to the low elevation gradient in the ARB and many riverine swamp systems, habitat variation is largely dependent on water flow patterns driven by these multi-scale processes. The resultant water exchange properties between the river and floodplain forest dictate the spatial and temporal extent of ephemeral habitat and habitat quality by driving the biological processes (i.e., photosynthesis, respiration) that alter water chemistry, ultimately affecting forest and aquatic resources (Junk et al. 1989).

This multi-scale flow regulation has altered flow in all of the waterways of the ARB floodway (Hupp et al. 2008). First, the discharge schedule of the Old River Control Structure, to adhere to the daily 70/30 latitudinal flow split between the Mississippi and Atchafalaya river basins, holds the ARB floodway in a quasi-natural flood cycle (Fig. 8.13). Although it follows the Mississippi River hydrograph, this cycle does not allow the full range of discharge conditions, especially at the low-flow end of the spectrum (Bryan et al. 1998, Haase 2010). Additionally, while the latitudinal flow split requirements have been maintained since the beginning of operations at the Old River Control Structure in 1963, there is evidence of significantly increased discharge for all portions of the hydrologic cycle at the entry point of the Atchafalaya River into the ARB floodway. Analysis of gauging records at Simmesport, Louisiana, demonstrates that the flow duration curve for mean daily flows for the period 1968–2009 is systematically greater than the flow duration curve for the period 1938–1961 (Fig. 8.14; Haase 2010). The relationship between the flow duration curves suggests that for the period 1968–2009, the quantity of water entering the basin via the Atchafalaya River at Simmesport, Louisiana, has been systematically greater for all flow conditions (low-flow events to flood events) when compared to the period 1938–1961. These flow increases mirror a similar increase in flows noted for the gauging station on the Mississippi River at Natchez, Mississippi (Haase 2010). Thus, because of the daily 70/30 latitudinal flow split, as flows across the hydrograph increase on the Mississippi River, immediately upstream of the

FIG. 8.13. The Old River Control Auxiliary Structure discharging into the ARB floodway. Structures in the Old River complex are used to adhere to daily latitudinal flow split requirements that divert 70% of the combined flow of the Mississippi and Red Rivers down the Mississippi channel and 30% down the Atchafalaya River channel. This flow split has been maintained since the beginning of operations at the Old River Control Structure in 1963 and hold the ARB floodway in a quasi-natural flood cycle. Photo by David Y. Lee © The Nature Conservancy.

Old River Control Structure, flows will correspondingly increase in the Atchafalaya River at the head of the ARB floodway. The observation of increased flows at the head of the basin is consistent with an observed increase in mean river stage at Morgan City, Louisiana (Fig. 8.15; Keim et al. 2006a).

The increased stage noted for the Morgan City, Louisiana, gauging station may be caused by a number of factors including: 1) increased Atchafalaya discharge due to increased overall discharge of the Mississippi and Red Rivers; 2) increased conveyance due to channel training; 3) increased river bottom elevation in the lower Atchafalaya, due to stored sediment; 4) decreased overall storage capacity in the ARB floodway due to infilling; 5) increased development of the Atchafalaya and Wax Lake deltas; and 6) increasing relative sea level rise (eustatic sea level rise plus land subsidence). Whatever the cause of the increased river stages, this overall increased stage suggests that, even within this quasi-natural fluctuation, flooding in the ARB floodway is increasing in magnitude, frequency, and duration, similar to the rest of Louisiana's coastal forest habitats (Chambers et al. 2005, Keim et al. 2006a, Shaffer et al. 2009).

The enclosure of the ARB floodway within levees, along with the combination of natural and anthropogenic development of the Atchafalaya River channel and distributary closure, has accelerated the natural transition of this system from a river-floodplain system that contained an interconnected series of lakes and bayous

Atchafalaya River at Simmesport, Louisiana

Flow Duration

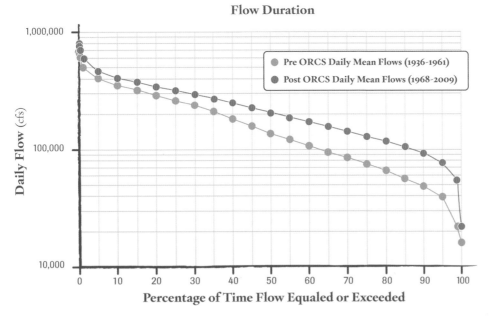

FIG. 8.14. Flow duration curves for the Atchafalaya River at Simmesport, Louisiana, comparing flow before and after construction of the Old River Control Structure complex. Curves show a pattern of increased flow duration after the Old River Control Structure began operation, particularly at the low end of the discharge range. Figure redrawn from Haase (2010).

Mean Atchafalaya River Stage at Morgan City, Louisiana
1963-2010

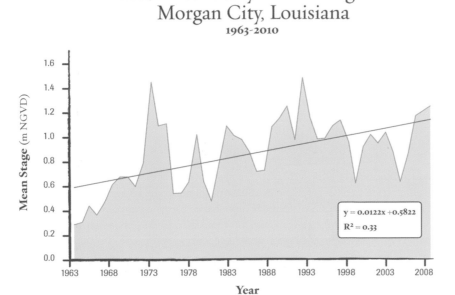

FIG. 8.15. Hydrograph showing mean annual stage in the Atchafalaya River at Morgan City, Louisiana after construction of the Old River Control Structure complex (1963–2010). Note that, despite the fact that the Atchafalaya River receives a set daily flow allotment, mean stage at Morgan City has increased through the time series. Means calculated from daily stage data from the US Army Corps of Engineers/US Geological Survey Gage 03780, Lower Atchafalaya River at Morgan City, Louisiana.

(low and high-energy) to an alternate state where a hydraulically efficient river channel bisects areas of adjacent swamps (Sabo et al. 1999b). Activities to allow access to the channel, pipeline construction, or levee maintenance, have also resulted in features (i.e., cuts and blockages) that are designed to divert water throughout the system (Hupp et al. 2008). While these features may have been designed to convey water through the system, there is evidence that their cumulative effects (natural sedimentation, static flow allocation at Old River, channel engineering, canalization, and local obstructions to flow) have converted the system. Historically, the river and distributaries drained the system from the higher land in the north toward the lower land in the south, and natural overbank flooding dispersed water, sediments, and nutrients across the flooded forest. Today, the system has been converted to one where some distributaries (natural and artificial) that remain on the landscape and anthropogenic alterations may be causing large-scale water stagnation by diverting water into the swamp in ways that set flow in opposing directions. The results of this conversion is affecting sedimentation patterns, water chemistry, and consequently, the distribution and productivity of biotic resources in the ARB (Bryan and Sabins 1979, Lambou 1990, Brunet 1997, Sabo et al. 1999a, b, Davidson et al. 1998, 2000, Aday 2000, Rutherford et al. 2001, Fontenot et al. 2001, Hupp et al. 2008).

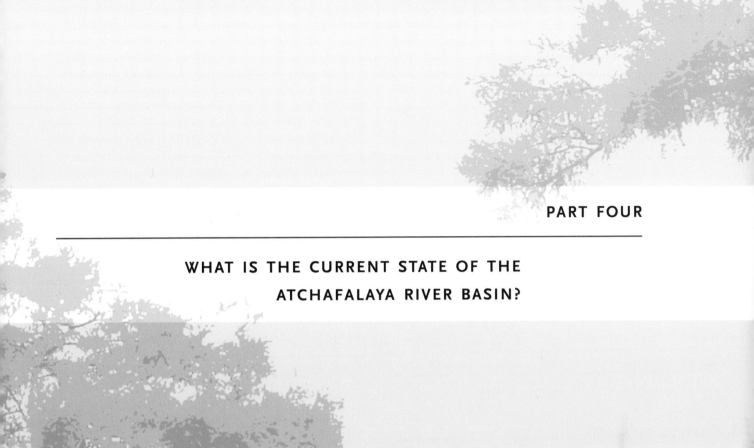

PART FOUR

WHAT IS THE CURRENT STATE OF THE ATCHAFALAYA RIVER BASIN?

THE EFFECTS OF WATER MANAGEMENT and human engineering features on aquatic resources in the ARB have been studied since the 1970s. Most studies have been conducted in water bodies (lakes, distributaries, low-energy bayous) within the two large swamp complexes on the eastern and western sides of the Atchafalaya River from Bayou Sorrel south to Duck and Flat Lakes just north of Morgan City (Cocodrie Swamp, Beau Bayou, Buffalo Cove, Pigeon Bay, Flat Lake, Upper Belle River, Six Mile Lake water management units; see Fig. 8.12). These areas are the lowest elevation areas in the middle region of the ARB floodway and, consequently, the majority of the cypress swamp habitat. These studies have described the relationship between water flow conditions and spatial and temporal variation in the characteristics of aquatic habitats and have shown that the prevailing water management system and other human engineering features have put aquatic and forested wetland habitats at risk. The reason they are at risk is because water quality is degraded over large areas for significant periods of time each year and sediment deposition is increased in some areas where sedimentation rates should be low.

WATER QUALITY IN THE ATCHAFALAYA RIVER BASIN FLOODWAY. In the 20 years between the 1970s and 1990s, incidence of hypoxia in the ARB floodway swamps doubled. Hypoxia events, which had always occurred periodically on the swamp floor, have become larger, longer, and also include waterways that received swamp runoff during or after floods when water temperatures are warm (> 16 °C [> 60 °F]; Bryan et al. 1998, Sabo et al. 1999a, b). Hypoxia was greatest during late flood pulses when air temperatures were highest (Bryan et al. 1998, Sabo et al. 1999a, b, Fontenot et al. 2001).

Today, hypoxia continues to occur most frequently during late flood pulses in hot weather due to natural sedimentation and channel engineering and other man-made features that have increased segmentation and isolation of the floodplain from the river, impeded drainage, and caused large-scale, persistent stagnation. The incidence and duration of hypoxia is greatest in low-energy waterways, principally canal/bayou junctions and east/west bayous and canals (Fig. 9.1). Lakes and distributaries generally

FIG. 9.1. Low oxygen water that has moved out of the swamp forest and into a bayou. The incidences of hypoxia have increased by more than a factor of two, and the duration is typically worse at bayou/canal junctions like the one pictured above. In this photo, low-oxygen water is flowing out of a bayou and mixing with high-oxygen water in the Gulf Intracoastal Waterway. Photo by Charles Demas, US Geological Survey Louisiana Water Science Center, Baton Rouge, Louisiana.

do not become hypoxic because primary production remains high, and wind circulation and riverine influences keep dissolved oxygen concentrations within the normoxic range (Bryan et al. 1998, Sabo et al. 1999a, b).

On the western side of the ARB floodway, there is almost no interaction between the river and the swamp when river levels are below about 2.7 m (9 ft) at Butte La Rose (Fig. 9.2). At these low to low-mid stages, most of the river water stays in the river and main channels (i.e, Bayou L' Embarras, Lake Fausse Pointe Cut). Some connectivity with the swamp begins to occur via Bayou Darby and the Buffalo Cove outlet when water levels reach 2.7 m (9 ft), and this pattern increases with increasing stage. As the water level drops, water drains from the area (through Bayou Darby and adjacent bayous and Buffalo Cove outlet) into the Lake Fausse Pointe Cut and out through Six Mile Lake and the river channel. Seasonal hypoxia west of the river results from reduced drainage, in some cases because water enters areas opposite from the direction of drainage (e.g., Buffalo Cove outlet), and is often present in flooded swamps in the Cocodrie Swamp, Beau Bayou, and Buffalo Cove water management units (Constant et al. 1999, Daniel E. Kroes, US Geological Survey, unpublished data).

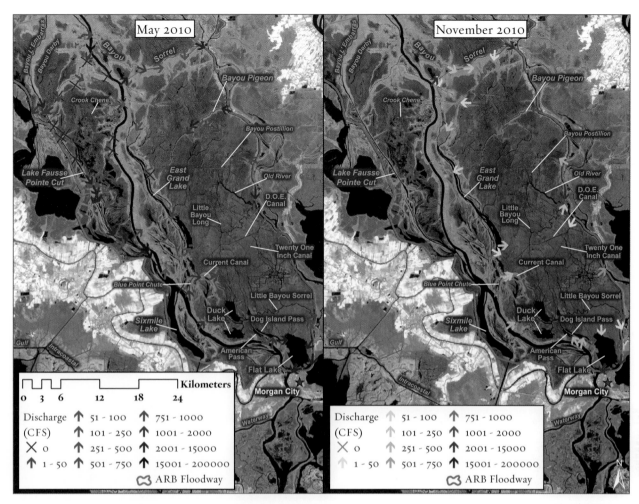

FIG. 9.2. Flow magnitude and direction in the ARB floodway during high flow (May 2010) and low flow (November 2010) periods on the Atchafalaya River. Discharge is provided in the inset legends and is represented in cubic feet per second (cfs). Image created from flow study data available on the Atchafalaya Basin Program Natural Resource Inventory and Assessment System (NRIAS; http://abp .cr.usgs.gov/).

The eastern side of the ARB floodway that extends from Bayou Pigeon south to Little Bayou Sorrel north of Flat Lake (Upper Belle River water management unit), an area of about 254 km² (98 mi²), is frequently hypoxic due to extended periods of persistent inundation (Fig. 9.3). This is the lowest area of the ARB floodway, and, consequently, as water rises in the river, this area acts like a vacuum sucking in water to balance the stage with the river. If the river and swamp were not separated throughout the floodway, this area would fill from overbank flooding, drainage from the north, and local distributary sources. However, because of the isolation of the river from the swamp throughout the floodway, flow into this area has to enter and exit indirectly.

Studies of water elevation, water quality, and flow across this area have related the indirect flow patterns to water quality degradation (Sabo et al. 1999a, b, Daniel E. Kroes, US Geological Survey, unpublished data). Sabo et al. (1999a, b) highlighted three distinct zones within the floodway where internal water flow properties are different but

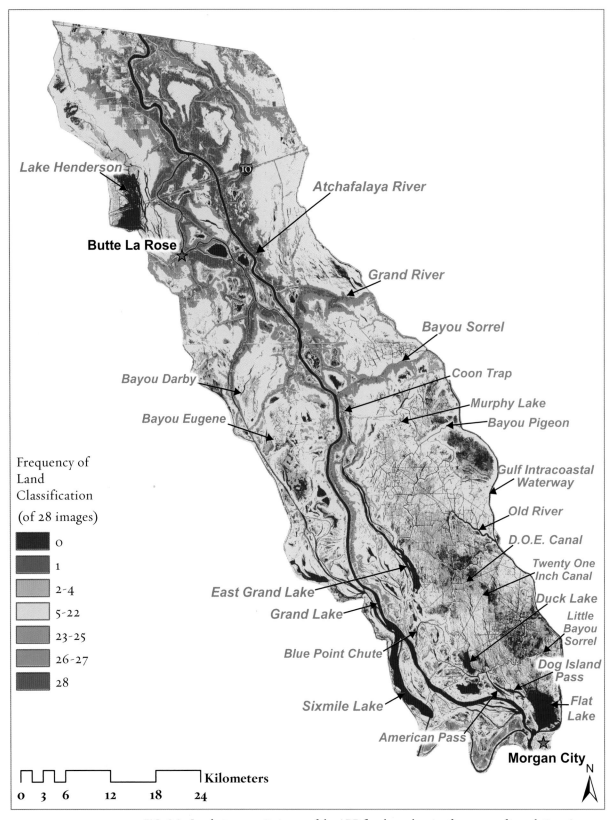

FIG. 9.3. Landsat composite image of the ARB floodway showing frequency of inundation. Areas colored in brown and green are mostly dry land. Areas colored in blues are mostly water and represent areas where flooding-induced hypoxia is persistent. Figure modified from Allen et al. (2008) and provided by Yvonne C. Allen, US Army Engineer Research and Development Center, Baton Rouge, Louisiana.

still directly related to the ultimate cause of hypoxia. New flow research done by the US Geological Survey at variable river stages, including the Mississippi River flood of 2011, generally corroborates the inferences made by Sabo and his coauthors but provides a great deal more understanding of current flow dynamics in this area (Daniel E. Kroes, US Geological Survey, unpublished data).

In the zone north of Old River (north zone), Atchafalaya River water enters the Bayou Sorrel distributary. Water in Bayou Sorrel flows into the Gulf Intracoastal Waterway and then enters interior swamps in a westward direction, via Bayou Pigeon and connected channels and eastward through Grand Lake and connected channels (Fig. 9.2). This water then drains from the area through Old River and the Gulf Intracoastal Waterway on the west and several waterways bordering Grand Lake. New flow research generally corroborates these findings and shows that flow into the interior through Bayou Pigeon is around 15% of the flow in the Gulf Intracoastal Waterway. Additionally, Bayou Postillion also serves as a distributary into this area, although its flow contribution is smaller than Bayou Pigeon (Daniel E. Kroes, US Geological Survey, unpublished data). Water flows out of this area through Old River, and, at low to low-mid stages, (> 2.7 m [9 ft] at Butte La Rose), outflow at Old River is about equal to distributary inflow. As flow in the Atchafalaya River increases, this same general flow pattern exists throughout this area; however, more water flows into the swamps than will drain from them.

Water in the central zone (south of Old River and north of the DOE Canal) enters from the south and moves northward through a northern branch of Blue Point Chute, passes through the Current Canal and into Little Bayou Long (Fig. 9.2). Blue Point Chute is a major feeder to the interior swamps in this area, even at stages as low as 0.9 m (3 ft) at Butte La Rose, and inputs of Atchafalaya River water through Blue Point Chute increase with river stage (Daniel E. Kroes, US Geological Survey, unpublished data). This water is sent northward and meets drainage from the north zone, stalling flow until eventually draining into the southern zone (south of Grand Lake).

The southern zone receives water through the southern branch of Blue Point Chute and Twenty One Inch Canal (Fig. 9.2). As river stage reaches or exceeds 2.7 m (9 ft) at Butte La Rose, water also enters this zone through Sixteen Inch Canal (Daniel E. Kroes, US Geological Survey, unpublished data). American Pass functions as both an inflow and a drain, depending on water level. During normal flows, across the hydrograph, water generally flows in through American Pass as water levels rise and drains through American Pass as water levels drop (Daniel E. Kroes, US Geological Survey, unpublished data). Inflowing water in this zone moves south and east until draining into the Gulf Intracoastal Waterway, Dog Island Pass, and Little Bayou Sorrel.

Hypoxia is most pervasive and persistent in the central zone, occurring early in the spring, as temperatures increase, typically on the rising limb of the Atchafalaya River flood pulse and remaining through the recession of the flood pulse (Sabo et al. 1999a, b). Hypoxia typically spreads northward from the central area into the northern portions of the zone during the falling limb of the pulse because drainage from this area is impeded by artificially elevated water levels in exit pathways that results from diverted water moving northward through Blue Point Chute/Current Canal, American Pass,

FIG. 9.4. An example of a hydraulic dam where high water levels in the Atchafalaya River prevent low-oxygen black water from draining out of the swamp. Photo by Charles Demas, US Geological Survey Louisiana Water Science Center, Baton Rouge, Louisiana.

Dog Island Pass, and the Gulf Intracoastal Waterway to the east, causing hydraulic damming.

Hypoxia events in the southern portion of the region occur mostly on the rising limb of the Atchafalaya River flood pulse, although infrequent hypoxic events were documented on the falling limb (Sabo et al. 1999a, b). Hypoxia in this zone has become more temporally and spatially pervasive in recent years. The cause is likely hydraulic damming that occurs when high water levels in the river and Gulf Intracoastal Waterway prevent low-oxygen water from draining out of the swamp through American Pass and Duck Lake (Fig. 9.4; Sabo et al. 1999a, b).

Hypoxia is generally absent in the northern zone by midsummer, after the flood pulse recedes and flow becomes more channelized; however, hypoxia persists in the central and southern zones, possibly due to the fact that the lower land elevation in this area allows persistent connection of the bayous and waterways to hypoxic swamp water. Hypoxia is particularly problematic when flood pulses over about 3 m (10 ft) at Butte La Rose persist into the summer or early fall (Sabo et al. 1999a, b).

Near the peak of the Mississippi River Flood of 2011 (~6.4 m [21 ft] at Butte La Rose, approximately 25,000 m^3 s^{-1} [883,000 cfs]), at least 60% of the flow in the ARB flood-

FIG. 9.5. During the Mississippi River Flood of 2011, the entire ARB floodway became a river, with water flowing rapidly through the floodplain forest. In this photo, the height of the tree canopy and the whitewater that can be seen in the swamp, indicate that this forest was inundated with at least 3 m (10 ft) of water moving rapidly downstream. Photo © CC Lockwood.

way was off-channel (outside the banks of the Atchafalaya River). On both the east and west sides of the river, the primary distributaries filled up and flowed overbank across the forested floodplain. At that point, the whole floodway became the river (Fig. 9.5; Kroes and Allen 2012). On the east side of the river, 50% of the outflow from the system just north of Morgan City, Louisiana, was coming through the river channel, and 50% of the flow was coming out of the swamp forest. At this stage, water was actually flowing out (in a southerly direction) from both Blue Point Chute and American Pass, two major inlets where water normally flows in the opposite direction to draining water (northerly flow) during lower stages.

WATER QUALITY OUTSIDE THE EAST AND WEST ATCHAFALAYA PROTECTION LEVEES. Swamps and water bodies outside the levees receive no riverine inflow. Scientific studies that specifically address water quality in the swamps outside the floodway levees are few. However, assessments of the state of coastal forests document that persistent inundation occurs annually in portions of the Verret Swamps (i.e., Keim and King 2006, Keim and Amos 2012), suggesting that water quality across portions of that system is compromised during at least part of the year.

One detailed evaluation was made of the Atchafalaya East Watershed, defined as a 2,118 km² (818 mi²) area of Iberville, Point Coupee, and West Baton Rouge Parishes located between the East Atchafalaya Flood Protection Levee and the Mississippi River Levee and spanning from the Morganza Floodway south to the Assumption Parish line, located just north of Lake Verret (Tri-Parish Partnership 2009). This evaluation divided the study area into 14 sub-watersheds and used Louisiana Department of Environmental Quality records from the Total Maximum Daily Loading Program and resource agency and stakeholder input to detail surface water quality problems throughout this area. Generally, this study documented poor water quality, including excessive sedimentation, surface water stagnation, reduced flow, algal growth, and large-scale fisheries decline across the study area. These problems were linked to reduced drainage capacity, increased flooding susceptibility, increased timber loss, increased closure of waterways to boat traffic, and limited recreational use of many waterways. Dissolved oxygen impairment was identified in 13 of the sub-watersheds. Eutrophic conditions (high concentrations of nitrogen and phosphorus) were documented in 10 (71%) of the sub-watersheds, and water quality impairment due to fecal coliform bacteria was documented in 9 (64%). Other water quality problems included chemical pollution (chloride and atrazine), and excessive amounts of total suspended solids.

SEDIMENTATION IN THE ATCHAFALAYA RIVER BASIN FLOODWAY. Currently, most of the sediments bypass the interior of the ARB floodway and are transported through the Atchafalaya River channel and its major distributaries to the delta and along the continental shelf (Hupp et al. 2008, Allen 2010, Xu 2010, Allison et al. 2012). Xu (2010) estimated that over a 30-year period (1975–2004), about 6 million metric tons (6.6 million short tons), or 9% of the suspended sediment in the Atchafalaya River was deposited in the floodway between Simmesport and Morgan City. However, Allison et al. (2012) estimated the sediment retention rate to be almost four times higher. During three flood years (2008, 2009, and 2010), the floodway sequestered 23.1×10^6 metric tons yr^{-1} (25.6×10^6 short tons yr^{-1}), representing 32.1% of total suspended sediment inputs from the Mississippi and Red Rivers. Most of the lost sediment (70%) was mud, 30% was sand, and channel bed profiles showed that most retained sediment was likely deposited on the floodplain rather than stored on the river bottom. Whatever the rate, it is clear that the floodway has likely reached a point where the hydraulic gradient in the river is declining, causing more sediment to be sent to Atchafalaya Bay (Xu 2010). Thus, sedimentation rates in the interior basin are relatively low compared to rates in the past. Deposited sediment is mostly in the fine-grained fraction, because most of the coarse bedload material never escapes the canal (Hupp et al. 2008). For example, sediment deposition from the river caused the conversion of open water to land at an average rate of 2.7 km^2 yr^{-1} from the 1800s to the early 1980s. These rates decreased to -0.04 km^2 yr^{-1} from the early 1980s into the early 2000s (Allen 2010).

Although large-scale delta building is not occurring in the floodway, there are localized areas that receive substantially higher rates of sediment deposition than they would naturally receive, and this accelerated deposition is largely due to anthropogenic factors. These localized depocenters are generally backswamp areas where very little sediment would normally accrete because they are disconnected from direct riverine input by levee ridges where the majority of the sediment is deposited (Fig. 9.6; Hupp et al. 2008). The prevailing water management system, as well as large-scale canalization of the ARB floodway, has created conditions where sediment is channeled

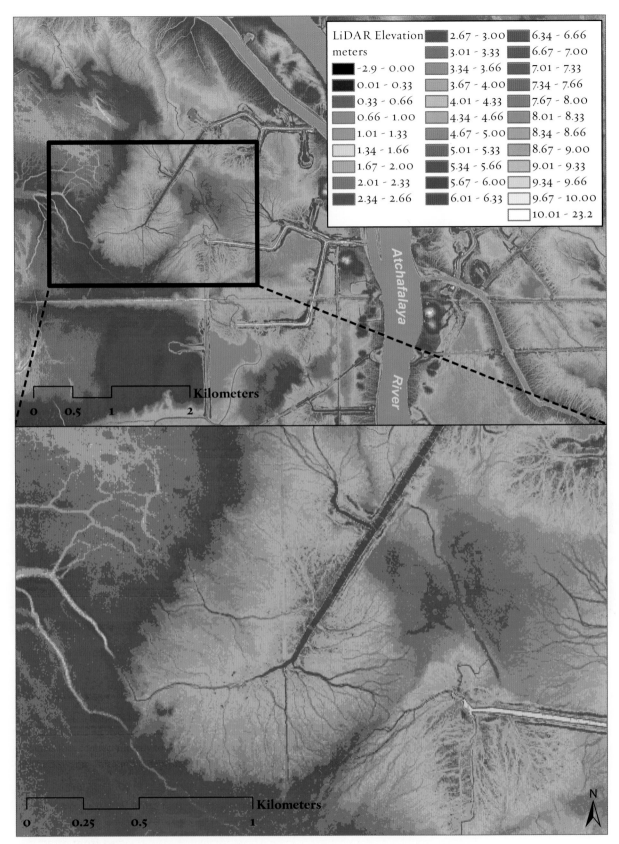

FIG. 9.6. LiDAR image showing an example of high sediment deposition in a backswamp area that normally would receive very low rates of sediment accretion. Exacerbated sedimentation is the result of canals that connect directly to the river or other distributaries. Sediment-laden water is delivered into backswamp areas and is deposited in a mini-delta splay that forms at the end of each canal.

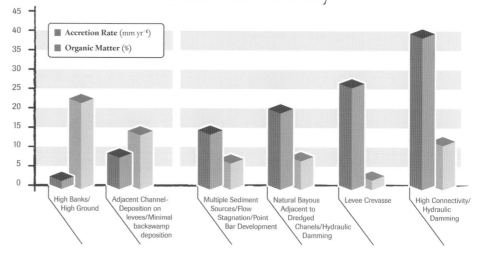

Sediment Accretion Rate and Composition in the ARB Floodway

FIG. 9.7. Histogram showing mean sediment accretion rate and percent organic matter of soils at sites across 20 transects in the Atchafalaya Basin Floodway System, ARB floodway. Site results clustered into 6 major categories, based on their location. The first 2 groups represent sites with sedimentation patterns that reflect natural overbank river-floodplain connectivity. The last 4 groups represent sites where sedimentation rates are elevated, which is indicative of water management problems. Figure created from data in Hupp et al. (2008).

into and deposited in these areas resulting in high rates of sediment deposition. This increased deposition can increase swamp segmentation and impoundment by creating levees along natural drainage pathways and increase elevation enough to favor higher-elevation trees (i.e., overcup oak, green ash, hackberry) and convert cypress forest to bottomland hardwood forest (Hupp et al. 2008, Kroes and Kraemer 2013).

A recent study illustrated the conditions that lead to exacerbated sedimentation in some areas (Hupp et al. 2008). Sedimentation patterns across a region spanning from Bayou Sorrel to Bayou Benoit were determined in the three major sedimentary environments in the middle ARB (22% levee, 34% transitional, 24% backswamp). Six groups of features were described, of which only two represented natural situations (Fig. 9.7). The first group (High Banks/High Ground) was areas with the highest levee banks and elevations. Sedimentation rates in these areas was likely high in the past but now the high elevations preclude long periods of inundation during a flood pulse, resulting in the lowest deposition rates of inorganic sediments. Soils at these sites contained the highest proportion of organic matter, suggesting that decomposing plant material is important in maintaining elevation. The second group was comprised of areas that are at equilibrium, where accretion from an adjacent channel is moderate on the levee ridges and decreases toward the backswamp. These areas support healthy trees and will likely not convert from swamp into bottomland hardwood forest.

The next four groups were areas where sedimentation rates were elevated, indicating problems related to water management (Fig. 9.7). These areas had one or a combination of the following characteristics: 1) high connectivity to distributaries (natural or

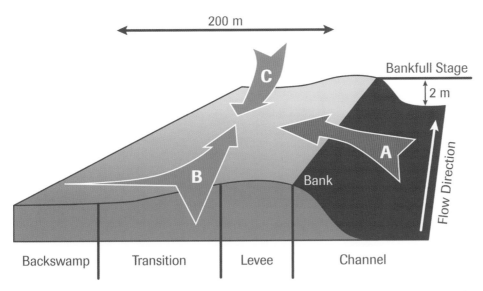

FIG. 9.8. Diagram showing inflow of sediment-laden river water from multiple sources—the mechanism that causes exacerbated sedimentation in ARB backswamps. Water containing high sediment concentration enters the backswamp from the adjacent channel (A), an upstream slough (B), and a downstream slough (C). When flows from B and C occur simultaneously, flow stagnation via hydraulic damming can occur, and retained water from all three sources unload their sediment into backswamp areas. Figure redrawn from Hupp et al. (2008) and provided by Daniel E. Kroes, US Geological Survey, Louisiana Water Science Center, Baton Rouge, Louisiana.

man-made); 2) multiple sediment sources; 3) proximity to dredged channels; 4) flow stagnation; and 5) levee crevasses (breach). These characteristics work in concert to inundate areas with sediment-laden water from multiple sources, and without proper drainage, a hydraulic dam occurs where flow stagnates and water is retained for extensive periods of time (Fig. 9.8). This hydraulic damming not only results in sedimentation rates that are 3 to 4 times the normal rate for riverine backswamps, it also causes extensive areas of hypoxia within the ARB floodway during summer months because periods of hydraulic damming typically occur when temperatures are high (Sabo et al. 1999a, b, Hupp 2000).

Currently, sedimentation rates are high in most lakes in the lower portion of the middle basin (northern end of Grand Lake, Murphy Lake, Six Mile Lake, Flat Lake), because of nearly constant inputs of waters carrying large sediment loads resulting from the large-scale channelization of water through the ARB floodway for flood control, navigation, and oil and gas extraction (Fig. 9.9; Sabo et al. 1999a, b, Hupp et al. 2008). In fact, sediment cores taken from open water areas of Buffalo Cove and Murphy Lake showed sedimentation rates of 30 mm yr^{-1} (1.2 in yr^{-1}) and 28 mm yr^{-1} (1.1 in yr^{-1}), respectively; this is similar to the highest rates found in some altered backswamp areas (Hupp et al. 2008). While it is important to understand that this lake sedimentation is a normal process in a deltaic system, there are lake areas where rates are greatly accelerated due to man-made features (Fig. 9.10). For example, the accretion rate at the northern end of East Grand Lake is up to 0.6 m yr^{-1} (2 ft yr^{-1}) due to the rock weir at the Coon Trap (Daniel E. Kroes, US Geological Survey, personal communication).

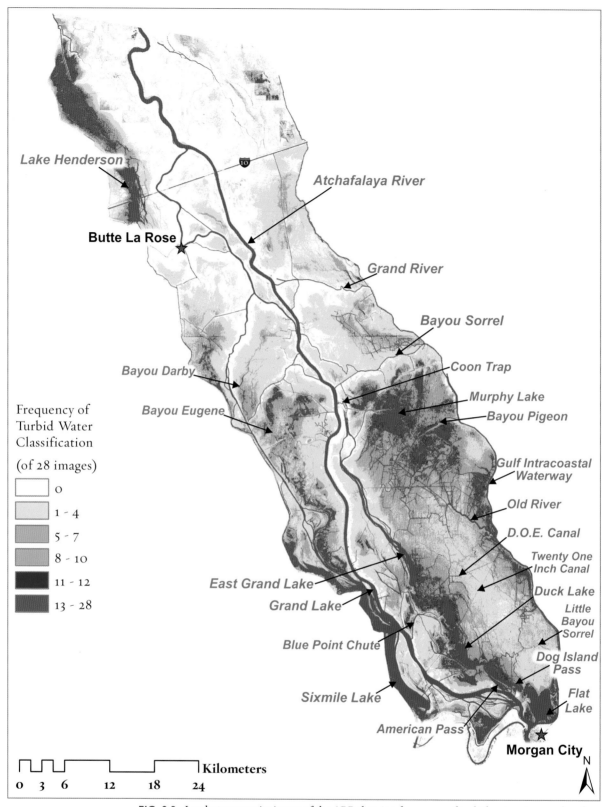

FIG. 9.9. Landsat composite image of the ARB showing frequency of turbid water. Areas colored in brown receive the most input of riverine input (sediment-laden water). Areas colored in gray and white receive the lowest input of sediment-laden water. Figure modified from Allen et al. (2008) and provided by Yvonne C. Allen, US Army Engineer Research and Development Center, Baton Rouge, Louisiana.

FIG. 9.10. Photo showing a lake that has largely filled in due to high sedimentation rates. Vegetation now grows on the filled lake bottom. For reference, the tree line marks the boundary of what was once open water. When lakes fill in, they can no longer provide deep water refuge for fish and other aquatic animals when dissolved oxygen levels drop. Photo by Anne Hayden, Southern Illinois University.

It is also important to note that the remaining lakes in the floodway are some of the only deep water refuge habitat left for fish when oxygen concentrations in the swamp and smaller bayous is reduced, and these areas are also important foraging and nesting habitat for bald eagles and osprey (Fig. 9.11). Deep water habitats provide cooler oxygenated water during periods of high temperatures and hypoxia, making these habitats critical to the life history of many fish species that live in the ARB.

Swamp areas east of the river that frequently receive sediment-laden water are found at the northern end of East Grand Lake and the area south of Bayou Pigeon, and southeast of Blue Point Chute (Fig. 9.9). West of the river, in the Lake Henderson and Buffalo Cove water management units and swamps immediately adjacent to bayous such as Eugene and Darby that deliver sediment-laden water directly from high-energy distributaries, sediment deposition rates are high. Many of these swamp areas may also experience long water retention times and, consequently, hypoxia (Fig. 9.3), if continued sedimentation builds significant barriers to flow (Allen et al. 2008, Hupp et al. 2008, Kroes and Kraemer 2013). These barriers to water flow may prevent continued sustainability of cypress-tupelo swamp and riverine wetland habitats throughout much of the ARB middle basin.

Interaction of fresh water and sediment in the ARB floodway represents a difficult balance. Riverine connectivity is needed for good water quality and continued sustain-

FIG. 9.11. Lakes that remain in the ARB floodway are some of the only deep water refuge habitat left for fish when oxygen concentrations in other water bodies drop. Lake habitat is critical to the maintenance of aquatic productivity in the floodway, consequently, it is also important for piscivores like osprey (*Pandion haliaetus*). Photo © Charles Bush. (www.charlesbushphoto.com)

ability of forests and aquatic habitats. However, that connectivity can also cause sedimentation problems in some areas, because fresh water and sediment delivery go hand in hand. Therefore, the challenge is to allow delivery of fresh water (and sediment) to floodplain areas while preventing barriers to flow.

SEDIMENTATION IN THE LOWER BASIN (DELTA PLAIN AND CONTINENTAL SHELF). Currently, the majority of the sediment entering the ARB at the head of the Atchafalaya River is transported through the floodway by the river (and other main flow paths) and deposited on bayhead deltas and the continental shelf. The increased conveyance of sediment seaward is due to the increasing isolation of the river and its floodplain (in the floodway), due to both natural and anthropogenic factors, and a steady decline in the hydraulic gradient of the Atchafalaya River between Simmesport, and Morgan City, Louisiana (Xu 2010).

Some of this sediment transported to the coast is deposited on the delta, and this rate is variable. The greatest sediment sequestration happens during large flood events, when large amounts of sediment are transported to the delta (Azure E. Bevington,

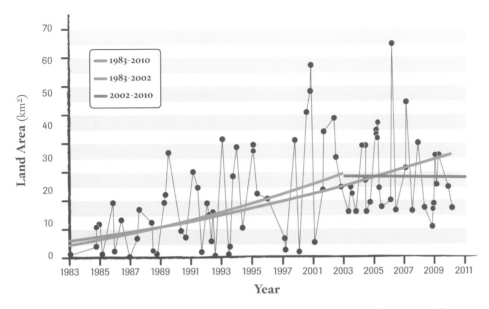

FIG. 9.12. Growth of the Wax Lake Delta from 1983–2010. Data are from Landsat imagery. Shown are the raw values and trendlines for the entire time series, as well as two series—1983–2002 and 2002–2010. Figure redrawn from Allen et al. (2011) and provided by Yvonne C. Allen, US Army Engineer Research and Development Center, Baton Rouge, Louisiana.

Guerry O. Holm, Charles E. Sasser, and Robert R. Twilley, Louisiana State University, unpublished data). Two large pulse events in 2008 caused the cumulative deposition of five million metric tons (5.5 million short tons) of sediment on the Wax Lake Delta, resulting in a vertical gain of 6.6 cm (2.6 in) across the delta. This vertical gain was a tenfold increase in the amount of accretion required for marshes to survive submergence. An estimated 23% of sediment load available to the delta was deposited in shallow island habitats. During the Mississippi River Flood of 2011, sediment accumulation on the Atchafalaya Delta was estimated at 1.6 ± 1.0 g cm^{-2}, a contribution of 88% of the annual average accumulation for the delta (Khan et al. 2013). Net growth of the Wax Lake delta was estimated at between 3 and 7 km^2 (1.1–2.7 mi^2), and over 85% of the existing delta either received an elevation boost or remained unchanged (Carle et al., in press).

Overall land gain has occurred since the deltas became subaerial, but there is evidence that growth rates of the deltas have slowed. Rouse et al. (1978) estimated growth of 6.5 km^2 yr^{-1} (2.5 mi^2 yr^{-1}) from 1973–1976. More recently, Roberts et al. (2003) estimated growth of only 3.0 km^2 yr^{-1} (1.2 mi^2 yr^{-1}) from 1989–1994. Analysis of long-term (1983–2010) growth of the Wax Lake Delta using Landsat imagery resulted in an average estimate of about 0.95 km^2 yr^{-1} (0.3 mi^2 yr^{-1}; Allen et al. 2011). The estimated land gain prior to 2002 was 1.1 km^2 yr^{-1} (0.4 mi^2 yr^{-1}) but from 2002–2010, land gain was essentially flat at 0.007 km^2 yr^{-1} (0.003 mi^2 yr^{-1}; Fig. 9.12). An analysis of satellite imagery of Atchafalaya and Wax Lake deltas also showed slower delta growth from 1989 to 2010 (Rosen and Xu 2013). Growth of both the Atchafalaya and Wax Lake deltas declined from 2.4 km^2 yr^{-1} to 1.6 km^2 yr^{-1} (0.9 to 0.6 mi^2 yr^{-1}) and 3.2 km^2 yr^{-1} to 0.6 km^2 yr^{-1} (1.2 to 0.2 mi^2 yr^{-1}), respectively. These observations do not necessarily mean that

Average Annual Suspended Sediment Discharge in the Atchafalaya River at Simmesport, Louisiana
1952-2008

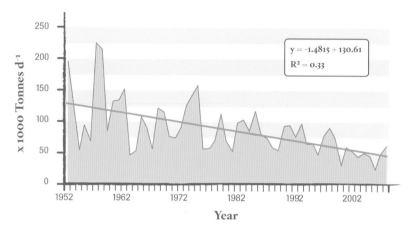

FIG. 9.13. Sediment hydrograph showing the suspended sediment load in the Atchafalaya River at Simmesport, Louisiana, 1952–2008. Also shown is a trendline showing long-term reductions in sediment load.

growth has stopped or that the early estimates were incorrect. However, it does appear that subaerial growth has slowed. This change may be a result of the natural slowing of growth that occurs as deltas mature and more coarse sediment is shunted seaward—the same process that occurred upstream in the ARB floodway (Wellner et al. 2005). However, there is still accommodation space available for bayhead and shelf expansion of the deltas; therefore, other factors may also be slowing deltaic growth and expansion. Factors that may be contributing to these lower growth rates include decreased overall sediment supply in the Mississippi River system, increased rate of relative sea-level rise, increased hurricane frequency, and more cold fronts (Barras 2009, Blum and Roberts 2009, Allen et al. 2011).

There is evidence that sediment supply in the Mississippi River, and consequently the Atchafalaya River, has declined (Fig. 9.13), largely due to damming on the upper Missouri River, soil erosion control, and other engineering features on the Mississippi River (cutoffs, flow-training structures, bank revetments; Swarzenski 2001, Blum and Roberts 2009, 2012, Meade and Moody 2010, Xu 2010, Tweel and Turner 2011). Estimates of suspended sediment loads have decreased by about half. Pre-dam estimates of suspended sediment loads were 400–500 million metric tons yr[-1] (441–551 million short tons yr[-1]). Post-dam estimates were about 205 million metric tons yr[-1] (226 million short tons yr[-1]). When combined with subsidence and sea-level rise, this sediment deficit may contribute to submergence-induced land loss of an estimated 10,000–13,500 km[2] (3,861–5,212 mi[2]) across the Mississippi River delta plain (Blum and Roberts 2009). Likewise, there is a relationship between historic changes in wetland area on the Mississippi River delta and suspended sediment concentrations in the Mississippi River (Tweel and Turner 2011). Variability in wetland area over the last 200 years reflects periods of increasing, stable, and decreasing sediment supply in the Mississippi River. These findings suggest that future decreases in suspended sediment may cause additional reductions in wetland area on the Mississippi River delta and may impose spatial limits on wetland restoration across the Mississippi River delta plain. Although there are no specific estimates of land change for the Atchafalaya and Wax Lake deltas, reduced sediment concentrations in the Mississippi River will likely affect their long-term growth.

Allison et al. (2012) documented that during 2008–2010 as much as 44% of the total suspended sediment load (80% of the sand load) in the Mississippi and Red Rivers was sequestered between Old River and the Gulf of Mexico. Storage mechanisms included river bottom aggregation, accretion in batture lands, and overbank storage. In addition to long-term declines in sediment availability, only slightly more than half of the sediment (and only about 20% of the suspended sand) at upstream stations (Tarbert Landing and Simmesport) is available for diversion into coastal wetlands for restoration.

Other research suggests, however, that even with the present sediment loads and relative rates of sea-level rise, restoration and maintenance of natural hydrological and geological processes may enable this region to adapt to future changes in relative sea level. The combination of land subsidence and sea-level rise (relative sea-level rise) in the central area of the coast, as measured by tide gauges at Grand Isle, Louisiana, has been estimated at 9.4 mm yr^{-1} (0.4 in yr^{-1}) which is among the highest rates recorded in North America. Additionally, when taking into account future climate uncertainties, this rate could increase by an additional 5.5 mm yr^{-1} (0.2 in yr^{-1}) or more (Gonzáles and Törnqvist 2006). However, this high rate of relative sea-level rise is a long-term (1947–2006) estimate that can be deconstructed into phases that correspond to patterns of subsurface fluid removal and patterns of wetland loss (Kolker et al. 2011). In this analysis, Louisiana is currently experiencing the lowest rates of relative sea-level rise in more than 50 years (1.0 ± 0.97 mm yr^{-1}) due to reduced onshore oil production. These results suggest that the volume of sediment needed to restore coastal wetlands in Louisiana may actually fall to within the low-end of some estimates. An estimated 700–1,200 km^2 (270–463 mi^2) of future land building is possible across the Mississippi River delta plain over the next century if constraints along the Mississippi and Atchafalaya Rivers are removed to allow for proper overbank deposition and large seasonal pulse events (Kim et al. 2009).

Transport and storage of sediment (> 10 million metric tons [11 million short tons]) out of Atchafalaya Bay and onto the continental shelf (~1.0–1.3 mm [0.04–0.05 in] deposition per cold front) opposite Atchafalaya Bay continues since 20–30 cold fronts affect the system each year. Accretion continues along the chenier plain when tropical storms distribute and rework sediments along the coast, delivering sediments to some nearshore marshes and eroding land in others (Allison et al. 2000, Turner et al. 2006, Bentley 2002, Bentley et al. 2003, Barras 2006, 2009, Jaramillo et al. 2009, and many others).

AQUATIC VEGETATION. Native aquatic plants create an important habitat type in riverine systems. Native macrophytes in the floodway include cabomba (*Cabomba* *caroliniana*), coontail (*Ceratophyllum demersum*) and sagittaria (*Sagittaria* spp.). These plants are typically distributed across the landscape in isolated, relatively sparse beds that create productive microhabitats for invertebrates, juvenile fish, and larger predators. However, like many aquatic systems worldwide, the ARB has experienced large-scale invasion of nonnative (exotic) aquatic plant species, and many of the water management features in the ARB have increased susceptibility of the system to invasion by nonnative species. Large-scale invasion of waterways by nonnative plants disrupts many natural processes (i.e., nutrient cycling), causes significant changes in water quality (pH, temperature, dissolved oxygen), and may physically block water movement. In the ARB, there is evidence that the invasion of nonnative plants has displaced native aquatic plant species and reduced their populations, exacerbated problems with water quality, and blocked waterways cost several million dollars to clear each year (Bryan et al. 1998, Colon-Gaud et al. 2004, Walley 2007). There is also evidence that these macrophytes may affect invertebrate and fish production in the ARB (Mason 2002, Colon-Gaud et al. 2004, Troutman et al. 2007).

Of the 20 aquatic plant species found in the ARB, 7 are nonnatives (Walley 2007). Nonnatives that have become particularly problematic in the ARB floodway include three floating plant species—water hyacinth (*Eichhornia crassipes*), common salvinia (*Salvinia minima*), and giant salvinia (*Salvinia molesta*)—and one submerged aquatic species, hydrilla (*Hydrilla verticillata*; Fig. 9.14). Common salvinia appears to be the most abundant since it was present in 97% of plant samples taken across middle basin swamps in 2005 and 2006 (Walley 2007). In that same study, water hyacinth (36%) and hydrilla (19%) were also commonly found. Percent cover of common salvinia was highest in canal (39%) and swamp (28%) habitats, and lowest in lakes (6%). Water hyacinth was most abundant in swamps (15%), where it was likely transported during the flood pulse, and bayous (7%). The location of hydrilla beds was variable, with greatest coverage in canals and bayous (4%) in 2005 and swamps (2%) in 2006. Other exotic plant species in the ARB include alligatorweed (*Alternanthera philoxeroides*), watermilfoil (*Myriophyllum spicatum*), and water lettuce (*Pistia stratiotes*), but none of these species cause significant ecological or economic problems (Walley 2007).

The four problem species (water hyacinth, common and giant salvinia, and hydrilla) are all able to reproduce and grow extremely rapidly. For example, water hyacinth can double its population size in less than 20 days, and a single hydrilla shoot can produce up to 6,000 new tubers per square meter (Mitchell 1976, Sutton et al. 1992). When compared with ecologically similar (similar life history, architecture, growth form) native counterparts in the ARB floodway, all four of the nonnative species accumulate more biomass per unit area than the native plant species (Walley 2007). It is extremely hard to control growth and spread of these plants, especially in the floodway, where spring pulses can transport plants far and wide, and chemical control agents are susceptible to dilution. In areas where water levels are controlled (e.g., Henderson Lake) or where floodplains can dewater, periodic drawdown can be moderately effective in controlling some species (i.e., hydrilla; Mason 2002). Consequently, these species are outcompeting native species and becoming dominant within the ARB plant community (Colon-Gaud et al. 2004, Walley 2007). Their excessive growth and reproduction results in

FIG. 9.14. A lake filled with water hyacinth (*Eichhornia crassipes*), a floating nonnative plant that has become particularly problematic in the ARB floodway. The opulation of water hyacinth can double in about 20 days, which means that water bodies can become totally covered by this species. Not only do these waterways become impassable, but these outbreaks of exotic plants can exacerbate hypoxia by restricting photosynthesis in the water column. Photo by Daniel E. Kroes, US Geological Survey Louisiana Water Science Center, Baton Rouge, Louisiana.

dense canopy mats at the surface of the water that can alter the quality of the water column below and exacerbate hypoxia during summer months (Colon-Gaud et al. 2004, Walley 2007).

Hypoxic conditions develop in water bodies where nonnative aquatic plants form dense surface canopies because they block light and prevent gas exchange between the water surface and the atmosphere, reducing photosynthesis by submerged native macrophytes and algae that reside under the mat (Caraco and Cole 2002, Colon-Gaud et al. 2004, Walley 2007). Dissolved oxygen is further reduced when native plants and algae die and decompose beneath the mat. It is important to note that decreased dissolved oxygen concentrations can be found beneath mats of native and nonnative plants. However, beds native aquatic species typically occur in isolated area with plenty of adjacent normoxic habitat and their density is relatively sparse, which allows more light to penetrate the water column, increasing dissolved oxygen concentrations beneath the mat (Fig. 9.15). The aggressive growth and rapid reproduction of invasive plants in the ARB creates large expanses with hypoxic conditions below the canopy. This is particularly problematic in impounded areas (e.g., Henderson Lake) and during late pulse events when high water temperatures and stagnation on the floodplain are already causing large-scale hypoxia. This alteration of water quality can reduce or eliminate habitat for native fishes and invertebrates that are sensitive to hypoxia (Gelwicks 1996, Davidson et al. 1998, 2000, Rutherford et al. 2001, Colon-Gaud et al. 2004).

FIG. 9.15. More light penetrates isolated, sparse beds of native aquatic vegetation, and the waters surrounding them are characterized by slightly higher concentrations of dissolved oxygen. These aquatic plant beds are important refuge areas for fish and other aquatic animals during periods of hypoxia. The extremely dense growth patterns of many nonnative vegetation species may actually decrease sub-canopy oxygen levels and exacerbate hypoxic conditions. Photo © Eric Engbretson, Engbretson Underwater Photography (www.underwaterfishphotos.com).

Native and nonnative aquatic plants provide important habitats for fish because they increase the structural complexity of the environment, reducing the risk of predation and supporting extensive food resources (Rozas and Odum 1988, Chick and McIvor 1994). Invertebrate prey, particularly those species that forage on periphyton (algae that grow on plant stems) comprise an important prey base for fish, creating an energetically valuable habitat, especially for juveniles (Castellanos and Rozas 2001, Mason 2002, Colon-Gaud et al. 2004).

In the ARB floodway, aquatic plants are important habitats for invertebrates and fish during both high and low water periods (Rutherford et al. 2001, Colon-Gaud et al. 2004, Troutman et al. 2007). In vegetated habitats of the ARB floodway 12 orders of invertebrates and 35 species of fish have been documented. For most invertebrate orders, there was no difference in density of animals found between hydrilla and the native coontail stands. However, hydrilla beds were associated with significantly lower densities of gastropods (aquatic snails), and this was likely due to less periphyton growth on the tightly packed hydrilla stems. Likewise, when compared to water hyacinth and the native emergent species bulltongue (*Sagittaria lancifolia*), hydrilla consistently contained the highest densities of fish (Fig. 9.16).

However, during low water periods, as the collapsed mats of hydrilla and hyacinth became extremely dense, fish tended to disperse, and the more structurally simple beds of bulltongue were used more heavily, due to more favorable dissolved oxygen levels

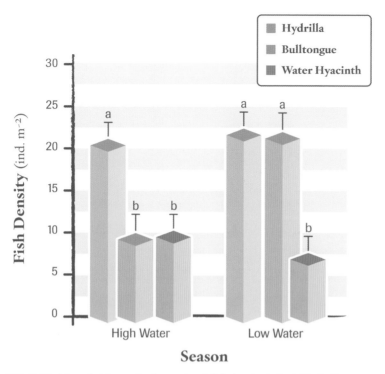

FIG. 9.16. Mean (with standard error bars) fish density in hydrilla, bull-tongue, and water hyacinth habitats during high and low water seasons in the ARB floodway. Bars with different letters are significantly different (p ≤ 0.05). Figure redrawn from Troutman et al. (2007).

and water depth. One study showed that the effects of hydrilla on fish may occur during early life stages (Mason 2002). Growth of age-0 largemouth bass was reduced in tightly packed hydrilla beds. These young fish had fewer prey fish in their diet when compared to fish in less dense hydrilla beds despite the fact that tightly packed hydrilla contained large numbers of prey fish. The young bass appeared unable to move effectively through the vegetation to catch prey fish. While studies show that invasive plants can provide comparable habitat for aquatic animals, an important question remains of whether the pervasive decline in habitat diversity resulting from the expansive spatial coverage of nonnatives will have a long-term effect on nekton, and, consequently, fisheries in the ARB.

10

Coastal Louisiana contains about 405,000 ha (1,000,000 ac) of wetland forest (Chambers et al. 2005). These coastal forests are extremely productive and valuable for flood and storm protection, fish and wildlife habitat, nutrient, carbon, and pollutant sequestration, as well as the economic and cultural values associated with both consumptive (e.g., hunting, fishing, timber production) and non-consumptive (e.g., ecotourism, birding) uses. The ARB contains about one-quarter of these coastal forests, and is a critical landscape to consider when assessing and addressing the future sustainability of Louisiana's coastal forests. Consequently, water management that promotes healthy and regenerating forests in the ARB is essential to promote the diverse and healthy wetland and aquatic habitats that are critical for keeping the ARB and the rest of Louisiana's coastal forests productive and valuable.

LOUISIANA COASTAL FORESTS. Since the 1980s the vigor and areal coverage of coastal cypress-tupelo forests has been declining across their range, but particularly in Louisiana (Conner et al. 1981, 1993, Conner and Toliver 1990, Pezeshki et al. 1990, Conner and Day 1988, 1992a, Thomson et al. 2002, Chambers et al. 2005, Keim et al. 2006a, Faulkner et al. 2007, 2009). In the last 70–80 years, the combination of subsidence, sea-level rise, and altered or engineered hydrology has increased duration of flooding as well as the depth of inundation in swamp forests, contributing to their decline.

This trend toward rising water levels has also been documented in the ARB floodway (see Fig. 8.15). Increased water levels in the ARB are also due to a combination of factors including land subsidence, sea-level rise, the prevailing water management system (i.e., static flow allocation at Old River, channel training, flood control levees), and natural and engineered hydrological barriers (oil field canals and levees; Sabo et al. 1999a, b, Keim et al. 2006a, Haase 2010). Mean monthly discharge from the Atchafalaya River increased after the Old River Control Structure began operating (Fig. 10.1; Haase 2010).

Although cypress and tupelo can tolerate extended flooding, persistent inundation can diminish swamp resources in two critical ways—by increasing mortality of

Atchafalaya River at Simmesport, Louisiana

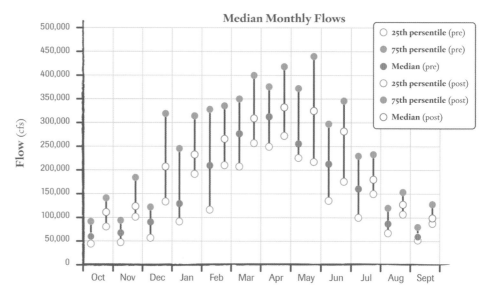

FIG. 10.1. Graph showing the median monthly discharge of the Atchafalaya River at Simmesport, Louisiana, from 1936–2009. Graph shows median flow both pre- (1936–1961) and post-operation (1968–2009) of the Old River Control Structure. Figure redrawn from Haase (2010).

adult trees and decreasing recruitment of young trees. Flooding stresses adult trees by reducing photosynthesis, which gradually leads to decreased growth and productivity, crown dieback, increased susceptibility to insects and pathogens, decreased root mass, and, ultimately, increased tree mortality (Fig. 10.2; Conner et al. 1981, Pezeshki 1990, King 1995, Keeland et al. 1997, Keim et al. 2006a). Flood tolerance in a tree species is dependent on the ability for individual trees to: 1) develop adventitious roots above the anoxic zone; 2) diffuse large enough quantities oxygen to the roots to both supply the roots and oxidize the rhizosphere (soil region adjacent to roots); 3) have a high capacity for anaerobic respiration; and 4) withstand high concentrations of carbon dioxide (Harms et al. 1980, Mitsch and Gosselink 1993). In continuously flooded conditions, water stagnates and toxins (including high concentrations of CO_2) accumulate to a point where even the most flood tolerant species cannot develop water roots. Pezeshki (1990) found that a CO_2 level of 180 ul L^{-1} was the threshold where photosynthesis was reduced in baldcypress.

Across the range of cypress swamps, including Louisiana and the ARB, growth and productivity of baldcypress decreases significantly in continuously flooded (impounded) conditions (Dicke and Toliver 1990, Conner and Toliver 1990, Conner et al. 1993, Megonigal et al. 1997, Keim and King 2006, Keim et al. 2006a, Shaffer et al. 2009, Keim and Amos 2012, and many others). Dicke and Toliver (1990) compared tree growth and mortality in seasonally flooded (SF) and continuously flooded (CF) cypress-tupelo stands near Bayou Pigeon in the ARB floodway. In this study, both stands had similar basal area (50 m² ha⁻¹ [11.5 ft² acre⁻¹]) and species composition (84% cypress, 16% tupelo). Continuous flooding did not reduce growth rates in

FIG. 10.2. An example of a baldcypress—water tupelo forest stand that is showing flood stress. Though flood tolerant, persistent flooding reduces stand productivity through reduced recruitment and growth and can even kill trees. Photo © Brittany App (www..appsphotography.com) for www .cyclingforwater.com.

tupelo or small-diameter cypress trees. In fact, mean diameter growth rate for water tupelo was slightly greater in the continuously flooded stand (1.3 mm yr^{-1} [0.05 in yr^{-1}]) than in the seasonally flooded stand (1.1 mm yr^{-1} [0.04 in yr^{-1}]), and, across all diameter classes, cypress and tupelo grew at a similar rate under continuous flooding. Conversely, continuous flooding reduced the growth of large cypress (25–35 cm dbh [10–14 in]) by 27–46%, compared to seasonal flooding conditions (CF 2.0 mm yr^{-1} [0.08 in yr^{-1}]; SF 3.2 mm yr^{-1} [0.1 in yr^{-1}]), suggesting that growth of larger trees was most affected by flooding. Basal area growth of both stands followed the same trends as diameter growth. While mortality of both cypress and tupelo trees was highest in seasonally flooded stands, the mortality rate of tupelo in seasonally flooded stands was three times higher than cypress. The continuously flooded stand contained fewer but larger trees (average 5 mm [0.2 in] or 24% larger) than the seasonally flooded stand. These patterns of growth and mortality suggest that water tupelo may compete better

FIG. 10.3. An example of a healthy baldcypress—water tupelo forest stand. Water marks on the trees show that this is a seasonally flooded stand that receives periodic influxes of nutrient-rich water that drains as river levels recede. Trees in these types of forest stands have greater survival, growth, and volume than those in continuously flooded stands. Additionally, the chance of successful baldcypress and water tupelo regeneration is greater in seasonally-flooded stands; however, overall regeneration of these species in the ARB floodway is low. Photo © CC Lockwood.

under continuous flooding, whereas baldcypress may be the preferred species under seasonally flooded conditions.

More than 15 years later, the same cypress-tupelo stands studied by Dicke and Tolivar (1990) were revisited to further understand the development, composition, and natural stand dynamics of established forest stands in relation to flood stress (Keim et al. 2013). In this study, the number of baldcypress and water tupelo trees in both stands had decreased. Tree growth was greater and the volume of baldcypress increased in the seasonally flooded stand (Fig. 10.3). Conversely, in the continuously flooded stand, tree growth was slower and there was a decrease in baldcypress volume. These results suggest that, while on the same overall development trajectory of self-thinning, the seasonally flooded stand was more productive, due simply to age, or to flooding, or a combination of the two. The slower growth rates at the continuously flooded stand appeared to be slowing its development. Interestingly, water tupelo was losing dom-

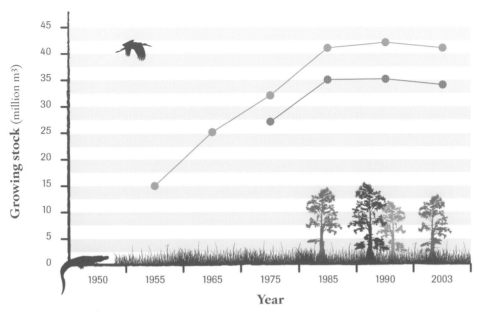

FIG. 10.4. Graph showing the temporal trend in growing stock of both Louisiana baldcypress and Louisiana coastal baldcypress only. Figure is modified from Keim et al. (2006 a) and is based on data from the US Forest Service.

inance to baldcypress at both stands. This was attributed to disturbance related to Hurricane Andrew, which broke the tops of over 60% of sampled tupelo trees in both stands (Keim et al. 2013).

Slower growth results in lower aboveground net primary production. In one study, aboveground net primary productivity on wet plots (675±271 g m^{-2} yr^{-1}) was significantly lower than on intermediate and dry plots (Megonigal et al. 1997). Additionally, when these authors analyzed their data in combination with other studies, they found that for every centimeter of flooding in wet plots, aboveground productivity of baldcypress decreased by 5 g m^{-2} yr^{-1}, and on sites where there was persistent flooding, this relationship was 5 times greater (-25 g m^{-2} yr^{-1} cm^{-1}). Another review of productivity reported in multiple investigations showed that persistent flooding was capable of decreasing aboveground productivity in cypress-tupelo swamps to less than 200 g m^{-2} yr^{-1} (Shaffer et al. 2009).

Ultimately, persistent flooding stress can kill cypress trees. After dam construction and impoundment (persistent flooding of 1–1.2 m [3.3–3.9 ft]) of the Oklawaha River in Florida, a healthy 1,620-ha mixed bottomland hardwood and cypress-tupelo swamp experienced 50% tree mortality, with 100% mortality within 3 years (Harms et al. 1980). In persistent flooding conditions, young trees exhibited the highest mortality rates, and trees in the larger size class (25–35 cm dbh [10–14 in]) that typically contribute most to stand growth, were usually the longest survivors (Harms et al. 1980, Conner et al. 1986, Dicke and Toliver 1990). An assessment of cypress growing stock in Louisiana from 1974–2003 showed that the mean diameter of cypress increased by more than 6 cm (2.3 in), but there was not a corresponding increase in total wood volume. Although cypress trees continue to grow, the lack of increases in total wood volume suggests that environmental stressors (i.e., persistent flooding, salinity) are causing increased mortality and decreased regeneration in baldcypress (Fig. 10.4; Keim et al. 2006a).

While flooding is an essential component of cypress swamps, all flooding is not equal. A number of studies have modeled long-term cypress tree growth (1895–2005) with historical hydrology (51 years) and climate (111 years) at sites that received riverine flooding (ARB floodway) and stagnant flooding (Lake Verret basin; Amos et al. 2005, Amos 2006, Keim and Amos 2012, Bohora 2012). Sites in the ARB floodway were flooded by headwater, riverine flooding, whereas sites in the Lake Verret basin were flooded by backwater inundation. Long-term radial growth rate in both systems averaged 1.1 mm yr^{-1} (0.04 in yr^{-1}), with a maximum radial growth rate of 11.4 mm yr^{-1} (0.45 in yr^{-1}). In both systems tree growth was determined by hydrology rather than climatic variables, and growth was directly or indirectly driven by river flooding—either direct flooding or river-influenced coastal water levels. Cypress at sites in the ARB floodway that received deeper headwater flooding grew more than those at stagnant sites in the Lake Verret basin that received backwater flooding. Hypoxic conditions due to persistent flooding may create stresses that decrease growth, whereas flooding by oxygen- and nutrient-rich water may create conditions conducive to growth.

Another study showed increased stem and diameter growth in cypress seedlings that were flooded by nutrient-rich water (Souther and Shaffer 2000). Discharge of sewage effluent can also greatly increase growth and productivity of cypress-tupelo forests, due to high attenuation rates (60–100%) for nitrogen and phosphorus (Zhang et al. 2000, Hunter et al. 2009). However, declining cypress growth was not reversed in chronically inundated forests that received wastewater effluent suggesting that the negative effects of prolonged inundation overrode the positive effects of nutrient addition (Keim et al. 2012).

Persistent flooding also decreases recruitment of cypress and tupelo due to its effects on seed germination and seedling growth (summary data for regeneration and growth in Appendix 4). Neither baldcypress nor water tupelo seeds can germinate in standing water; but once the water recedes, seeds of both these species can sprout (Fig. 10.5; Demaree 1932, DeBell and Naylor 1972, Hook 1984, Kozlowski 1997). In contrast, young seedlings are extremely vulnerable to flooding, and mortality can be expected in as few as 2–3 days if seedlings are completely submerged (Conner et al. 1986). Seed germination densities vary widely depending on inundation patterns (range 0–950 acre^{-1}; Appendix 4) with seasonal flooding regimes producing more than 40 times more seedlings than continuously flooded sites; newly germinated seeds can drown in 2–3 days. For the first two years after seedlings germinate they must grow rapidly to ensure that at least the terminal portion of their stem remains above flood height for all but a few days (Demaree 1932, DeBell and Naylor 1972, Conner et al. 1986, Souther and Shaffer 2000). Newly germinated cypress seedlings began dying after at about 45 days of complete inundation, with 100% mortality after 57 days (Souther and Shaffer 2000). In contrast, completely submerged 1-year old seedlings survived up to 100 days (75% survival at 100 days flooded). Long-term flooding that is deeper than the foliage of young seedlings results in high mortality. Therefore, the limiting hydrological factor for cypress and tupelo regeneration is the number of consecutive unflooded days during the grow-

FIG. 10.5. A baldcypress seedling in the ARB floodway. Both baldcypress and water tupelo are capable of regenerating from seeds once water level recedes. However, due to the depth and duration of flooding in most areas of the floodway, most seedlings succumb to mortality in their first year. Therefore, the future sustainability of cypress-tupelo forests in the ARB is dependent on achieving water conditions that promote the germination and survival of young trees. Photo by Richard F. Keim, Louisiana State University.

ing season, which in the ARB is approximately 214 days (April 1–October 31; Keim et al. 2006a). It is important to note that these survival estimates represent maximum survival times under ideal conditions and that turbid water may reduce the time that a seedling can survive submergence (Richard F. Keim, Louisiana State University Agricultural Center, personal communication).

Both baldcypress and water tupelo are capable of regenerating through stump sprouts after a tree is harvested ("coppicing"; Fig. 10.6). This type of regeneration has been studied extensively to determine whether coppicing can sustain swamps that have continuous flooding regimes. Initial stump sprouting in cypress is extensive (≥ 80% of stumps) after tree harvest; however, survival of sprouts is low and ranges from 0–23%

FIG. 10.6. Both baldcypress and water tupelo are capable of regenerating from cut stumps (coppicing). These stump sprouts regenerate quickly after the initial harvest, but mortality in regenerated sprouts is high. Therefore, coppicing is not considered a viable regeneration mechanism in cut-over cypress-tupelo forests. Photo by Richard F. Keim, Louisiana State University.

between 2 and 7 years after harvest (Conner et al. 1986, Conner 1988, Ewel 1996, Keim et al. 2006b). In southeastern Louisiana swamps, survival of stump sprouts 10–50 years after harvest ranged from 0–72% (Keim et al. 2006b). This is similar to other studies that have reported vigorous sprouting and long survival (> 30 years) in some areas and low sprouting rates and high mortality in others (e.g., Hook et al. 1967, Hook and DeBell 1970, Brinson 1977, Kennedy 1982, Aust et al. 1997, 2006). Thus, coppicing is not considered a reliable regeneration practice for baldcypress.

In southern deepwater swamps with seasonal flooding regimes, flood waters often remain too deep to expose mineral/organic soils long enough to stimulate seed germination. In fact, natural drawdowns may only occur once every 3–5 years and correspond with drought cycles (Messina and Conner 1998). Because baldcypress has such exacting requirements for germination and seedling survival, cypress swamps are typically comprised of a matrix of even-age cohorts.

The body of research on survival and growth of cypress-tupelo forests has several management implications, with respect to water management and restoration of freshwater flow. To optimize the sustainability of cypress-tupelo forests, flood waters should be nutrient rich, and water management should mimic natural seasonal flooding conditions. Water levels during the growing season do not need to be low for extended periods of time every year. Rather, management should promote extended periods of drawdown that expose forest soils every 3–5 years to allow seeds to germinate and grow. After trees reach 1 year of age, they can tolerate extended flooding; however to assure optimum conditions for seedling survival, water levels should be managed during the next 2–3 growing seasons (Demaree 1932, Souther and Shaffer 2000).

FORESTS IN THE ATCHAFALAYA RIVER BASIN. Assessments of tree health inside the ARB floodway are largely lacking, but two stands, 1 near Bayou Pigeon and 1 near Bayou Mallet, have been studied since 1990. Remote sensing and ground-truthing resulted in a classification of the Bayou Pigeon stand as healthy, mostly due to the influence of riverine flooding. The Mallet stand, further south and more separated from riverine flow, was characterized as a mix of healthy and intermediate, defined as between healthy and degraded in their analysis (Keim and King 2006). This classification is provisional, because regeneration potential at the sites was not assessed and the mapping using remote sensing may have resulted in scrub-shrub species in the Bayou Pigeon being classified healthy cypress. More on-the-ground studies of this type are critically needed throughout the range of cypress-tupelo swamps inside the floodway levees.

This same study assessed forest condition outside the floodway levees in the Verret Swamps. Forests at the southern end of the basin, surrounding the Gulf Intracoastal Waterway were degraded, largely due to inundation and their position at the forest/marsh interface. Other degraded areas included three locally impounded areas: 1) east of Bayou Corne; 2) southeast of Stephensville; and 3) east of Elm Hall Wildlife Management Area. These degraded forests encompassed about 10% of the 40,000 ha (98,842 acres) study area. The majority of the rest of the area was classified as intermediate health, and due to persistent flooding found in the area and the lack of riverine pulsing (Keim and King 2006).

Cypress forest encompasses 43% of the ARB floodway, and regeneration data show that the current water management regime is not adequate to sustain these forests. Using National Wetland Inventory maps and remote sensing (Landsat Thematic Mapper imagery) to classify the regeneration condition of the entire cypress swamp area in the ARB showed that of the 106,000 ha (261,931 acres) of cypress-tupelo forest in the ARB floodway, more than 23% (24,525 ha [60,600 acres]) was unable to regenerate either naturally (from seed) or artificially (planting of cypress seedlings), because it was permanently flooded (Faulkner et al. 2009). Only 5.8% (6,175 ha [15,258 acres]) of cypress swamp was classified as capable of natural regeneration. These areas were located in the middle and upper portion of the cypress range in the ARB floodway and

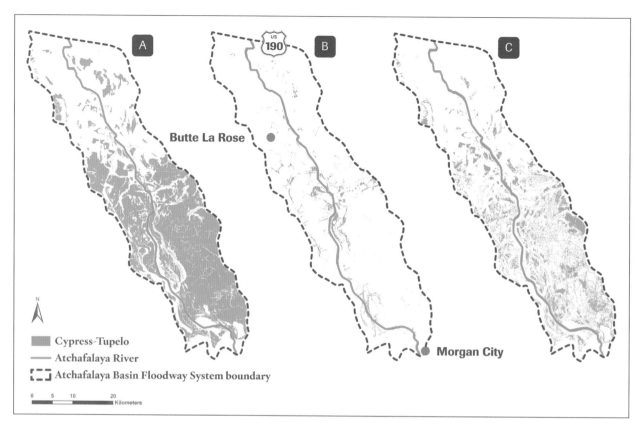

FIG. 10.7. Results of remote-sensing analyses showing the A) extent of cypress-tupelo forest in the Atchafalaya Basin Floodway System; B) areas classified as having hydrological characteristics favorable for natural regeneration of baldcypress and water tupelo trees; and C) areas classified as having hydrological characteristics unfavorable for either natural or artificial (planting) regeneration of baldcypress and water tupelo trees. Figure modified from Faulkner et al. (2009).

were small, linear features, likely along the backslope of small distributary levee ridges (Fig. 10.7; Faulkner et al. 2009). Results from this assessment show that over 25,000 ha (61,776 acres) of cypress swamp in the ARB floodway are not sustainable and are at risk of converting to marsh or open water.

More recently, a land classification method that used a combination of hydrograph and remote sensing data to identify the healthiest and most sustainable cypress swamp areas in the floodway was developed (Piazza et al., in review). These were areas that supported apparently healthy cypress-tupelo stands and were influenced by hydrographic cycles conducive to a wet-dry cycle that supports natural cypress regeneration. Areas considered most sustainable from a cypress regeneration perspective were flood-free for a minimum of 100 days and a maximum of 160 continuous days during the growing season, providing ideal conditions for cypress seed germination and establishment, based on values provided in the literature (e.g., see Keim et al. 2006a). Additionally, areas that were connected to the main Atchafalaya River received frequent pulses of riverine floodwater and provided water quality conditions (high oxygen and nutri-

TABLE 10.1. Results of a landscape analysis of the Atchafalaya River Basin Floodway

Area	Total area (ha)	Proportion suitable for baldcypress regeneration (%)	Proportion suitable for regeneration and exhibiting high water quality index (%)
Beau Bayou	9,160	33	2
Flat Lake	17,375	31	15
Buffalo Cove	12,469	35	21
Lake Henderson	11,738	38	2
Upper Belle River	11,658	29	9
Pigeon Bay	14,683	21	12
All areas combined	77,083	31	11

In the analysis, 6 priority conservation areas were identified where long-term hydrographic records showed that hydrologic conditions and riverine connectivity during the growing season were potentially favorable for baldcypress regeneration. Shown are the proportions of each area that had hydrology and turbidity characteristics (water quality index) that were potentially suitable for baldcypress regeneration and growth. The analysis was completed by The Nature Conservancy and the Engineer Research and Development Center, US Army Corps of Engineers.

ents) that correspond with increased growth and vigor in baldcypress and water tupelo. Therefore, areas that contained suitable flooding conditions and exhibited high levels of turbidity in at least seven (25%) of the 28 remote images available for the 25-year period of study were classified as having hydrological characteristics most suitable for baldcypress and water tupelo regeneration and growth.

From this analysis, six areas of cypress-dominated forest were identified within the floodway (77,083 ha [190,478 acres]) as the most sustainable under current hydrological conditions (Table 10.1; Fig 10.8). The six areas comprise approximately 23% of the forested wetlands, bayous and lakes that are contained within the 339,000 ha (838,000 acre) floodway. Although these areas were selected because they were generally better suited to sustain cypress-tupelo forest than the floodway in general, in none of the areas did the proportion of land considered ideal for cypress regeneration exceed 38% (range: 21–38 %; Table 10.1). To illustrate the widespread effects of poor water quality within the floodway, the proportion of area within each of the identified areas that was both ideal from a cypress regeneration perspective and received significant main channel pulses ranged from 2–21%, generally agreeing with the results presented by Faulkner et al. (2009).

Field-based and site-specific data are critically needed to refine the assessment of regeneration condition in the ARB and allow for future forecasting in light of changing hydrological conditions. For example, at a site located near Bayou Pigeon that Faulkner

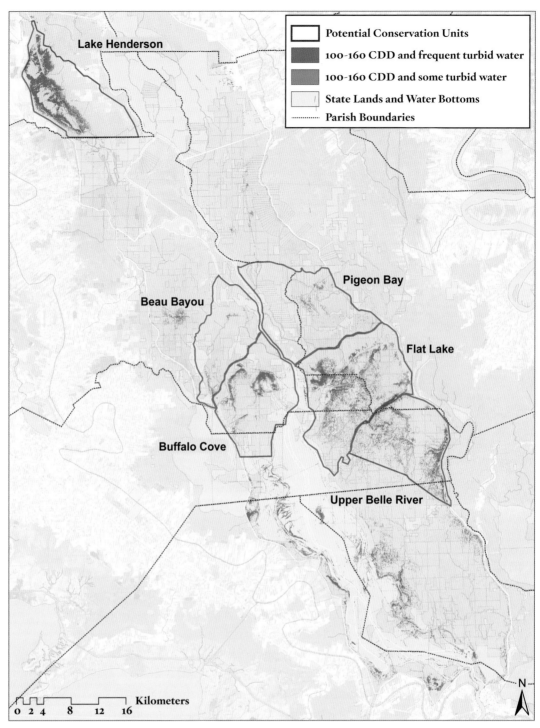

FIG. 10.8. Map showing six priority land conservation areas, determined by classification of cypress-tupelo swamp forests, hydrological conditions, potential growing days, and frequency of turbid water within the Atchafalaya Basin Floodway System. Areas colored in blue exhibited flooding conditions conducive for 100–160 growing day durations during the growing season and received relatively frequent observations of turbid water (7 of 28 images over 25 year time series). Areas colored in green exhibited the same flooding conditions and received the most frequent observations of turbid water (10 of 28 images analyzed over 25 year time series; Piazza et al., in review). Areas were named for the water management unit they encompassed.

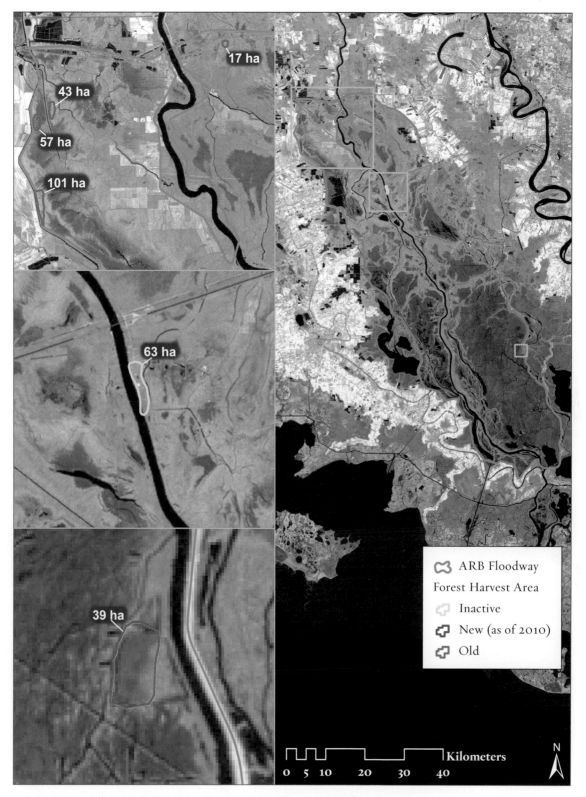

FIG. 10.9. Image showing the location of logging sites within the Atchafalaya Basin Floodway System, ARB floodway. Image is based on overflight data from the Louisiana Department of Agriculture and Forestry, Office of Forestry in spring 2010. Overflight data provided by Mike Thomas, Louisiana Department of Agriculture and Forestry, Office of Forestry, Baton Rouge, Louisiana.

FIG. 10.10. While many of the forests in the ARB floodway are not managed, some, particularly in the upper portion of the floodway are managed and harvested for wood production. Therefore, active logging does take place, though at a much lower rate than it once did. Photo © Alison M. Jones for www.nowater-nolife.org.

et al. (2009) reported was incapable of regeneration, a field-based study showed that there was actually an average of 116 potential growing days for a seedling, making that site capable of natural regeneration (Keim et al. 2006a). However, at a site near Bayou Mallet, a seedling has only 35 growing days on average before it is flooded, making the site incapable of natural regeneration. Flooding patterns at the Bayou Mallet site may provide a limited, but potentially available, window for planting cypress seedlings (artificial regeneration; Keim et al. 2006a). Additionally, because of the dynamic sedimentary nature of the ARB floodway, studies that rely on remote sensing alone may inadvertently classify one forest type as another due to errors that result from an incomplete understanding of the critical flooding dimensions for baldcypress and water tupelo regeneration in the floodway. Therefore, research that combines remote sensing and field investigation to better understand the interaction between flooding and forest sustainability in the ARB is critical.

FOREST HARVEST. According to the Louisiana Department of Agriculture and Forestry, forest harvest in the ARB floodway has slowed substantially, and very little harvest in the cypress-tupelo swamps of the floodway has occurred in recent years (Mike Thomas, Louisiana Department of Agriculture and Forestry, personal communication). Overflight data from spring 2010 showed only 201 ha (497 acres) of new harvest, and these tracts were located in bottomland hardwood forest habitat, not cypress-tupelo, immediately adjacent to the West Atchafalaya Protection Levee north of Indian Bayou Wildlife Management Area (Fig. 10.9). Recent reports confirmed the harvest of about 421 ha (1,040 acres) of cypress-tupelo in the Cocodrie Swamp area during December 2011 (Fig. 10.10; Yvonne C. Allen, US Army Engineer Research and Development Center, unpublished data). It is important to note that, with current hydrologic conditions both inside and outside the floodway levees, harvesting of cypress-tupelo swamp in the ARB can result in the loss of this forest type, likely to a scrub-shrub community (i.e., buttonbush, swamp-privet, water elm). Because baldcypress and water tupelo cannot regenerate under continuous flooding conditions during the growing season, the scrub-shrub ecotype may be favored (see Fig. 3.8; Keim and King 2006, Faulkner et al. 2007, 2009).

STATE OF THE FISHERIES AND WILDLIFE RESOURCES 11

THE ARB SUPPORTS LOCALLY, continentally, and globally important populations of fisheries and wildlife species (Fig. 11.1). Like the rest of coastal Louisiana, the ARB boasts abundant and diverse nekton resources and rich fisheries that are supported by fresh water and nutrients transported by the Mississippi and Atchafalaya Rivers and a diversity of wetland habitats. These fisheries not only provide livelihoods and economic value to the state, but they also provide a local food source that is affordable, widely accessible, and central to the culture of Louisiana—a culture that is appreciated worldwide. The forests and wetlands of the ARB are also important to migratory and resident populations of game and non-game species: black bear, waterfowl, shorebirds and wading birds, and neotropical migrants. Habitat loss or alteration and/or improper water management negatively affects the quantity and quality of economically and culturally important fisheries species, as well as overall biodiversity. Water and forest management that promotes diverse and healthy forests, wetlands, and water bodies is critical to ensure that the important ecologic and economic services provided by the ARB continue into the future.

PHYTOPLANKTON AND ZOOPLANKTON. The quality of aquatic habitat affects phytoplankton densities. Green-water habitats are extremely productive and typically contain very high densities of phytoplankton that generate large amounts of dissolved oxygen and provide ample food resources for higher-level consumers (Sager and Bryan 1981, Gelwicks 1996, Davidson et al. 1998, 2000, Rutherford et al. 2001). Brown-water habitats are very turbid and contain relatively low phytoplankton densities. Black-water habitats also contain low numbers of phytoplankton, likely due to shading (Sager and Bryan 1981).

In the ARB swamp zooplankton density ranged from 0.04–1,754 ind m⁻³, with green-water habitats supporting densities of zooplankton that were more than 10 times higher than brown- or black-water habitats, probably because densities of phytoplankton were greatest in green-water habitats (Fig. 11.2; Davidson et al. 2000). Although zooplankton densities in black-water habitats were much lower, species

FIG. 11.1. Adult and juvenile white ibis in the ARB floodway. The ARB supports local, regional, national, and global biodiversity by providing large blocks of forested wetland, marsh, and aquatic habitats that are important for many fisheries and wildlife species. Photo © Brittany App (www.appsphotography.com) for www.cyclingforwater.com.

richness of cladocerans (9 spp.) and copepods (12 spp.) was similar to both brown- and green-water habitats (Davidson et al. 2000). Species that were found in highest densities in black-water habitats likely employed a physiological mechanism (i.e., hemoglobin synthesis) to tolerate the reduced dissolved oxygen concentrations typical of hypoxic habitats, and these habitats may offer a refuge because planktivorous fishes are largely absent (Gelwicks 1996, Davidson et al. 1998, 2000, Rutherford et al. 2001).

CRAWFISH IN THE ATCHAFALAYA RIVER BASIN FLOODWAY. Despite their ecological, cultural, and economic importance, relatively little is known about the current status of crawfish populations in the ARB floodway. Crawfish stocks in flooded swamps in the ARB have been estimated only in the Henderson Lake area, where up to 47 animals m^{-2} were found (Pollard et al. 1983). In addition, little is known about the effects of harvest and water quality on crawfish stocks in the ARB floodway.

There are few regulations on crawfish harvest in the ARB floodway. Commercial harvesters are only required to possess a valid freshwater fishing license and a commer-

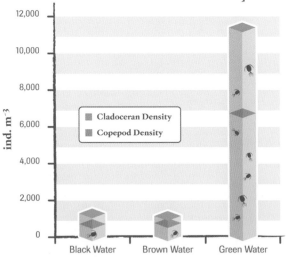

Zooplankton Density in the ARB Floodway

FIG. 11.2. Histogram showing the mean density of cladoceran and copepod zooplankton among three floodplain habitats in the Atchafalaya Basin Floodway System, ARB floodway.
Black water refers to water bodies that are clear, with little
water movement and low dissolved oxygen concentrations;
black-water habitats are typically found in overflow habitats
especially as water temperatures increase. Brown-water habitats are high energy waterways connected to the main river
channel with large amounts of water discharge, sediment, and
high concentrations of dissolved oxygen. Green-water habitats
are waterways where water temperatures are high and phytoplankton are abundant, that are often supersaturated with dissolved oxygen. Figure modified from Davidson et al. (2000).

cial crawfishing license. Trap dimensions and mesh size are regulated (flues ≤ 5.1 cm
[2 in], mesh size ≥ 1.9 cm by 1.7 cm [0.75 in by 0.67 in]) but not crawfish size or take.
There are also no regulations on entry into the fishery or the numbers of traps used by
fishermen. The lack of regulations and recordkeeping requirements, coupled with the
fact that the wild fishery is a source of supplemental income for many who participate
in it, results in limited information on harvest pressure and its effect on stocks. Some
studies have shown that harvest pressure can depress crawfish stocks in localized areas
of the ARB floodway. However, other studies suggest that, because of their life cycle
and tremendous impact of environmental factors on recruitment and survival, crawfish in the ARB floodway are an annual crop that cannot be overfished (Bell 1986, Dellenbarger and Luzar 1988, Bonvillain 2012).

The effects of water management and water quality in the ARB floodway may have
effects on crawfish stocks because of the importance of water flow characteristics and
environmental factors on crawfish recruitment and growth. From 1955–1981 pulse timing during the harvest year and the previous year affected yield of crawfish in the ARB

FIG. 11.3. Crawfish employ a number of mechanisms to cope with hypoxia. One of those mechanisms is to crawl out of the water to breathe air. This red swamp crawfish has crawled onto a water hyacinth to escape anoxic conditions. Photo by Christopher P. Bonvillain, Louisiana State University.

(Bell 1986). Similarly Dellenbarger and Luzar (1988) showed that water level change in the floodway, not simply water level, affected crawfish yield from 1970–1980. The effects of the pulse timing and duration on 1) flooding of backswamps where crawfish reside, 2) dewatering of backswamps so that crawfish spawning occurred in protected burrows rather than in predator-laden waters, and 3) water temperature and dissolved oxygen concentrations appear to determine crawfish yield. Bonvillain (2012) showed interannual variability in model parameters that explained crawfish catch and growth in the area just north of Flat Lake. In 2008 catch was a function of water temperature and turbidity, as well as the presence of aquatic macrophytes, whereas crawfish growth was a function of water temperature. In 2009, crawfish catch and growth was a function of dissolved oxygen concentrations only.

Crawfish have evolved mechanisms to cope with low oxygen conditions (Fig. 11.3). These mechanisms include breathing air at the air-water interface or by moving onto land, as well as physiological modifications that allow them to alter metabolism, modulate breathing and heart rate, and regulate oxygen transport through the hemolymph (see reviews in Bonvillain 2012, Bonvillain et al. 2012). While red swamp crawfish are more tolerant of low oxygen concentrations than many other crawfish species, the effect of chronic hypoxia on crawfish populations is not well understood. Bonvillain et al. (2012) found that moderate hypoxia (2 mg L^{-1}) did not stimulate either field-caught or laboratory reared red swamp crawfish to switch to anaerobic respiration, suggesting that the species can tolerate extended periods of moderate hypoxia, likely by breathing air. However, mean hemolymph protein concentrations were significantly reduced in field-caught crawfish subjected to chronic hypoxia, signifying a switch from active foraging to using protein reserves. This switch to protein reserves causes slowed juvenile growth and can affect molting (McClain and Romaire 2007). Conversely, in acute severe hypoxia, mean hemolymph lactate and hemolymph glucose levels in experimental crawfish (wild-caught and laboratory-reared) were significantly higher than crawfish in normoxic conditions, but their hemolymph protein was not reduced. These physiological effects signified a switch to anaerobic respiration pathways to maintain homeostasis under the stressful conditions. Crawfish will die in severely hypoxic waters, both in the laboratory and in traps in the field, when they are unable to reach the surface (Pollard et al. 1983, Bonvillain et al. 2012).

FISH IN THE ATCHAFALAYA RIVER BASIN FLOODWAY. Effects of water management and human engineering features on fishes have been studied since the 1970s, mostly in the water bodies (lakes, distributaries, low energy bayous) within the two large swamp complexes on the eastern and western sides of the Atchafalaya River from Bayou Sorrel south to Duck and Flat Lakes just north of Morgan City. These areas are some of the most affected by increased sedimentation and seasonal hypoxia that has resulted from the way that water is currently managed, both of which represent potential threats to nekton and commercial and recreational fisheries in the ARB.

Hypoxia affects fish physiologically. The partial pressure of oxygen (PO_2) in fish blood ranges from 50–110 mm Hg (2–4 in Hg; Davis 1975). At the range of water temperatures that occur in the ARB (5–35 °C [41–95 °F]), the PO_2 in fish blood falls below 45 mm Hg (1.8 in Hg) at a dissolved oxygen concentration of 2.0 mg L^{-1} (2 ppm). At this PO_2, normal respiration is impaired and fish must alter their behavior to compensate for the lack of oxygen (i.e., anaerobic respiration, mouth breathing, decreased activity; Davis 1975). Within the basin, hypoxia occurs in critical nursery habitats on the floodplain during peak spawning of many native fishes. This not only affects juvenile fish abundance on the floodplain but also can cause direct effects on predator species that use the floodplain as foraging habitat (Bryan and Sabins 1979, Snedden et al. 1999, Bonvillain 2006).

In the ARB floodway, the effect of the current water management model on fish has been assessed by documenting species presence-absence and comparing fish abundance and diversity to more complex measurements of community dynamics and production-related necessities (i.e., survival and growth). Most of these studies were conducted in the 900 km² (347 mi²) swamp area from Bayou Sorrel south to Morgan City. Fish abundance, species richness, and diversity were significantly greater in normoxic swamp habitats than in hypoxic habitats, with the greatest abundance of fishes found in phytoplankton-rich green-water habitats (Fig. 11.4; Gelwicks 1996). Community analysis showed that species composition of the fish assembly was affected most by dissolved oxygen concentrations, specific conductance, and current velocity, and that most adult game and non-game fish species in this area were positively associated with higher dissolved oxygen concentrations (Gelwicks 1996). Interestingly, there

Fish Abundance and Diversity
in the ARB Floodway

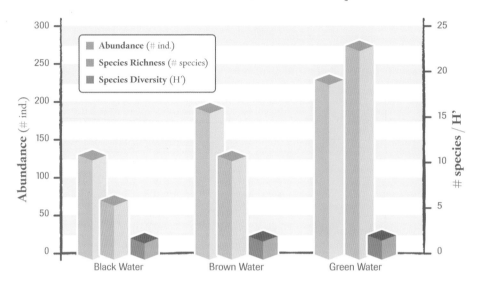

FIG. 11.4. Histogram showing the mean abundance, species richness, and species diversity of fishes among three floodplain habitats in the ARB. See Fig. 11.2 for definitions of black-, brown-, and green-water habitats. Figure redrawn from Gelwicks (1996).

was a limited suite of species (i.e., warmouth [*Lepomis gulosus*], spotted gar [*Lepisosteus oculatus*], pirate perch [*Aphredoderus sayanus*]) that was typically found in highest abundances in hypoxic, black-water habitats (Rutherford et al. 2001). Studies of fish growth in the ARB floodway are largely lacking, but one study showed that growth of bluegill (*Lepomis macrochirus*), an important recreational species, was significantly lower in hypoxic habitats, and another found growth of juvenile bass found in tightly packed hydrilla beds was low (Aday et al. 2000, Mason 2002).

Extended hypoxia also affects recruitment of fishes in the ARB. Brunet (1997) documented that in normoxic habitats the percentage of females with developed ova was significantly higher than in hypoxic habitats for bluegill, longear sunfish (*Lepomis megalotis*), and redspotted sunfish (*Lepomis punctatus*). Gonadosomatic index values, defined as the ratio of fish gonad weight to body weight, were three times higher in bluegills and longear sunfish in normoxic habitat than hypoxic areas, and, although egg abundance was similar, all three species produced significantly smaller eggs in hypoxic swamp habitats. In addition, the abundance of larval fish from 10 families was significantly lower in hypoxic habitats, and species-specific abundance was related to the timing and duration of flooding (Fontenot et al. 2001, Engel 2003). For example, abundance of larval sunfish during a late pulse event in 1995 was 81% less than the abundance at the same sites during an earlier spring pulse in 1994 (Fontenot et al. 2001). Similarly, a delayed pulse event in 2002 resulted in half the abundance of larval fish and 1.5 times less juvenile fish (at the same sites) than during a normal (earlier) pulse in 2001 (Engel 2003). Decreased abundance during later pulses was related to the increased scale and

FIG. 11.5. A wood stork (*Mycteria americana*) eating a spotted gar (*Lepisosteus oculatus*) on the floodplain. Like many fishes, spotted gar are closely tied to flooded habitats, rapidly moving onto the floodplain to feed. Flooplains are also hotspots for trophic transfer, as predators can quickly become prey. Photo © iStockphoto.com/milehightraveler.

intensity of hypoxia that occurs during delayed seasonal pulses. Although less abundant in hypoxic habitats, larval fish of some species tolerated hypoxia by using macrophyte beds (i.e., *Hydrilla verticillata*) as oxygen refuges until water levels receded and oxygen concentrations recovered; however, this presented a greater risk for predation loss (Fontenot et al. 2001, Engel 2003). Hypoxia appears to have great potential to affect the size of fish populations in the ARB, by reducing recruitment and survival.

Studies have also quantified the importance of floodplain connectivity for fishes in the ARB floodway. Both Snedden et al. (1999) and Bonvillain (2006) showed that flooded forest was important foraging habitat for spotted gar (Fig. 11.5). Alford and Walker (2011) found a distinct relationship between the number of days of spring flooding (optimized at 124–157 days yr^{-1}) and fall relative abundance of largemouth bass, crappie, blue catfish, and black buffalo, and that recruitment was tied to flooding conditions 1–2 years earlier (see Table 5.2). Flooding that lasts for more than 4–5 months results in lower recruitment and abundance of some species. In addition, water management and altered hydrology that limits the healthy connectivity and drainage of the floodplain can also have negative effects on the fish in the ARB floodway, by reducing fish recruitment and abundance. Effects on fish of commercial importance were mixed. Monthly crawfish harvest was positively related to flood magnitude and duration. Conversely, the gizzard shad, an important commercial fish species and forage fish in the ARB floodway was not dependent on the flood pulse, and their abundance was optimized during low-flow and drought years, likely due to the abundance of plankton (Alford and Walker 2011).

Halloran (2010) found that flooded forest habitat was less important than expected for fish recruitment, citing a mismatch between the onset of reproduction and flooding for a number of species. Instead, there was a stronger relationship between the onset of reproduction for fish and the increase in zooplankton populations that typically coincides with falling water levels. In this same study, densities of juvenile fishes using the floodplain varied; densities of centrarchids (crappies, bass, bluegills) were greatest. It is important to note that Halloran (2010) did not assess the importance of floodplain connectivity for poeciliid reproduction, foraging, reproduction and provisioning of food resources which are related to flooding timing and duration for these and other resident wetland nekton species in other wetland systems (Kneib 1997, Piazza and La Peyre 2007, 2010, and others).

The available research targets species that maintain much of the ecological and commercial productivity of the ARB floodway and has management implications with regard to large-scale water management and restoration of local hydrology in the ARB floodway. The combination of a balanced and variable flow regime and local hydrological management, both of which promote riverine connectivity and proper drainage, would potentially benefit the fisheries of the ARB floodway by increasing nekton recruitment and growth.

The available research also suggests an inundation period that benefits nekton could be combined with the inundation patterns necessary for healthy forests to benefit biotic resources as a whole. Research on nekton species provides an opening to begin discussing and formulating an adaptive water management program in the ARB floodway that uses both the Old River Control Structure and local hydrologic restoration projects to benefit both fisheries and forestry resources. It is important to note, however, that variability in flow regime is critical to balancing the needs of the suite of nekton species. Some species prefer deeper water and more persistent flooding, while others prefer shallower water and shorter flood durations; thus managing water in concert with climatic cycles could be a way to achieve balance in the system.

NEKTON IN THE ATCHAFALAYA AND WAX LAKE DELTAS. In the Atchafalaya and Wax Lake deltas 101 species of nekton (finfish and decapods crustaceans) have been documented in waterways and flooded tidal marsh habitats (Castellanos and Rozas 2001, Thompson and Peterson 2001, 2003, 2006, Peterson et al. 2008). A common theme is the fact that the deltas support a full suite of species (freshwater to marine). Also, this growing delta system provides heterogeneous and healthy wetland and aquatic habitats, making it excellent nursery habitat for a wealth of wetland-dependent nekton species. One study that compared nekton communities on the Atchafalaya and Wax Lake deltas demonstrated that the two deltas supported essentially similar nekton communities, and slightly increased nekton diversity within the Atchafalaya Delta suggested that anthropogenic alteration of the main Atchafalaya River delta, in the form of beneficial use of dredge disposal for marsh creation, had not negatively affected fish communities (Peterson et al. 2008). Artificial acceleration of the delta cycle resulting from strategic deposition of dredge spoil may have actually increased aquatic habitat diversity, as compared to the largely unaltered Wax Lake Delta (Peterson et al. 2008). As long as this delta system remains healthy, and if management continues to promote healthy wetland and aquatic habitats on the deltas, nekton populations will continue to flourish.

WILDLIFE IN THE ATCHAFALAYA RIVER BASIN FLOODWAY. Management of high quality bottomland hardwood forest habitat for timber stands of oaks and other hard mast producing species is also good for producing optimum black bear foraging habitat (Black Bear Conservation Coalition 2011). Desired future conditions include restoration and maintenance of naturally diverse forest stands that include a significant hard mast component and a structurally complex forest with trees of all age classes.

Because large baldcypress and water tupelo trees are important escape and denning habitats for Louisiana black bears, retention of both small isolated groups of cypress trees within bottomland hardwoods, as well as large stands of cypress and tupelo is important for bears. In fact, under the Endangered Species Act, all large baldcypress and water tupelo trees adjacent to water that are ≥ 91 cm dbh (36 in), with visible signs of defects (i.e., cavities, broken tops), must be protected within the range of Louisiana black bear (US Fish and Wildlife Service 1995, Black Bear Conservation Coalition, 2011).

Additionally, because canebrakes were historically important habitats for Louisiana black bears, as well as many wildlife species, including the critically endangered, but likely extinct, Bachman's warbler, native cane habitat should be favored in hardwood forest stands and restored wherever appropriate (i.e., abandoned agricultural fields, utility rights-of-way; Remsen 1986, Black Bear Conservation Coalition 2011).

Across the United States, the populations of roughly one-third of all forest bird species have declined, largely due to habitat loss, and their vulnerability is expected to increase with future climate change (State of the Birds 2009, 2010). Therefore, large expanses of contiguous forest, like the ARB, are critical to supporting large, sustainable populations of birds, including many of the most imperiled species (State of the Birds 2010). Management that threatens the health and availability of forested wetlands of the ARB can reduce the diversity and abundance of birds. Altered hydrology at local (canals and levee banks) and basin-wide (water management system) scales, affects bird habitat in many ways, including effects on cypress-tupelo forest regeneration, distribution of forest types, individual tree mortality, and distribution and abundance of aquatic prey species. Forest management that results in extensive tracts of even-age and young forest (i.e., clearcutting) causes habitat fragmentation, selects against those bird species that rely on contiguous stands of older forest types, and provides colonization sites for invasive tree species, such as Chinese tallow, which provide poor habitat for most native birds (Baldwin 2005).

12

RECENT STUDIES HAVE USED targeted surveys, stakeholder observation, and extensive interviewing of stakeholder groups (members of stakeholder groups are listed in chapter 7) to determine: 1) how different stakeholders view the ARB, based on their perspectives and values; 2) where conflicts or potential conflicts reside; 3) how different stakeholders feel about the current state of the ARB and its resources; and 4) the perception of the roles of science and stakeholders in management and restoration of the ARB.

Stakeholders see and define the ARB quite differently, depending upon background, values, livelihoods, perspectives, and mandates (Fig. 12.1; van Maasakkers 2009). These differing viewpoints not only define how stakeholders view issues in the ARB but also affect the way they define the ARB and view its boundaries. What has resulted from these differing definitions is an array of maps and separate plans with different and shrinking definitions of the ARB boundaries, and this has created confusion and frustration. For example, the US Army Corps of Engineers (1982) defined the ARB as the historical extent of the basin, incorporating areas both inside and outside the floodway protection levees. The Louisiana State Master Plan (Louisiana Department of Natural Resources 1998) restricted that definition to the area inside the protection levees between Simmesport and Morgan City, and the US Army Corps of Engineers (2000) further restricted the boundaries to the Atchafalaya Basin Floodway System (see Fig. 1.2). The cultural definition of the ARB again broadens the geographical scope to include all 14 parishes that compose the Atchafalaya National Heritage Area (Atchafalaya Trace Commission 2011). Coastal plans have differed in their treatment of the ARB with regard to its relationship with and its inclusion in the coastal zone (e.g., see State of Louisiana 2012). Adding to this confusion is the fact that physical processes (i.e., sedimentation) can cause relatively rapid habitat change that can also change legal boundaries in the ARB, and this phenomenon has led to both confusion over who is defined as a legitimate stakeholder and litigious conflicts over rights (van Maasakkers 2009).

FIG. 12.1. Stakeholders view and define the basin differently, depending on their perspectives and values. Differing viewpoints can serve as a source of frustration and conflict and illustrate the importance of a clear, well-defined decision-making process that accounts for differences in perspectives and values (see Kozak and Piazza, in review). For example, while consumptive use like hunting (top photo) and fishing (commercial and recreational) have traditionally been major uses of the ARB, there are a growing number of non-consumptive stakeholders, like this paddler (bottom photo). These new stakeholders may have different viewpoints than traditional users of the basin. Photos © Bob Marshall.

FIG. 12.2. Land access is a continuing source of conflict in the ARB floodway. This conflict stems from questions about the locations of legal boundaries. All stakeholder groups feel that the resolution of this conflict is paramount because they view it as a hindrance to restoration. Photo by Kristopher Simon, University of Louisiana at Lafayette.

These differing viewpoints, definitions, and mandates have led to broad stakeholder frustration over planning, management, and allocation of restoration funding in the ARB. The most commonly cited reasons for stakeholder frustration are tied to the lack of a clear and scale-appropriate definition of the ARB: 1) the lack of clear management objectives; 2) inappropriate consideration of geographic scale; and 3) funding challenges (Louisiana Department of Natural Resources 2002, van Maasakkers 2009, Consensus Building Institute 2010a).

Stakeholder conflict and its ramifications for restoration are also sources of concern for many of the stakeholders in the ARB. There has been a long-standing and bitter conflict over access rights in the ARB floodway (Fig. 12.2). Landowners feel that private property rights are in jeopardy, and commercial and recreational stakeholders cite fear and confusion over future use of the ARB. Some of this hostility is a relic from lack of landowner inclusion in decision-making processes of early years, and much of it results from questions over the location of legal boundaries. As a result, litigation over the conflict between public and commercial use and private property rights has gone to the highest courts, and both parties can cite court victories that have advanced their cause. Stakeholders universally name this particular dispute as a source of confusion, angst, and a hindrance to restoration of the ARB floodway (specifically project implementation), and express a strong desire for a solution (Reuss 2004, van Maasakkers 2009, Consensus Building Institute 2010a).

Despite differing viewpoints and conflicts, studies show that stakeholders are in broad agreement over key points (Louisiana Department of Natural Resources 2002,

Consensus Building Institute 2010a, State of Louisiana 2012). Most stakeholders agree that the ARB is an important ecosystem locally, regionally, nationally, and globally. They are also worried about the future trajectory of the system, because all groups envision a different ideal ecosystem from what currently exists. Stakeholders also agree with the flood control objective of the ARB floodway; however, they desire a system where flood control is implemented in a manner that considers what is best for the natural resources. For example, there is broad agreement that the prevailing water management model (i.e., 70/30 flow split) is negatively affecting basin resources in the floodway. Stakeholders independently identified their interests as being negatively affected (i.e., crawfish, timber, and fishing) and feel that problems with water levels and flood/drawdown timing are driving those effects. There is a broad desire to allow for slight changes in the 70/30 flow allocation to alleviate these problems; however, some stakeholders worry about change to the prevailing model. Stakeholders also showed broad agreement that coastal issues are important and that the Atchafalaya River has a potential role in helping restore coastal wetlands, particularly in the Terrebonne basin; however, there is uncertainty over using the river to meet coastal objectives.

Stakeholders generally agree with the science that the ARB is not on a good ecological trajectory; however, they have an almost universal frustration with past planning and management efforts in the ARB and differing amounts of confidence in current restoration efforts (van Maasakkers 2009, Consensus Building Institute 2010a). This frustration is rooted in: 1) the lack of a clear and universal definition of the ARB; 2) the lack of cohesion among different programmatic management and restoration goals and mandates; 3) the limited process for stakeholder involvement, including limited use of local knowledge; and 4) limited cooperation among government agencies responsible for management and restoration.

While stakeholders generally agree with the science and with its use in planning and restoration, they feel that the current planning process does not employ enough local knowledge to bolster the science and that stakeholders are brought in too late, after objectives are set, information is gathered, and interventions have been formulated. This limited and late-stage role is unacceptable to stakeholders whose lives will be affected by changes to the system. Stakeholders expressed a desire for a more integrated planning process that engages them throughout the process, from defining system boundaries and relevant problems, to setting objectives, gathering information, and making decisions (van Maasakkers 2009).

Land protection is an important issue in the ARB floodway because programs to enroll landowners in land conservation programs have been underfunded and remain incomplete. Under the authorization of the US Congress, the US Army Corps of Engineers has purchased 20,234 ha (50,000 acres) of land in the ARB floodway under fee ownership for public access. This land includes Indian Bayou Wildlife Management Area, purchased in 2001, as well as inholdings purchased in Sherburne and Attakapas Island wildlife management areas. In 2007, the US Congress again authorized the Corps to protect land in the ARB floodway, this time by authorizing the purchase of an additional 8,093 ha (20,000 acres) in fee ownership from willing sellers, as well as to enroll 135,560 ha (335,000 acres) into developmental control and environmental protection easements. However, due to lack of appropriations, only about 36,421 ha (90,000 acres) have been enrolled in easements, and none of the additional fee land has been acquired.

As a land protection vehicle, US Army Corps of Engineers developmental control and environmental easements are minimally restrictive and include only the following prohibitions: 1) land cannot be converted from existing use or otherwise developed; 2) new permanently habitable structures cannot be placed or constructed; 3) any other new structures cannot be placed or constructed without written permission from the Corps, with the exception of structures used for mineral exploration/production or to rebuild an existing structure. Timber harvest is allowed, except for the following: 1) baldcypress >107 cm dbh (42 in); 2) oaks (*Quercus* spp.), ashes (*Fraxinus* spp.), and sweet pecan (*Carya illinoinensis*) < 51 cm dbh (20 in); and 3) water tupelo (*Nyssa aquatica*) and baldcypress < 61 cm dbh (24 in), unless 9.1 m^2 ha^{-1} (2.1 ft^2 $acre^{-1}$) of basal area is maintained in any combination of these species.

The results of a recent survey of landowner stakeholders in the ARB floodway indicated that there was a potential opportunity for a prioritized and comprehensive land protection program. Of landowners contacted, 80% were interested in participating in land protection programs, and 26% already had land enrolled in the US Army Corps of Engineers developmental control and environmental protection easement program. Those already enrolled in the Corps program were universally satisfied with it. All landowners who were interested in participating in land protection programs were also willing to accept more restrictions for more compensation (Piazza et al., in review).

PART FIVE

HOW DO WE CREATE A NEW FUTURE FOR THE ATCHAFALAYA RIVER BASIN?

COMMON PATTERNS ACROSS SCIENTIFIC DISCIPLINES 13

T HE BODY OF RESEARCH ON THE ARB highlights the role of the river basin as a provider of living resources that are not only critical for maintaining local, regional, national, and international biodiversity but are also critical for people who benefit both directly and indirectly from its goods and services. This review of the research has also highlighted some consistent patterns that have emerged as long-term threats to continued, sustainable viability of the ARB.

THE WATER MANAGEMENT MODEL. The prevailing water management model in the ARB is damaging the living resources of the ARB. It is degrading cypress forests—via excessive accretion and conversion to bottomland hardwood in some areas inside the floodway and through flood-induced stress and mortality both inside and outside the floodway levees. Additionally, excessive flooding both inside and outside the East and West Atchafalaya Protection Levees has caused almost complete failure of baldcypress and water tupelo regeneration throughout the ARB. These forests are critical for maintaining the healthy fish and wildlife resources that many Louisianans rely on for their livelihoods and recreation.

Poor water circulation and excessive, stagnant flooding are the result of the prevailing water management system and localized obstructions to water flow. These conditions harm aquatic resources because they degrade water quality, especially by reducing dissolved oxygen concentrations. Currently, spatially and temporally extensive dead zones (hypoxia) both inside and outside the floodway levees recur every summer and persist for months, causing habitat quality to deteriorate and fish kills. In addition, the system is unable to reset properly after tropical storms.

HYDROLOGIC RESTORATION. The distribution and management of quality water throughout the ARB is paramount to maintaining sustainable ecosystems for wildlife and people. The key to sustainable management of water quality is the restoration of natural flooding and drying cycles, in forested wetlands within the floodway and outside the protection levees. When the magnitude, timing, frequency, and duration of floods are within the natural range of their variability, biological production is enhanced and key ecological processes are functional. Natural water cycling in many riverine systems relies on a late-winter/early-spring flood pulse to inundate and purge the swamp of organic debris that collects throughout the year. Flooding typically peaks around mid-April and floodwater remains into June. The drainage cycle that follows the pulse is equally important to maintaining healthy wetland habitats and typically occurs in summer and fall when water temperatures peak. When this natural flood-drain cycle is broken and flooding becomes excessive and persistent (i.e., flood waters do not drain properly), excessive organic matter collects on the swamp floor. If water temperatures are high, decomposition depletes dissolved oxygen in the water column causing widespread hypoxia in the floodway, as well as in swamp areas outside the protection levees.

Restoration of the natural flooding and drying cycle within the ARB floodway requires a combination of strategies. At the basin scale, an appropriate water management system that incorporates a balanced and variable seasonal flow regime, to which native species have adapted, may require reevaluation of congressionally mandated water distribution at the Old River Control Structure (Fig. 13.1). Flows entering the ARB floodway have increased over the past 50 years mirroring a similar trend in the Mississippi River. This fact, along with corresponding evidence of excessive duration and magnitude of flooding, indicate that the 70/30 longitudinal split of daily discharge between the Mississippi and Atchafalaya Rivers may need to be modified to reestablish flooding patterns within the ARB that are within the historical range of natural variability for the basin. Deviation of only a few percentage points during the flood cycle (depending on river volume) and across the entire hydrograph (high and low flow) may significantly improve water depth and flood conditions in the swamp and allow proper large-scale conditions for drainage (van Beek et al. 1979, Bryan et al. 1998).

At the local scale, appropriate hydrology management projects need to address channel engineering features (dredging, bank-stabilization levees, and distributary closure) that resulted from efforts to control flooding and barriers to flow caused by obstructions (canals and levees) to assure proper throughflow and drainage of water in floodplain wetland environments. Channel engineering has separated many swamps from natural overbank river flooding except during the highest flood stages by increasing water conveyance in the Atchafalaya River (Alford and Walker 2011). The river-floodplain interaction that included natural overbank flooding from the rivers and natural distributaries that drained from higher land in the north toward the lower land in the south has been replaced by a system where the remaining distributaries (natural and artificial) and anthropogenic alterations such as channelization are diverting water into the swamp in ways that set flow in the opposing directions to drainage causing large-scale water stagnation. This stagnation affects sedimentation patterns and water chemistry and, consequently, the distribution and productivity of biotic resources.

Canals and levees, particularly for oil and gas exploration, have separated some swamps from overbank flooding while channeling sediment-laden water directly into others, causing increased segmentation and isolation of habitats and, in some cases, transition to a higher-elevation forest community (Fig. 13.2). Isolation of Atchafalaya

FIG. 13.1. The Old River Control Structure ensures that the ARB floodway provides essential flood control for the lower Mississippi River. This structure is also central to restoring a flood cycle that will promote the health and sustainability of the floodway's natural resources. Photo by David Y. Lee © The Nature Conservancy.

River water from habitats on the floodplain diminishes the ability of the ARB to remove nutrients from the water, resulting in higher loading rates to the Gulf of Mexico.

Because of the extremely low water flow rates found in the swamp, even small obstructions can have large effects on the direction of flow. Therefore, local flow obstructions, particularly in the lowest areas of the floodway, impede interaction between the river and the swamp and prevent proper swamp drainage. These obstructions create flooding and water quality conditions that are unhealthy for aquatic species and that are not conducive to cypress and tupelo regeneration. For example, even when flood stages were at their highest during the Mississippi River flood in 2011, there were areas of the upper Belle River water management unit that were not adequately flushed, due to the myriad of flow obstructions in the area (Daniel E. Kroes, US Geological Survey, personal communication).

Based on a review of studies covering over 20 years of water management in the ARB, an "ideal" large-scale water management model would provide flood control (when it is needed) and navigation but would also allow an annual flooding/drying cycle that would benefit the ecological resources of the ARB floodway by promoting

FIG. 13.2. An example of a canal dredged and maintained for oil and gas exploration in the ARB floodway. This canal, connected to a distributary, channels sediment-laden water into interior water bodies and swamps. Additionally, the associated levees disrupt local hydrology and separate swamps from natural overbank flooding. Basin-wide water management strategies coupled with local strategies and projects that are designed to minimize canalization and promote hydrologic reconnection are key to the long-term viability of swamp forests in the ARB floodway. Photo by Yvonne C. Allen, US Army Engineer Research and Development Center, Baton Rouge, Louisiana.

the highest diversity of aquatic habitats and healthy forests. This model would not be static, but would be adaptive, allowing slight deviations at critical times. It would account for the vital interests and engineering features that have been put in place but at the same time, approximate the natural water cycle that probably existed prior to the levees when flooding depths were significantly lower (0.5–0.6 m [1.6–2 ft]); Bryan et al. 1998, Sabo et al. 1999a), such that livestock were routinely grazed in the ARB (Jim Delahoussaye, Friends of the Atchafalaya, personal communication).

One specific alternative for managing water with this type of model, proposed by Bryan et al. (1998), has been adapted to report stages at Butte La Rose, instead of Bayou Sorrel Lock. These authors stated that an "ideal" water management model would promote a cycle where the stage would exceed 3 m (9 ft) at the Butte La Rose gauge by early January. Stages would increase to ≥ 4 m (14 ft) by mid-April. Water levels would then be reduced by mid-June to ≤ 1.5 m (5 ft), to allow the swamp to drain and detritus to oxidize. At this stage, most of the swamp on the west (except Beau Bayou area) and east (north of Old River) sides of the river would drain (Daniel E. Kroes, US Geological Survey, unpublished data). This type of management, where swamps are allowed to drain periodically, is also critical for managing nonnative species, particularly water hyacinth, because many plants would be stranded on the floodplain and die (Bryan et al. 1998). Sabo et al. (1999a) recommended a similar flow regime but also suggested that management strategies consider climatic conditions, so that high discharge levels would occur when air temperatures are below the median.

Climate should also be considered when timing periodic prolonged dry periods that are necessary for baldcypress and water tupelo health and regeneration. Drawdowns should be timed to coincide with natural climatic drought cycles that occur every 3–5 years (Bryan et al. 1998). Both of these strategies that focus on seasonal and meso-scale cycles would maximize the time when low discharge and high water temperatures co-occur. This low stage-high temperature combination would optimize the diversity of habitat characteristics in the ARB floodway and is considered exceedingly important for aquatic and forest productivity (Bryan and Sabins 1979, Brunet 1997, Lambou 1990, Sabo et al. 1999a, b, Davidson et al. 1998, 2000, Aday et al. 2000, Rutherford et al. 2001, Fontenot et al. 2001, Keim and King 2006, Keim et al. 2006a).

To further ensure optimal habitat diversity, it has been suggested that remaining lakes in the ARB be protected from unnaturally accelerated rates of sedimentation. Lakes share many habitat characteristics with high-energy distributaries and stay normoxic throughout the year, providing critical refuges for aquatic organisms during periods of hypoxia elsewhere in the ARB (Sabo et al. 1999a). Conduits that allow direct delivery of sediment-laden water into open-water habitats should be discouraged. While intact cypress-tupelo stands may benefit from the nutrients and sediments supplied by well-engineered diversion projects, open-water habitats subject to similar inputs will quickly convert to land, be colonized by willow trees, and serve to further isolate and segment the habitat.

With respect to the coast, the Atchafalaya River drives delta formation and growth, and it must continue to do so into the future. The continued viability of the Atchafalaya and Wax Lake deltas depends on the fresh water and sediments delivered by the river, and significant changes in flow and the transport of sediments may have significant and potentially detrimental effects on these deltas. When developing competing management scenarios for the ARB, including the potential role of the Atchafalaya River in the future of coastal restoration, it is clear that future management strategies must not only see the Atchafalaya and Mississippi Rivers in their role as fresh water and sediment delivery systems to the coast but must also remember their role in restoring and maintaining the forested wetland and emerging deltaic systems within the ARB.

TRADEOFFS. The scientific literature provides guidance for different management and restoration objectives in the ARB, based on individual ecosystem services. However, when considering management based on the interaction of ecosystem services (and interrelationships between human uses), clear tradeoffs emerge. For example, management actions that restore riverine connectivity into swamp forests for nutrient removal, forest health, or water quality improvement must also contend with the sediment that accompanies that water. Access to areas in the floodway and the basin's resources also represent tradeoffs, because access routes (i.e., canals) may also alter hydrology, channeling sediment into lakes or creating water quality problems. Water management that allows forest regeneration may affect commercial and recreational fisheries. The list goes on.

The ARB is a complicated place, and sorting out these tradeoffs will also be complicated and a likely source of conflict. While the science is there to provide guidance, decisions about these tradeoffs will ultimately be made by the basin's stakeholders, the people whom the management will affect. Making hard decisions that balance stakeholder values with objective science will result in an ARB that provides ecological, cultural, and economic benefits well into the future, but the science also shows that making science-based decisions is hard work.

STAKEHOLDER VISIONING AND DECISION MAKING. A collective stakeholder vision of a healthy and sustainable ARB and an understanding of how this system would interact with an imperiled coast are essential for achieving the goal of ecosystem restoration (Kozak and Piazza, in review). There are many stakeholders to whom the ARB provides livelihood, recreation, and cultural and aesthetic value. Therefore, a collective vision is essential for developing a stakeholder-driven decision-making process and adaptive management framework that promotes objective-driven and learning-based management. In this framework, stakeholders can weigh the current scientific information with their values to make informed decisions on how current and future generations of people will interact with the ARB.

The science clearly shows that the ARB is a dynamic system and historic anthropogenic alterations as well as current management regimes will continue to alter the composition and structure of terrestrial and aquatic habitats, force changes in the distribution and abundance of wildlife that depends on those habitats, and greatly influence the commercial and recreational uses by a wide variety of stakeholders. In the face of inaction, the science clearly shows that there will likely be continued degradation of terrestrial and aquatic resources that form the centerpiece for the southern Louisiana way of life. If we rise to the challenge, we have a unique opportunity to ensure that viable human and natural communities persist well into the future.

14

SCIENTISTS HAVE CONTRIBUTED a great deal of knowledge about the geological and ecological processes that have built and maintained the Atchafalaya River basin, and the common patterns that have emerged across disciplines clearly show that the prevailing water management model is damaging the living resources in the system. However, while we already know a great deal about the current state of the Atchafalaya River Basin, there are a number of questions, that if answered, will elevate the knowledge of the system greatly, and this knowledge will aid the ongoing and future conservation and restoration of the ARB. A few examples of these questions are provided below, and this list should not be considered exhaustive. Some of these scientific efforts are already under way, and some have not yet been undertaken.

It is important to note, however, that the science is really ahead of the game in the ARB. It is already influencing local-scale hydrology restoration projects, focused on individual or multiple water management units, through a process that engages scientific expertise and uses rigorous analysis tools (i.e., Atchafalaya Basin Program Natural Resource Inventory and Assessment Tool). Therefore, future studies should be done with the goal of improving and scaling up the restoration program and being relevant to decision making, and not used as an excuse for inaction. It should be the goal that these questions be answered in a framework for designing and informing science-based decision making and management systems through empiricism, rigorous monitoring, and improved system modeling, all of which will inform and improve management through improved stakeholder involvement, continued learning, and process adaptation.

Hydrology and Sedimentation

▶ Balancing freshwater delivery, water quality, and sediment accretion in the ARB floodway has been an ongoing challenge to both project management and habitat restoration. There is a need to compare accretion rates over a wide variety of factors (i.e., habitats, landscape position, sediment source, sediment composition, delivery route—

natural vs. cuts and gaps—delivery times, etc.) and to determine a range of strategies for balancing water quality improvement, via riverine connectivity, with the goal of reducing sediment accretion in some areas.

► There is a good amount of basic information on inundation frequency in the ARB floodway, including how changes in water level impact the horizontal extent of inundation. However, what is lacking is a clear understanding of how water levels impact water depths throughout the floodway. This information is critical to understanding water flow patterns and planning restoration at the scale of water management units, multiple water management units, and system wide. Accurate elevation and subsidence measurements are needed throughout the floodway, as well as bathymetry information across the floodway.

► There is a need for study of water flow patterns and subsidence outside the East and West Atchafalaya Protection Levees, in the Verret Swamps and Lake Fausse Point. These areas also represent large blocks of the remaining coastal forests in the lower Mississippi Valley, and there are virtually no scientific data on how this system will respond to future stressors, namely sea-level rise. The future of these forests may require some restoration of riverine connectivity, but there is no information on how the systems may react to such inputs.

► There is a tremendous wealth of information on delta growth and development on the Atchafalaya and Wax Lake deltas. However, there is conflicting information on the present status and future trajectory of this system, including predictions of how changes in flow management could either benefit or potentially slow growth and evolution of the deltas.

► There is a need for hydrodynamic and sediment modeling in the ARB. Models need to incorporate the ARB floodway, the delta plain, and the areas outside the protection levees. Models also need to be scalable so that they can inform both large-scale and local water management decisions, as well as local hydrology improvement projects, stakeholder visioning, and decision making. Models are also needed to assess the ability of the Atchafalaya River to sustain its deltas, as well as expand its contribution to coastal restoration, particularly with respect to the western Terrebonne marshes. Lastly, models are also needed for areas outside the East and West Atchafalaya Protection Levees to understand water flow patterns and subsidence to help sustain these large forest blocks.

Biogeochemistry

► Understanding the role of the Atchafalaya River and its floodplain habitats in nutrient and carbon cycling is vital, and there have been several good baseline studies addressing this need. Additional studies are needed to quantify the future sequestration and denitrification potential of the floodway and coastal portions of the system and the effects of hydrology restoration on this potential. This information can be used to assess the value of restoration alternatives that reconnect isolated areas to riverine processes. Studies of the potential for nutrient and carbon sequestration of forested floodplains outside the floodway levees (Lake Fausse Point, Verret Swamps) would provide further information on the costs of keeping those systems disconnected in terms of Gulf hypoxia.

▶ There is a need for a source study of mercury in the ARB to determine how if local conditions are adding to mercury concentrations. This information is also important because the lability of mercury, or its ability to break down or methylize and become bioavailable, varies with source.

It is important to understand the complex role of oxidation-reduction and dissolved organic matter in the microbial mercury methylization, and consequently, bioavailability of mercury from sediments. In some studies freshwater systems are characterized as excellent mercury sinks, because methylization rates are low due to lack of sulfate. Freshwater systems have also been characterized as sinks for methyl-mercury because of the combination of highly reduced conditions and the high concentrations of dissolved organic matter in the soil. For example, in salt marshes where soils are mostly mineral and periodically dewater, rates of methylization are less than expected. It may be that seasonal (and tidal) drying of soils may get rid of mercury when the soils become aerobic, but that is yet unknown. This research is potentially important for water management in the ARB floodway because it may be that proper water management can mitigate high mercury levels through the aerobic effects of drawdowns, thus providing value-added benefits to water management projects at multiple scales.

Forests

▶ Field-based and site-specific data are critically needed to refine assessments of cypress-tupelo regeneration conditions from remote sensing data, both inside the ARB floodway and outside the protection levees, to allow for future forecasting of actual and potential regeneration rates in light of changing hydrological conditions and different management scenarios. For example, field-based studies in small areas of cypress-tupelo forest that were thought incapable regeneration based on assessments made using remote sensing showed that there may be enough potential growing days for a seedling to survive submergence during the growing season. Field studies have also identified specific locations where flooding would preclude seedling establishment, but where there may be a limited window when artificial regeneration with larger cypress seedlings could be viable. Additionally, because of the dynamic sedimentary nature of the ARB floodway, studies that rely on remote sensing alone may classify forest types incorrectly simply due to parameterization errors. Such forest classification errors result from an incomplete understanding of the critical flooding dynamics for baldcypress and water tupelo regeneration in the floodway. Research that combines remote sensing and field investigation to better understand the interaction between flooding and forest type maintenance and regeneration will facilitate development of better forest restoration tools and management options for the ARB. Without an accurate inventory of regeneration, it is impossible to understand what it will take, both ecologically and socioeconomically, to achieve meaningful forest sustainability in the ARB.

Fisheries Resources

▶ While the effects of large-scale water management, channel engineering, and local hydrology obstructions on water quality and nekton abundance and recruitment have been well described, the little information available on nekton stocks in the ARB is old and likely obsolete due to the effects of these factors on habitat variability in the basin. There is a need to assess nekton stocks (abundance, biomass, energy density) to modernize the information and quantify the effects of the current water management paradigm in terms of nekton stocks and their economic value. These types of estimates are not only critical for further defining the value of the ARB's ecological services, but also for providing an objective determination of the future costs and value of the status quo versus future proposed actions on commercial and recreational fisheries and to create measures of restoration success. For example, there is currently no quantitative information on how water management and harvest pressure affect red swamp and white river crawfish populations, and these species support a fishery that is a major economic driver in the ARB.

▶ Fisheries research is needed to quantify the value of flooded forest habitat in terms of secondary production and food web support for commercially and recreationally important species in the ARB. These types of estimates are also necessary to assess the cost and value of no-action to the fisheries economy versus future proposed actions, as well as create measures of restoration success.

▶ Halloran (2010) identified three fish research needs. First, there is a need to further understand the linkages between floodplain inundation and recruitment of fishes. To do this he identified the need to more accurately estimate the relationship between timing of reproduction and peak fitness to determine the role of fish "health" in the relationship (or lack thereof) between floodplain inundation and recruitment. Second, he identified a need to quantify primary productivity (phytoplankton) and its relationship to flood pulses to further understand the importance of periodic flooding to fish recruitment. Lastly, he called for a robust ichthyoplankton survey in the Atchafalaya River to determine the degree to which the ARB is an ecological sink for juveniles of river fishes.

▶ More work is needed to determine how to control invasive aquatic plants. These invasive species are competitively superior to native species and can cause deleterious impacts to water quality in the ARB. However, there is no apparent solution to the invasive aquatic plant problem in the ARB at this time.

▶ Assessments of the effects of river engineering and water management structures on river shrimp are necessary. This species is vital to the food web in the Atchafalaya River, and this information will be vital to designing large-scale water and sediment management plans for the Atchafalaya River.

Wildlife Resources

▶ There is a need for a more rigorous scientific assessment of the importance of the ARB to birds and other wildlife species. The information currently available, while helpful, is not at a resolution that allows assessment of no-action on wildlife resources or their response to future restoration.

▶ While a general understanding of habitat use by birds in the ARB is available via the literature, a paucity of data exists regarding overall importance of the ARB to neotropical species. Given the continental and global importance of the ARB to birds, there is a need for field investigations that document the value of the ARB to avian species. Additionally, there is a great deal of remotely sensed bird data (Doppler radar imagery) that is housed at the US Geological Survey National Wetlands Research Center in Lafayette, Louisiana. To date, these data have not been reviewed or analyzed to any great extent, and investigation of these data may provide a useful way to more accurately quantify the current and future value of the ARB for birds as well as a way to determine where to develop and prioritize land conservation projects.

Multiple-Use Planning and Decision Making

▶ The future of the ARB will require land-use planning. Balancing the need for freshwater input for sustainable forests and fisheries with management of the sediment that comes with it creates a need to identify criteria for balancing multiple uses (i.e., commercial fishing, crawfish harvest, recreational fishing, ecotourism, forest management, flood storage, carbon sequestration, etc.). For example, there may be ways to balance habitat-use patterns that already exist in the Atchafalaya Basin with water and sediment management activities. However, there is currently no blueprint that investigates land-use for the ARB, which will be driven by a myriad of interdisciplinary factors and issues.

▶ The future ARB needs to be conceptualized and designed by its stakeholders. However, to engage stakeholders in this exercise, there needs to be a strong decision-making process in place that combines values-based objective-setting with rigorous scientific assessment (e.g., see Kozak and Piazza, in review). Decision making is a blend of art and science. Stakeholders develop the objectives—the fundamentally important factors—based on their values. Science is used to explicitly try to find a solution that maximizes the objectives. There is already a limited framework for engaging stakeholders, and there are also rigorous scientific tools to support a decision-making process. The only component missing is an inclusive process that focuses on identifying the "right" problem and finding optimal solutions that are values-based and technically sound. There are a number of models that have been proposed and used for consensus building, decision-making and adaptive management on river systems (e.g., see Gregory et al. 2001, Irwin and Freeman 2002, Consensus Building Institute 2010b), that can be used as a framework for developing a process for the ARB.

APPENDIXES

APPENDIX 1 Annual Harvest of Hardwood Sawtimber and Pulpwood and Severance Tax Revenue from Hardwood Harvest in Atchafalaya Basin Parishes

TABLE 1

Year	Assumption Sawtimber (bd. ft)	Assumption Pulpwood (cords)	Avoyelles Sawtimber (bd. ft)	Avoyelles Pulpwood (cords)	Iberia Sawtimber (bd. ft)	Iberia Pulpwood (cords)	Iberville Sawtimber (bd. ft)	Iberville Pulpwood (cords)
1980	1,234,218	74	5,912,900	7,185	0	66	12,366,733	2,425
1981	1,841,994	316	1,066,948	8,391	0	0	8,738,548	8,388
1982	363,408	52	157,833	9,733	0	30	7,551,459	12,822
1983	1,016,268	0	137,635	8,051	0	378	4,885,960	7,033
1984	3,306,633	83	931,187	25,830	0	18	8,264,869	5,897
1985	578,364	1,370	1,728,403	16,854	0	303	10,075,195	9,134
1986	1,300,867	8,282	4,915,665	26,492	0	0	3,536,270	1,241
1987	64,881	34	4,382,000	13,974	0	0	6,884,577	3,949
1988	379,656	1,214	3,474,686	3,673	15,627	0	6,078,727	3,559
1989	437,165	1,287	9,671,524	17,425	108,919	29	6,171,309	10,170
1990	405,995	4,170	2,590,414	2,324	0	26	9,217,502	7,924
1991	68,688	1,479	5,888,163	9,014	0	4,176	2,707,802	2,394
1992	56,732	2,658	6,039,629	19,363	5,700,579	8,469	3,333,630	58,172
1993	139,774	7,452	9,055,717	17,651	98,526	0	8,959,341	42,078
1994	0	0	5,088,582	11,246	0	29	4,162,323	32,168
1995	12,528	484	2,725,112	16,528	0	1,119	7,453,297	30,999
1996	0	1,282	6,689,464	25,516	201,223	22	10,772,212	47,945
1997	10,698	718	7,285,762	19,898	45,416	26	4,164,497	15,008
1998	6,443	3,386	8,094,420	27,210	21,260	0	4,892,926	47,369
1999	1,638,069	1,009	2,985,455	11,349	20,840	19	11,284,143	64,006
2000	7,528	154	1,053,772	7,551	212,113	1,219	15,563,459	85,905
2001	58,338	1,201	7,161,697	35,988	260,071	7,096	3,085,408	29,825
2002	226,559	4,301	4,825,325	17,256	4,393	269	1,854,954	21,714
2003	127,382	1,284	7,100,601	33,956	382,492	65	9,486,974	28,269
2004	2,925	1,261	4,798,427	22,431	451,156	165	4,864,966	28,282
2005	0	401	7,210,964	30,956	420,159	193	5,952,069	39,581
2006	33,116	2,148	4,506,819	16,473	117,505	227	5,534,321	28,751
2007	58,338	1,201	7,161,697	35,988	260,071	7,096	3,085,408	29,825
2008	0	1,794	5,993,542	14,330	11,975	79	1,341,092	10,032
2009	313,361	7,647	6,775,247	16,874	0	156	4,904,988	27,669

Year	Pointe Coupee		St. Landry		St. Martin		St. Mary	
	Sawtimber (bd. ft)	Pulpwood (cords)	Sawtimber (bd. ft)	Pulpwood (cords)	Sawtimber (bd. ft)	Pulpwood (cords)	Sawtimber (bd. ft)	Pulpwood (cords)
1980	11,380,894	27,840	10,312,369	34,019	701,694	158	900,454	9
1981	10,575,950	26,811	4,179,894	10,509	973,732	386	768,124	0
1982	3,653,605	10,400	7,074,430	5,376	58,793	1,153	790,048	0
1983	7,842,466	12,521	4,588,901	5,672	86,151	281	990,910	0
1984	2,801,908	18,921	4,761,847	38,945	193,684	1,931	1,247,214	0
1985	5,414,400	28,028	9,870,045	21,671	957,057	2,782	198,823	407
1986	3,149,799	3,391	12,112,692	23,435	330,266	1,110	0	0
1987	9,494,769	18,103	10,827,230	18,388	125,247	796	944,846	0
1988	4,430,992	3,393	16,987,303	22,705	0	399	147,768	0
1989	5,896,674	8,511	15,539,018	22,422	239,027	5,843	35,060	0
1990	8,674,839	2,737	19,032,691	14,635	323,356	19,800	24,109	0
1991	7,351,026	6,882	16,373,816	25,422	729,719	28,841	112,336	5,454
1992	6,964,639	15,157	15,564,245	37,165	695,376	21,935	12,371	0
1993	6,138,258	7,306	11,614,324	48,399	303,462	32,330	0	0
1994	40,334	0	7,765,765	26,015	1,219,269	4,482	0	0
1995	13,509,865	19,161	10,676,558	31,577	1,451,366	7,031	4,800	16
1996	8,649,979	19,346	6,471,115	11,490	796,621	2,724	305,981	2,323
1997	7,114,697	14,624	11,687,809	18,649	2,698,298	9,969	0	39
1998	9,913,008	27,070	8,492,965	16,078	3,950,805	9,629	25,331	19
1999	7,347,623	24,835	6,915,271	13,823	1,821,538	5,038	5,967	87
2000	5,577,920	13,533	5,739,089	11,998	191,003	2,552	267,007	964
2001	7,567,486	33,911	9,515,544	31,238	537,235	6,354	1,906	8
2002	8,348,162	43,198	8,800,364	34,397	73,248	1,230	18,951	62
2003	9,653,569	49,200	7,765,598	24,291	1,727,111	3,241	11,677	9
2004	6,544,525	33,753	6,716,583	13,840	681,573	5,461	11,763	2,209
2005	7,738,318	39,715	6,237,868	22,705	1,571,622	7,855	3,128	0
2006	5,997,533	24,533	6,504,981	24,044	5,935,917	14,545	127,868	363
2007	7,567,486	33,911	9,515,544	31,238	537,235	6,354	1,906	8
2008	5,245,224	17,388	14,011,848	22,755	3,040,599	10,822	3,052	158
2009	8,374,705	24,895	11,297,238	26,924	1,294,955	5,325	19,100	1,936

Note that these data are collected parishwide and, therefore, information specific to the Atchafalaya Basin Floodway is included in these numbers but is not recorded separately.

APPENDIX 1 Annual Harvest of Hardwood Sawtimber and Pulpwood and Severance Tax Revenue
from Hardwood Harvest in Atchafalaya Basin Parishes

TABLE 2

| | Landowner Income ($) | | | | | | | |
| | Assumption | | Avoyelles | | Iberia | | Iberville | |
Year	Sawtimber	Pulpwood	Sawtimber	Pulpwood	Sawtimber	Pulpwood	Sawtimber	Pulpwood
1980	74,053	294	354,774	28,740	-	264	742,004	9,701
1981	110,520	1,391	64,017	36,921	-	-	524,313	36,907
1982	21,804	207	9,470	38,933	-	119	453,088	51,288
1983	60,976	-	8,258	32,206	-	1,511	293,158	28,130
1984	198,398	332	55,871	103,321	-	73	495,892	23,587
1985	41,064	5,479	122,717	67,414	-	1,212	715,339	36,536
1986	87,158	33,128	329,350	105,9680	-	-	236,930	4,964
1987	4,671	170	315,504	69,871		-	495,690	197,747
1988	27,335	6,068	250,177	18,361	1,099	-	437,668	17,796
1989	33,002	7,465	730,103	101,067	8,222	170	465,872	58,988
1990	14,183	23,937	240,183	13,338	-	149	854,647	45,486
1991	6,176	13,484	529,404	82,212	-	38,084	243,458	21,835
1992	5,918	21,979	630,054	160,136	-	83	594,684	70,041
1993	17,564	76,156	1,137,941	180,398	12,381	-	1,125,831	430,034
1994	-	-	922,865	116,962	-	301	754,879	334,546
1995	2,272	5,029	494,226	171,886	-	11,637	1,351,730	322,386
1996	-	19,294	1,261,499	384,016	37,947	331	2,031,424	721,572
1997	2,017	10,806	1,373,949	299,465	8,565	391	785,341	225,870
1998	1,585	67,178	1,991,794	539,846	5,231	-	1,204,002	939,801
1999	399,509	14,842	728,123	166,944	5,083	279	2,752,090	941,528
2000	1,719	1,960	240,671	96,124	48,444	15,518	3,554,538	1,093,571
2001	14,379	14,160	1,765,143	424,299	64,100	83,662	760,461	351,637
2002	65,337	59,999	1,391,575	240,721	1,267	3,753	534,950	302,910
2003	37,988	20,660	2,117,541	546,352	114,067	1,046	2,829,205	454,848
2004	872.00	20,289	1,430,987	360,915	134,544	2,655	1,450,830	455,057
2005	-	5,015	2,260,493	464,340	131,711	2,895	1,865,855	593,715
2006	11,250	41,177	1,531,057	315,787	39,919	4,352	1,880,120	551,157
2007	14,379	14,160	1,765,143	424,299	64,100	83,662	760,461	351,637
2008	-	29,655	1,602,493	236,875	3,202	1,306	358,568	165,829
2009	92,326	130,152	1,996,191	287,195	-	2,655	1,445,157	470,926

	Landowner Income ($)							
	Pointe Coupee		St. Landry		St. Martin		St. Mary	
Year	*Sawtimber*	*Pulpwood*	*Sawtimber*	*Pulpwood*	*Sawtimber*	*Pulpwood*	*Sawtimber*	*Pulpwood*
1980	682,854	111,359	618,742	136,074	42,102	630.00	54,027	38
1981	634,557	117,968	250,794	46,241	58,424	1,699	46,087	-
1982	219,216	41,599	424,466	21,503	3,528	4,611	47,403	-
1983	50,086	275,334	22,689	5,169	1,126	59,455	-	
1984	75,685	285,711	155,779	11,621	7,725	74,833	-	
1985	384,422	112,113	700,773	86,685	67,951	11,130	14,116	1,627
1986	211,037	13,563	811,550	93,740	22,128	4,438	-	-
1987	683,623	90,514	779,561	91,938	9,018	3,982	68,029	-
1988	319,031	16,967	1,223,086	113,526	-	1,996	10,639	-
1989	445,140	49,365	1,173,040	130,047	18,044	33,888	2,647	-
1990	804,331	15,713	1,764,711	84,004	29,982	113,654	2,235	-
1991	660,931	62,766	1,472,170	231,851	65,609	263,033	10,100	49,745
1992	726,551	125,345	1,623,662	307,355	72,542	181,406	1,290	-
1993	771,334	74,665	1,459,456	494,642	38,133	330,410	-	-
1994	1,420,661	144,765	1,408,399	270,555	221,127	46,608	-	-
1995	2,450,149	199,274	1,936,301	328,400	263,220	73,127	871	166
1996	1,361,213	291,157	1,220,323	172,925	150,227	40,996	57,702	34,961
1997	1,341,690	220,091	2,204,087	280,667	508,845	150,033	-	587
1998	2,439,294	537,069	2,089,864	318,988	972,175	191,039	6,233	377
1999	1,792,012	365,323	1,686,565	203,336	444,255	74,109	1,455	1,280
2000	1,273,941	172,275	1,310,751	152,735	43,623	32,487	60,982	12,272
2001	1,865,158	399,811	2,345,296	368,296	132,412	74,914	470	94
2002	2,407,526	602,612	2,537,937	479,838	21,124	17,159	5,465	865
2003	2,878,887	791,628	2,315,857	390,842	515,059	52,148	3,482	145
2004	1,951,708	543,086	2,003,019	222,686	203,259	87,867	3,508	35,543
2005	2,425,808	595,725	1,955,447	340,575	586,716	117,826	981	-
2006	2,037,482	470,298	2,209,872	460,923	2,016,550	278,828	43,439	6,959
2007	1,865,158	399,811	2,345,296	368,296	132,412	74,914	470	94
2008	1,402,416	287,424	3,746,348	376,140	812,965	178,888	816	2,612
2009	2,467,439	423,713	3,328,505	458,246	381,533	90,632	5,627	32,951

Note that these data are collected parishwide and, therefore, information specific to the Atchafalaya Basin Floodway is included in these numbers but is not recorded separately.

TABLE 1

MAMMALS	Common Name	Latin Name
Deer	Whitetailed deer	*Odocoileus virginianus*
Rodents	Eastern fox squirrel	*Sciurus niger*
	Eastern gray squirrel	*Sciurus carolinensis*
	Beaver	*Castor canadensis*
	Cotton mouse	*Peromyscus gossypinus*
	White-footed mouse	*Peromyscus leucopus*
	Cotton rat	*Sigmodon hispidus*
	Eastern wood rat	*Neotoma floridana*
	Nutria (exotic)	*Myocastor coypus*
	Muskrat	*Ondatra zibethicus*
Rabbits	Eastern cottontail	
	Swamp rabbit	*Sylvilagus aquaticus*
Carnivores	River otter	*Lontra canadensis*
	American mink	*Neovison vison*
	Louisiana black bear	*Ursus americanus luteolus*
	Striped skunk	*Mephitis mephitis*
	Opposum	*Didelphis virginiana*
	Bobcat	*Lynx rufus*
	Cougar	*Puma concolor*
	Florida panther	*Felis concolor coryi*
	Gray fox	*Urocyon cinereoargenteus*
	Red fox	*Vulpes vulpes*
	Coyote	*Canis latrans*
	Racoon	*Procyon lotor*
Bats	Rafinesque's big-eared bat	*Corynorhinus rafinesquii*
	Southeastern myotis	*Myotis austroriparius*
	Big brown bat	*Eptesicus fuscus*
	Eastern red bat	*Lasiurus borealis*
	Hoary bat	*Lasiurus cinereus*
	Northern yellow bat	*Lasiurus intermedius*
	Seminole bat	*Lasiurus seminolus*
	Evening bat	*Nycticeius humeralis*
	Mexican free-tailed bat	*Tadarida brasiliensis*
Armadillos	Nine-banded armadillo	*Dasypus novemcinctus*

Data sources are Lowery (1974b), US Army Corps of Engineers (2000), and US Geological Survey (2011).

TABLE 2

BIRDS	Common Name	Latin Name	Conservation Concern*
Grebes	Pied-billed grebe	*Podilymbus podiceps*	
Cormorants	Double-crested cormorant	*Phalacrocorax auritus*	
Anhingas	Anhinga	*Anhinga anhinga*	
Herons and Bitterns	Great blue heron	*Ardea herodias*	
	Great egret	*Ardea alba*	
	Snowy egret	*Egretta thula*	
	Little blue heron	*Egretta caerulea*	USFWS
	Tricolored heron	*Egretta tricolor*	USFWS
	Cattle egret	*Bubulcus ibis*	
	Green heron	*Butorides virescens*	
	Black-crowned night-heron	*Nycticorax nycticorax*	USFWS
	Yellow-crowned night-heron	*Nyctanassa violacea*	
Ibises	White ibis	*Eudocimus albus*	
	Glossy ibis	*Plegadis falcinellus*	
	White-faced ibis	*Plegadis chihi*	
	Roseate spoonbill	*Platalea ajaja*	
Storks	Wood stork	*Mycteria americana*	USFWS
American Vultures	Black vulture	*Coragyps atratus*	
	Turkey vulture	*Cathartes aura*	
Waterfowl	Greater white-fronted goose ("specklebelly")	*Anser albifrons*	
	Snow goose (including "blue goose")	*Chen caerulescens*	
	Wood duck	*Aix sponsa*	
	Gadwall	*Anas strepera*	
	American widgeon ("baldpate")	*Anas americana*	
	Mallard	*Anas platyrhynchos*	
	Blue-winged teal	*Anas discors*	
	Northern shoveler	*Anas clypeata*	
	Northern pintail	*Anas acuta*	

*USFWS denotes that the species is listed as a species of concern due to declining populations in the official US Fish and Wildlife Service bird list for the Atchafalaya National Wildlife Refuge (US Fish and Wildlife Service 2006). ABC denotes that the species has been listed on the American Bird Conservancy United States Watchlist of Birds of Conservation Concern (American Bird Conservancy 2007). This watchlist was developed in partnership with the National Audubon Society and includes the National Audubon Society Watchlist. Data come from a number of scientific sources, including Partners in Flight. All but one species (Mottled duck) are Yellow List species, denoting declining or rare continental species. Mottled duck is a red list species, signifying that it is a species of highest continental concern.

(continued next page)

BIRDS	Common Name	Latin Name	Conservation Concern
Waterfowl, continued	Green-winged teal	*Anas carolinensis*	
	Canvasback	*Aythya valisineria*	
	Redhead	*Aythya americana*	
	Ring-necked duck	*Aythya collaris*	
	Lesser scaup	*Aythya affinis*	
	Bufflehead	*Bucephala albeola*	
	Hooded merganser	*Lophodytes cucullatus*	
	Red-breasted merganser	*Mergus serrator*	
	Ruddy duck	*Oxyura jamaicensis*	
	Mottled duck	*Anas fulvigula*	ABC (Red List)
Hawks	Osprey	*Pandion haliaetus*	
	Swallow-tailed kite	*Elanoides forficatus*	USFWS, ABC
	Mississippi kite	*Ictinia mississippiensis*	USFWS
	Bald eagle	*Haliaeetus leucocephalus*	USFWS
	Northern harrier	*Circus cyaneus*	USFWS
	Sharp-shinned hawk	*Accipiter striatus*	USFWS
	Cooper's hawk	*Accipiter cooperii*	USFWS
	Red-shouldered hawk	*Buteo lineatus*	
	Broad-winged hawk	*Buteo platypterus*	
	Red-tailed hawk	*Buteo jamaicensis*	
Falcons	American kestrel	*Falco sparverius*	USFWS
	Merlin	*Falco columbarius*	
	Peregrine falcon	Falco peregrinus	USFWS
Turkey	Wild turkey	*Meleagris gallopavo*	
Rails	King rail	*Rallus elegans*	ABC
	Virginia rail	*Rallus limicola*	
	Sora	*Porzana carolina*	
	Purple gallinule	*Porphyrio martinica*	
	Common moorhen (gallinule)	*Gallinula chloropus*	
	American coot	*Fulica americana*	
Plovers	Black-bellied plover	*Pluvialis squatarola*	
	Semipalmated plover	*Charadrius semipalmatus*	
	Killdeer	*Charadrius vociferus*	
Stilts	Black-necked stilt	*Himantopus mexicanus*	
Sandpipers	Greater yellowlegs	*Tringa melanoleuca*	
	Lesser yellowlegs	*Tringa flavipes*	
	Solitary sandpiper	*Tringa solitaria*	USFWS
	Spotted sandpiper	*Actitis macularia*	
	Semipalmated sandpiper	Calidris pusilla	USFWS, ABC
	Western sandpiper	*Calidris mauri*	ABC
	Least sandpiper	*Calidris minutilla*	USFWS
	Pectoral sandpiper	*Calidris melanotos*	USFWS

BIRDS	Common Name	Latin Name	Conservation Concern
Sandpipers, continued	Dunlin	*Calidris alpina*	USFWS
	Stilt sandpiper	*Calidris himantopus*	USFWS
	Wilson's snipe	*Gallinago delicata*	
	American woodcock	*Scolopax minor*	
	Long-billed dowitcher	*Limnodromus scolopaceus*	
Doves	Rock dove	*Columba livia*	
	Eurasian collared-dove (introduced)	*Streptopelia decaocto*	
	Mourning dove	*Zenaida macroura*	
Cuckoos	Black-billed cuckoo	*Coccyzus erythropthalmus*	
	Yellow-billed cuckoo	*Coccyzus americanus*	USFWS
Barn Owls	Barn owl	*Tyto alba*	
Owls	Eastern screech-owl	*Megascops asio*	
	Great horned owl	*Bubo virginianus*	
	Barred owl	*Strix varia*	
Nightjars	Common nighthawk	*Chordeiles minor*	
	Chuck-will's-widow	*Caprimulgus carolinensis*	USFWS
	Whip-poor-will	*Caprimulgus vociferus*	USFWS
Swifts	Chimney swift	*Chaetura pelagica*	
Hummingbirds	Ruby-throated hummingbird	*Archilochus colubris*	USFWS
	Rufous hummingbird	*Selasphorus rufus*	
Kingfishers	Belted kingfisher	*Megaceryle alcyon*	
Woodpeckers	Red-headed woodpecker	*Melanerpes erythrocephalus*	USFWS, ABC
	Red-bellied woodpecker	*Melanerpes carolinus*	
	Yellow-bellied sapsucker	*Sphyrapicus varius*	
	Downy woodpecker	*Picoides pubescens*	
	Hairy woodpecker	*Picoides villosus*	
	Northern flicker	*Colaptes auratus*	
	Pileated woodpecker	*Dryocopus pileatus*	
Flycatchers	Olive-sided flycatcher	*Contopus cooperi*	ABC
	Eastern wood-pewee	*Contopus virens*	USFWS
	Yellow-bellied flycatcher	*Empidonax flaviventris*	
	Acadian flycatcher	*Empidonax virescens*	
	Alder flycatcher	*Empidonax alnorum*	
	Least flycatcher	*Empidonax minimus*	
	Eastern phoebe	*Sayornis phoebe*	

(continued next page)

BIRDS	Common Name	Latin Name	Conservation Concern
Flycatchers, continued	Vermilion flycatcher	*Pyrocephalus rubinus*	
	Great-crested flycatcher	*Myiarchus crinitus*	
	Eastern kingbird	*Tyrannus tyrannus*	USFWS
	Scissor-tailed flycatcher	*Tyrannus forficatus*	
Vireos	White-eyed vireo	*Vireo griseus*	USFWS
	Yellow-throated vireo	*Vireo flavifrons*	
	Blue-headed vireo	*Vireo solitarius*	
	Warbling vireo	*Vireo gilvus*	
	Philadelphia vireo	*Vireo philadelphicus*	
	Red-eyed vireo	*Vireo olivaceus*	
Jays and Crows	Blue jay	*Cyanocitta cristata*	
	American crow	*Corvus brachyrhynchos*	
	Fish crow	*Corvus ossifragus*	
Swallows	Purple martin	*Progne subis*	
	Tree swallow	*Tachycineta bicolor*	
	Northern rough-winged swallow	*Hirundo rustica*	USFWS
	Bank swallow	*Riparia riparia*	
	Cliff swallow	*Petrochelidon pyrrhonota*	
	Barn swallow	*Hirundo rustica*	
Titmice	Carolina chickadee	*Poecile carolinensis*	USFWS
	Tufted titmouse	*Baeolophus bicolor*	
Nuthatches	Red-breasted nuthatch	*Sitta canadensis*	
Creepers	Brown creeper	*Certhia americana*	
Wrens	Carolina wren	*Thryothorus ludovicianus*	
	House wren	*Troglodytes aedon*	
	Winter wren	*Troglodytes troglodytes*	
Kinglets	Golden-crowned kinglet	*Regulus satrapa*	
	Ruby-crowned kinglet	*Regulus calendula*	
Gnatcatchers	Blue-gray gnatcatcher	*Polioptila caerulea*	
Thrushes	Eastern bluebird	*Sialia sialis*	
	Veery	*Catharus fuscescens*	
	Gray-cheeked thrush	*Catharus minimus*	
	Swainson's thrush	*Catharus ustulatus*	
	Hermit thrush	*Catharus guttatus*	
	Wood thrush	*Hylocichla mustelina*	ABC
	American robin	*Turdus migratorius*	
Shrike	Loggerhead shrike	*Lanius ludovicianus*	USFWS

BIRDS	Common Name	Latin Name	Conservation Concern
Mockingbirds and Thrashers	Gray catbird	*Dumetella carolinensis*	USFWS
	Northern mockingbird	*Mimus polyglottos*	
	Brown thrasher	*Toxostoma rufum*	
Starlings	European starling (introduced)	*Sturnus vulgaris*	
Pipits	American pipit	*Anthus rubescens*	
Waxwings	Cedar waxwing	*Bombycilla cedrorum*	
Wood Warblers	Blue-winged warbler	*Vermivora cyanoptera*	
	Golden-winged warbler	*Vermivora chrysoptera*	
	Tennessee warbler	*Oreothlypis peregrina*	
	Orange-crowned warbler	*Vermivora celata*	
	Nashville warbler	*Vermivora ruficapilla*	
	Northern parula	*Parula americana*	USFWS
	Yellow warbler	*Dendroica petechia*	
	Yellow-throated warbler	*Dendroica dominica*	
	Chestnut-sided warbler	*Dendroica pensylvanica*	
	Magnolia warbler	*Dendroica magnolia*	
	Yellow-rumped warbler	*Dendroica coronata*	
	Black-throated green warbler	*Dendroica virens*	
	Blackburnian warbler	*Dendroica fusca*	
	Pine warbler	*Dendroica pinus*	
	Prairie warbler	*Dendroica discolor*	ABC
	Palm warbler	*Dendroica palmarum*	USFWS
	Bay-breasted warbler	*Dendroica castanea*	
	Cerulean warbler	*Dendroica cerulea*	USFWS, ABC
	Black-and-white warbler	*Mniotilta varia*	
	American redstart	*Setophaga ruticilla*	
	Prothonotary warbler	*Protonotaria citrea*	USFWS, ABC
	Worm-eating warbler	*Helmitheros vermivorus*	USFWS
	Swainson's warbler	*Limnothlypis swainsonii*	USFWS, ABC
	Ovenbird	*Seiurus aurocapillus*	
	Northern waterthrush	*Parkesia noveboracensis*	
	Louisiana waterthrush	*Parkesia motacilla*	USFWS
	Kentucky warbler	*Oporornis formosus*	USFWS, ABC
	Mourning warbler	*Oporornis philadelphia*	
	Common yellowthroat	*Geothlypis trichas*	
	Hooded warbler	*Wilsonia citrina*	
	Wilson's warbler	*Wilsonia pusilla*	
	Canada warbler	*Wilsonia canadensis*	
	Yellow-breasted chat	*Icteria virens*	USFWS
Tanagers	Summer tanager	*Piranga rubra*	
	Scarlet tanager	*Piranga olivacea*	

(continued next page)

BIRDS	Common Name	Latin Name	Conservation Concern
Grosbeaks,	Eastern Towhee	*Pipilo erythrophthalmus*	
Sparrows,	Fox sparrow	*Passerella iliaca*	
Buntings	Song sparrow	*Melospiza melodia*	
	Lincoln's sparrow	*Melospiza lincolnii*	
	Swamp sparrow	*Melospiza georgiana*	
	White-throated sparrow	*Zonotrichia albicollis*	
	White-crowned sparrow	*Zonotrichia leucophrys*	
	Dark-eyed junco	*Junco hyemalis*	
	Northern cardinal	*Cardinalis cardinalis*	
	Rose-breasted grosbeak	*Pheucticus ludovicianus*	
	Blue grosbeak	*Passerina caerulea*	
	Indigo bunting	*Passerina cyanea*	
	Painted bunting	*Passerina ciris*	USFWS, ABC
	Dickcissel	*Spiza americana*	USFWS
Blackbirds	Bobolink	*Dolichonyx oryzivorus*	
and Orioles	Red-winged blackbird	*Agelaius phoeniceus*	
	Eastern meadowlark	*Sturnella magna*	USFWS
	Rusty blackbird	*Euphagus carolinus*	ABC
	Common grackle	*Quiscalus quiscula*	
	Brown-headed cowbird	*Molothrus ater*	
	Orchard oriole	*Icterus spurius*	USFWS
	Baltimore oriole	*Icterus galbula*	USFWS
	Purple finch	*Carpodacus purpureus*	
	House finch	*Carpodacus mexicanus*	
	American goldfinch	*Spinus tristis*	
Terns and	Black skimmer	*Rynchops niger*	ABC
Skimmers	Least tern	*Sternula antillarum*	ABC
	Royal tern	*Thalasseus maximus*	
	Sandwich tern	*Thalasseus sandvicensis*	
	Forster's tern	*Sterna forsteri*	
	Black tern	*Chlidonias niger*	
Gulls			
	Laughing gull	*Leucophaeus atricilla*	
	Ring-billed gull	*Larus delawarensis*	

Data sources are US Fish and Wildlife Service (2006), American Bird Conservancy (2007), National Audubon Society (2011a, b).

TABLE 3

NEKTON (Fish and Crustaceans	Common Name	Latin Name	Ecological Affinity*
Fishes			
Bass and Sunfish	Largemouth bass	*Micropterus salmoides*	FW
	Spotted bass	*Micropterus punctulatus*	FW
	Warmouth	*Lepomis gulosus*	FW
	Bluegill	*Lepomis macrochirus*	FW
	Longear sunfish	*Lepomis megalotis*	FW
	Spotted sunfish	*Lepomis punctatus*	FW
	Orangespotted sunfish	*Lepomis humulis*	FW
	Redear sunfish	*Lepomis microlophus*	FW
	Redspotted sunfish	*Lepomis miniatus*	FW
	Bantam sunfish	*Lepomis symmetricus*	FW
	Green sunfish	*Lepomis cyanellus*	FW
	Dollar sunfish	*Lepomis marginatus*	FW
	Flier	*Centrarchus macropterus*	FW
	Black crappie	*Pomoxis nigromaculatus*	FW
	White crappie	*Pomoxis annularis*	FW
Striped Basses	Yellow bass	*Morone mississippiensis*	FW
	White bass	*Morone chrysops*	FW
	Striped bass	*Morone saxatilis*	AN
North American Catfish	Channel catfish	*Ictalurus punctatus*	FW
	Blue catfish	*Ictalurus furcatus*	FW
	Flathead catfish	*Pylodictis olivaris*	FW
	Yellow bullhead	*Ameirus natalis*	FW
	Black bullhead	*Ameirus melas*	FW
	Tadpole madtom	*Noturus gyrinus*	FW
	Brown bullhead	*Ameiurus nebulosus*	FW
Sea Catfishes	Hardhead catfish	*Ariopsis felis*	ESM
	Gafftopsail catfish	*Bagre marinus*	ESM
Suckers	Smallmouth buffalo	*Ictiobus bubalus*	FW
	Bigmouth buffalo	*Ictiobus cyprinellus*	FW
	Black buffalo	*Ictiobus niger*	FW
	Lake chubsucker	*Erimyzon sucetta*	FW
	Blue sucker	*Cycleptus elongatus*	FW
	River carpsucker	*Carpiodes carpio*	FW
	Quillback	*Carpiodes cyprinus*	FW
	Creek chubsucker	*Erimyzon oblongus*	FW
	Spotted Sucker	*Minytrema melanops*	FW
	Blacktail redhorse	*Moxostoma poecilurum*	FW

*Ecological affinity is reported from Thompson and Deegan (1983) and Thompson and Peterson (2003, 2006) as: FW—freshwater, ES—estuarine, ESM—estuarine/marine, AN—anadromous, CA—catadromous.

(continued next page)

NEKTON (Fish and Crustaceans	Common Name	Latin Name	Ecological Affinity
Carps and Minnows	Common carp (exotic)	*Cyprinus carpio*	FW
	Grass carp (exotic)	*Ctenopharyngodon idella*	FW
	Silver carp (exotic)	*Hypopthalmichthys molitrix*	FW
	Bighead carp (exotic)	*Hypopthalmichthys nobilis*	FW
	Sicklefin chub	*Macrhybopsis meeki*	FW
	Golden shiner	*Notemigonus chryoleucas*	FW
	Blacktail shiner	*Cyprinella venusta*	FW
	Silverband shiner	*Notropis shumardi*	FW
	Red shiner	*Cyprinella lutrensis*	FW
	Redfin shiner	*Lythrurus umbratilis*	FW
	River shiner	*Notropis blennius*	FW
	Bullhead minnow	*Pimephales vigilax*	FW
	Pugnose minnow	*Opsopoeodus emiliae*	FW
	Cypress minnow	*Hybognathus hayi*	FW
	Mississippi silvery minnow	*Hybognathus nuchalis*	FW
	Striped shiner	*Luxilus chrysocephalus*	FW
	Ribbon shiner	*Lythrurus fumeus*	FW
	Redfin shiner	*Lythrurus umbratilis*	FW
Herrings and Shads	Gulf menhaden	*Brevoortia patronus*	ESM
	Gizzard shad	*Dorosoma cepedianum*	FW
	Threadfin shad	*Dorosoma petenense*	FW
	Skipjack herring	*Alosa chrysochloris*	FW
Gars	Spotted gar	*Lepisosteus oculatus*	FW
	Longnose gar	*Lepisosteus osseus*	FW
	Alligator gar	*Atractosteus spatula*	FW
	Shortnose gar	*Lepisosteus platostomus*	FW
Bowfin	Bowfin	*Amia calva*	FW
Mullets	Striped mullet	*Mugil cephalus*	ESM
	White mullet	*Mugil curema*	ESM
Pikes	Grass pickerel	*Esox americanus*	FW
	Chain pickerel	*Esox niger*	FW
Pirate Perch	Pirate perch	*Aphredoderus sayanus*	FW
Needlefish	Atlantic needlefish	*Strongylura marina*	ESM
New World Silversides	Inland silverside	*Menidia beryllina*	ESM
	Rough silverside	*Membras martinica*	ESM
	Brook silverside	*Labidesthes sicculus*	FW
	Mississippi silverside	*Menidia audens*	FW
Livebearers	Sailfin molly	*Poecillia latipinna*	FW
	Western mosquitofish	*Gambusia affinis*	FW
	Least killifish	*Heterandria formosa*	FW

NEKTON (Fish and Crustaceans	Common Name	Latin Name	Ecological Affinity
Freshwater Sturgeon	Pallid sturgeon	*Scaphirhynchus albus*	FW
	Shovelnose sturgeon	*Scaphirhynchus platorynchus*	FW
Sharks	Bull shark	*Carcharinus leucas*	ESM
Whiptail Stingrays	Atlantic stingray	*Dasyatis sabina*	MA
Paddlefishes	Paddlefish	*Polyodon spathula*	FW
Mooneyes	Mooneye	*Hiodon tergisus*	FW
	Goldeye	*Hiodon alosoides*	FW
Ladyfish	Ladyfish	*Elops saurus*	ESM
Freshwater Eels	American eel	*Anguilla rostrata*	CA
Snake Eels	Speckled worm eel	*Myrophis punctatus*	ESM
Anchovies	Bay anchovy	*Anchoa mitchilli*	ESM
Toadfish	Atlantic midshipman	*Porichthys plectrodon*	MA
Pupfish	Sheepshead minnow	*Cyprinodon variegatus*	ESM
Pipefish and Seahorses	Gulf pipefish	*Syngnathus scovelli*	ESM
	Chain pipefish	*Syngnathus louisianae*	ESM
Killifish and Topminnows	Diamond killifish	*Adinia xenica*	
	Gulf killifish	*Fundulus grandis*	ESM
	Bayou killifish	*Fundulus pulvereus*	ESM
	Rainwater killifish	*Lucania parva*	ESM
	Golden topminnow	*Fundulus chrysotus*	FW
	Saltmarsh topminnow	*Fundulus jenkinsi*	ESM
	Blackspotted topminnow	*Fundulus olivaceus*	FW
Searobins	Bighead searobin	*Prionotus tribulus*	MA
Perches and Darters	Slough darter	*Etheostoma gracile*	FW
	Mud darter	*Etheostoma asprigene*	FW
	Bluntnose darter	*Etheostoma chlorosomum*	FW
	Swamp darter	*Etheostoma fusiforme*	FW
	Slough darter	*Etheostoma gracile*	FW
	Cypress darter	*Etheostoma proeliare*	FW
	Logperch	*Percina caprodes*	FW
	Sauger	*Sander canadense*	FW
Bluefish	Bluefish	*Pomatomus saltatrix*	ESM
Jacks	Crevalle jack	*Caranx hippos*	ESM
	Leatherjacket	*Oligoplites saurus*	ESM
Snappers	Gray snapper	*Lutjanus griseus*	MA

(continued next page)

NEKTON (Fish and Crustaceans	Common Name	Latin Name	Ecological Affinity
Mojarras	Spotfin mojarra	*Eucinostomus argenteus*	ESM
Grunts	Sheepshead	*Archosargus probatocephalus*	ESM
	Pinfish	*Lagodon rhomboides*	ESM
Drum	Freshwater drum	*Aplodinotus grunniens*	FW
	Red drum	*Sciaenops ocellatus*	ESM
	Black drum	*Pogonias cromis*	ESM
	Star drum	*Stellifer lanceolatus*	ESM
	Spotted seatrout	*Cynoscion nebulosus*	ESM
	Sand seatrout	*Cynoscion arenarius*	ESM
	Silver perch	*Bairdiella chrysoura*	ESM
	Spot	*Leiostomus xanthurus*	ESM
	Atlantic croaker	*Micropogonias undulatus*	ESM
Pygmy Sunfish	Banded pygmy sunfish	*Elassoma zonatum*	FW
Sleepers	Fat sleeper	*Domitator maculatus*	ES
	Largescaled spinycheek sleeper	*Eleotris amblyopsis*	ES
	Spinycheek Sleeper	*Eleotris pisonis*	ES
Gobies	Darter goby	*Ctenogobius boleosoma*	ES
	Freshwater goby	*Ctenogobius shufeldti*	ES
	Lyre goby	*Evorthodus lyricus*	ES
	Highfin goby	*Gobionellus oceanicus*	ES
	Naked goby	*Gobiosoma bosc*	ES
	Clown goby	*Microgobius gulosus*	ES
Butterfishes	Harvestfish	*Peprilus paru*	MA
Sand Flounders	Bay whiff	*Citharichthys spilopterus*	ESM
	Southern flounder	*Paralichthys lethostigma*	ESM
American Soles	Lined sole	*Achirus lineatus*	ESM
	Hogchoker	*Trinectes maculatus*	ESM
Tonguefishes	Blackcheek tonguefish	*Symphurus plagiusa*	ESM
Crustaceans			
Crayfish	Red swamp crayfish	*Procambarus clarkii*	FW
	White River crayfish	*Procambarus zonangulus*	FW
	Cambarellus crayfish species	*Cambarellus* spp.	FW
Crabs	Blue crab	*Callinectes sapidus*	ESM

NEKTON (Fish and Crustaceans	Common Name	Latin Name	Ecological Affinity
Shrimp	River shrimp	*Macrobrachium ohione*	CA
	Riverine grass shrimp	*Palaemonetes paludosus*	FW
	White shrimp	*Litopenaeus setiferus*	ESM
	Northern brown shrimp	*Farfantepenaeus aztecus*	ESM
	Daggerblade grass shrimp	*Palaemonetes pugio*	ES
	Mississippi grass shrimp	*Palaemonetes kadiakensis*	FW
	Mysid shrimp	*Taphromysis louisianae*	FW

This list includes nekton collected in both the ARB floodway and the Atchafalaya and Wax Lake deltas and was compiled from a variety of sources including: Lambou (1959, 1963, 1990), Chabreck et al. (1982), Gelwicks (1996), Constant et al. (1997), Castellanos and Rozas (2001), US Geological Survey (2001), Thompson and Deegan (1983), Thompson and Peterson (2006), and Bauer and Delahoussaye (2008).

TABLE 4

HERPETOFAUNA	Common Name	Scientific Name
Amphibians		
Mole salamanders	Marbled salamander	*Ambystoma opacum*
Amphiumas	Three-toed amphiuma	*Amphiuma tridactylum*
Woodland Salamanders	Dusky salamander	*Desmognathus fuscus*
	Southern dusky salamander	*Desmognathus auriculatus*
	Dwarf salamander	*Eurycea quadridigitata*
Newts	Eastern newt	*Notophthalmus viridescens*
Sirens	Lesser siren	*Siren intermedia*
Toads	Gulf Coast toad	*Bufo valliceps*
	Woodhouse's toad	*Bufo woodhousei*
	Fowler's toad	*Bufo fowleri*
Treefrogs and other Hylids	Northern cricket frog	*Acris crepitans*
	Gray treefrog	*Hyla versicolor*
	Cope's gray treefrog	Hyla chrysoscelis
	Green treefrog	*Hyla cinerea*
	Spring peeper	*Pseudacris crucifer*
	Squirrel treefrog	*Hyla squirella*
	Striped chorus frog	*Pseudacris triseriata*
	Cajun chorus frog	*Pseudacris fouquettei*
Narrow-Mouthed Toads	Eastern narrow-mouthed toad	*Gastrophryne carolinensis*
True Frogs	American bullfrog	*Rana catesbeiana*
	Green frog (also Bronze frog)	*Rana clamitans*
	Pig frog	*Rana grylio*
	Southern leopard frog	*Rana sphenocephala*
Reptiles		
Snapping Turtles	Snapping turtle	*Chelydra serpentina*
	Alligator snapping turtle	*Macroclemys temminckii*
Land and Freshwater Turtles	Painted turtle	*Chrysemys picta*
	Chicken turtle	*Deirochelys reticularia*
	Mississippi map turtle	*Graptemys kohnii*
	False map turtle	*Graptemys pseudogeographica*
	Diamond-backed terrapin	*Malaclemys terrapin*
	River cooter (also Slider, Mobilian)	*Pseudemys concinna*
	Cooter	*Pseudemys floridana*
	Eastern box turtle	*Terrapene carolina*
	Slider	*Trachemys scripta*

HERPETOFAUNA	Common Name	Scientific Name
Mud and Musk Turtles	Eastern mud turtle (also Stinkpot)	*Kinosternon subrubrum*
	Razor-backed musk turtle	*Sternotherus carinatus*
	Stinkpot (also Common musk turtle)	*Sternotherus odoratus*
Soft-Shelled Turtles	Spiny softshell	*Apalone spinifera*
Lizards	Green anole	*Anolis carolinensis*
	Five-lined skink	*Eumeces fasciatus*
	Broad-headed skink	*Eumeces laticeps*
	Ground skink	*Scincella lateralis*
Snakes	Racer	*Coluber constrictor*
	Ring-necked snake	*Diadophis punctatus*
	Corn snake	*Elaphe guttata*
	Rat snake	*Elaphe obsoleta*
	Mud snake	*Farancia abacura*
	Eastern hog-nosed snake	*Heterodon platyrhinos*
	Common kingsnake	*Lampropeltis getulus*
	Milk snake	*Lampropeltis triangulum*
	Western green water snake	*Nerodia cyclopion*
	Plain-bellied water snake	*Nerodia erythrogaster*
	Southern water snake	*Nerodia fasciata*
	Diamond-backed water snake	*Nerodia rhombifera*
	Rough green snake	*Opheodrys aestivus*
	Graham's crayfish snake	*Regina grahamii*
	Glossy crayfish snake	*Regina rigida*
	Brown snake	*Storeria dekayi*
	Western ribbon snake	*Thamnophis proximus*
	Common garter snake	*Thamnophis sirtalis*
	Copperhead	*Agkistrodon contortrix*
	Cottonmouth	*Agkistrodon piscivorus*
	Timber rattlesnake	*Crotalus horridus*
Crocodilians	American alligator	*Alligator mississippiensis*

This list was compiled with Dundee and Rossman (1989) and Walls et al. (2011).

COL#	LOC.	BRPE	CORM	ANHI	GBHE	GREG	SNEG	LBHE	TRHE	REEG	CAEG	GNHE	BCHN
61	MARSH	-	-	-	-	800	150	-	-	-	-	-	-
87	VER	-	-	-	-	-	40	70	-	-	-	-	-
116	ABFS	-	550	-	30	10	-	-	-	-	-	-	-
117	ABFS	-	-	-	10	25	145	140	5	-	-	-	-
126	VER	-	-	-	5	40	-	-	-	-	-	-	-
135	ABFS	-	-	-	30	86	-	-	-	-	-	-	-
136	ABFS	-	-	25	165	550	40	20	-	-	-	-	-
158	VER	-	50	-	70	105	-	-	-	-	-	-	-
251	ABFS	-	-	10	5	100	600	450	-	-	200	-	-
263	A. DEL	-	-	-	-	-	15	85	10	-	-	-	10
265	A. DEL	-	-	-	40	225	-	-	-	-	-	-	-
290	ABFS	-	-	-	-	20	60	50	-	-	-	-	-
333	ABFS	-	-	5	120	186	700	350	-	-	-	-	-
339	A. DEL	-	-	10	145	175	-	-	-	-	-	-	-
365	ABFS	-	-	15	40	180	325	335	10	-	-	-	-
366	VER	-	-	20	145	40	-	-	-	-	-	-	-
414	VER	-	-	20	30	264	-	-	-	-	-	-	-
415	VER	-	-	5	225	-	-	-	-	-	-	-	-
439	MARSH	-	-	-	-	-	-	-	-	-	-	-	-
463	WAX	-	-	-	-	100	100	-	-	-	-	-	-
500	VER	-	-	-	5	-	-	-	-	-	-	-	-
503	ABFS	-	-	-	-	-	1700	250	-	-	-	-	-
512	ABFS	-	-	-	-	-	1900	450	-	-	-	-	-
531	ABFS	-	-	5	-	-	200	100	-	-	600	-	-
532	ABFS	-	-	-	-	-	60	200	-	-	-	-	-
534	VER	-	-	-	55	650	-	-	-	-	-	-	-
614	ABFS	-	-	-	-	60	1190	130	-	-	-	-	1
617	VER	-	-	3	5	-	1950	200	-	-	-	-	-
618	VER	-	-	-	50	325	170	40	-	-	-	-	-
620	ABFS	-	-	-	-	-	750	270	-	-	50	-	-
621	ABFS	-	-	-	-	10	-	-	-	-	200	-	-
624	VER	-	-	-	11	90	-	-	-	-	-	-	-
625	ABFS	-	-	1	-	-	-	-	-	-	400	-	-
626	ABFS	-	-	-	-	5	90	10	-	-	-	-	1
628	ABFS	-	-	-	-	25	500	10	-	-	200	-	-
629	ABFS	-	-	-	-	10	145	70	-	-	-	-	-
632	ABFS	-	130	15	82	137	30	70	-	-	-	-	1
633	ABFS	-	-	-	45	-	125	75	-	-	-	-	-
634	ABFS	-	60	1	5	-	1060	70	-	-	-	-	-
635	VER	-	-	1	20	220	405	115	-	-	-	-	-
636	ABFS	-	-	-	-	40	200	200	-	-	-	-	-
638	ABFS	-	-	15	286	334	30	155	-	-	-	-	-
675	W. COTE	-	-	-	5	81	-	-	-	-	-	-	-
678	W. COTE	-	-	-	20	-	-	-	-	-	-	-	-
680	MARSH	-	-	-	-	-	-	-	-	-	-	-	-
682	W. COTE	-	-	-	-	-	20	30	-	-	-	-	-
686	A. DEL	-	-	-	-	20	100	150	-	-	-	-	-
687	A. DEL	-	-	-	-	-	-	-	-	-	-	-	-
AVG		0	198	10	63	164	441	152	8	0	275	0	3
N		0	4	15	26	30	29	27	3	0	6	0	4
TOT		0	790	151	1649	4913	12800	4095	25	0	1650	0	13

YCNH	WHIB	DAIB	ROSP	LAGU	HERG	CATE	ROYT	SATE	FOTE	LETE	BLSK	Total pairs/colony
-	-	-	-	-	-	-	-	-	-	-	-	950
-	-	-	-	-	-	-	-	-	-	-	-	110
-	-	-	-	-	-	-	-	-	-	-	-	590
-	-	-	4	-	-	-	-	-	-	-	-	329
-	-	-	-	-	-	-	-	-	-	-	-	45
-	-	-	-	-	-	-	-	-	-	-	-	116
-	-	-	-	-	-	-	-	-	-	-	-	800
-	-	-	-	-	-	-	-	-	-	-	-	225
-	-	-	-	-	-	-	-	-	-	-	-	1365
-	2000	-	15	-	-	-	-	-	-	-	-	2135
-	-	-	5	-	-	-	-	-	-	-	-	270
-	-	-	-	-	-	-	-	-	-	-	-	130
-	-	-	-	-	-	-	-	-	-	-	-	1361
-	-	-	-	-	-	-	-	-	-	-	-	330
-	-	-	25	-	-	-	-	-	-	-	-	930
-	-	-	-	-	-	-	-	-	-	-	-	205
-	-	-	8	-	-	-	-	-	-	-	-	322
-	-	-	-	-	-	-	-	-	-	-	-	230
-	-	-	-	40	-	-	-	-	600	-	450	1090
-	20	-	45	-	-	-	-	-	-	-	-	265
-	-	-	-	-	-	-	-	-	-	-	-	5
-	-	-	-	-	-	-	-	-	-	-	-	1950
1	-	-	-	-	-	-	-	-	-	-	-	2351
-	-	-	-	-	-	-	-	-	-	-	-	905
-	-	-	-	-	-	-	-	-	-	-	-	260
-	-	-	-	-	-	-	-	-	-	-	-	705
2	-	-	-	-	-	-	-	-	-	-	-	1383
-	-	-	-	-	-	-	-	-	-	-	-	2158
-	-	-	-	-	-	-	-	-	-	-	-	585
-	-	-	-	-	-	-	-	-	-	-	-	1070
-	-	-	-	-	-	-	-	-	-	-	-	210
-	-	-	-	-	-	-	-	-	-	-	-	101
1	-	-	-	-	-	-	-	-	-	-	-	402
-	-	-	-	-	-	-	-	-	-	-	-	106
-	-	-	-	-	-	-	-	-	-	-	-	735
-	-	-	-	-	-	-	-	-	-	-	-	225
-	-	-	-	-	-	-	-	-	-	-	-	465
-	-	-	-	-	-	-	-	-	-	-	-	245
-	-	-	-	-	-	-	-	-	-	-	-	1196
10	-	-	-	-	-	-	-	-	-	-	-	771
-	-	-	-	-	-	-	-	-	-	-	-	440
-	-	-	-	-	-	-	-	-	-	-	-	820
-	-	-	-	-	-	-	-	-	-	-	-	86
-	-	-	-	-	-	-	-	-	-	-	-	20
-	-	-	-	-	-	-	-	-	300	-	-	300
-	-	-	-	-	-	-	-	-	-	-	-	50
-	6900	-	20	-	-	-	-	-	-	-	-	7190
-	-	-	-	-	-	-	-	-	10	-	350	360
4	2973	0	17	0	0	0	0	0	303	0	400	769
4	3	0	7	1	0	0	0	0	3	0	2	48
14	8920	0	122	40	0	0	0	0	910	0	800	36892

Shown are colony number, colony location (ABFS=Atchafalaya Basin Floodway System, A. DEL = Atchafalaya Delta, WAX = Wax Lake Delta, MARSH = Marsh Island, VER = Vermilion Bay, W. COTE = West Cote Blanche Bay), and the number of pairs per colony for each species (AOU abbreviation).

APPENDIX 4 Summary of Studies Investigating Effects of Nutrients and Flooding on Regeneration and Seedling Growth for Baldcypress, Water Tupelo, and Red Maple

Species	Treatment	Water properties	Regeneration rate (num. acre⁻¹)	Growth rate (cm yr⁻¹)	Diameter growth rate (cm yr⁻¹)	Basal area growth rate (cm² yr⁻¹)
Baldcypress	Hydrology	Natural flooding		25		-
Baldcypress			-	-	-	22.2
Baldcypress			-	-	-	27.1
Baldcypress			-	-	0.02	-
Baldcypress			-	-	0.33	-
Baldcypress			-	-	0.3	-
Baldcypress				90	0.48	-
Baldcypress			-	-	-	41
Baldcypress	Nutrients		-	-	-	1.3
Baldcypress	Nutrients		-	-	-	20.2
Baldcypress	Nutrients		-	-	-	5
Baldcypress	Nutrients		-	-	-	18
Baldcypress	Nutrients		-	-	-	6.7
Baldcypress	Nutrients		-	-	-	15.5
Baldcypress			-	-	0.122	-
Baldcypress	Hydrology	Impounded	0	0	-	-
Baldcypress	Hydrology	Natural flooding	950	22	-	-
Baldcypress	Hydrology	Natural flooding	25	25	-	-
Baldcypress	Hydrology	Natural flooding	25	55	-	-
Baldcypress	Hydrology	Natural flooding	622	24.4	-	-
Baldcypress	Hydrology	Impounded	15	47	0.93	-
Baldcypress	Hydrology	Natural flooding	5	50	0.6	-
Baldcypress	Hydrology	Natural flooding	24	60	1.35	-
Baldcypress	Hydrology	Natural flooding	102	130	2.2	-
Baldcypress	Hydrology	Natural flooding	67	100	1.5	-
Baldcypress	Hydrology	Continuous flooding	-	-	0.158	-
Baldcypress	Hydrology	Seasonal flooding	-	-	0.164	-
Baldcypress	Hydrology	Continuous flooding	-	-	0.2	-
Baldcypress	Hydrology	Seasonal flooding	-	-	0.32	-
Baldcypress	Hydrology	Control	-	20	-	-
Baldcypress	Hydrology	Flooded	-	20	-	-
Baldcypress	Hydrology	Natural flooding	-	-	-	17.2
Baldcypress	Hydrology	Impounded	-	-	-	24.5
Baldcypress	Hydrology	Spoil Break	-	-	-	21.3
Baldcypress	Hydrology	Crawfish pond	-	-	-	35.1
Baldcypress	Hydrology	Natural flooding	-	-	0.163	7
Baldcypress	-	-	-	90		-
Baldcypress	Hydrology	Variable flooding	-	25	-	-

Sampling unit	Observation type (field/lab)	Location	Reference
Seed	-	-	Matoon (1915)
-	-	-	McClurkin (1965)
-	-	-	McClurkin (1965)
-	-	-	Mitsch and Ewel (1979)
-	-	-	Mitsch and Ewel (1979)
-	-	-	Williston et al. (1980)
-	-	-	Williston et al. (1980)
-	-	-	Brown (1981)
-	-	-	Brown (1981)
-	-	-	Brown (1981)
-	-	-	Lemlich and Ewel (1984)
-	-	-	Lemlich and Ewel (1984)
-	-	-	Nessel and Bayley (1984)
-	-	-	Nessel and Bayley (1984)
-	-	Great Dismal Swamp	Day (1985)
Seed	Field	Barataria Basin, Louisiana	Conner et al. (1986)
Seed	Field	Barataria Basin, Louisiana	Conner et al. (1986)
Seed	Field	Barataria Basin, Louisiana	Conner et al. (1986)
Seed	Field	Barataria Basin, Louisiana	Conner et al. (1986)
Seed	Field	Barataria Basin, Louisiana	Conner et al. (1986)
Coppice Sprout	Field	Barataria Basin, Louisiana	Conner et al. (1986)
Coppice Sprout	Field	Barataria Basin, Louisiana	Conner et al. (1986)
Coppice Sprout	Field	Barataria Basin, Louisiana	Conner et al. (1986)
Coppice Sprout	Field	Barataria Basin, Louisiana	Conner et al. (1986)
Coppice Sprout	Field	Barataria Basin, Louisiana	Conner et al. (1986)
All diameter classes (5–55 cm dbh)	Field	Bayou Pigeon, Atchafalaya Basin	Dicke and Toliver (1990)
All diameter classes (5–55 cm dbh)	Field	Bayou Pigeon, Atchafalaya Basin	Dicke and Toliver (1990)
25–35 cm dbh	Field	Bayou Pigeon, Atchafalaya Basin	Dicke and Toliver (1990)
25–35 cm dbh	Field	Bayou Pigeon, Atchafalaya Basin	Dicke and Toliver (1990)
Seedling to 6 mo. old	Lab	Louisiana State University	Pezeshki (1990)
Seedling to 6 mo. old	Lab	Louisiana State University	Pezeshki (1990)
-	Field	Barataria Basin, Louisiana	Conner and Day (1992)
-	Field	Barataria Basin, Louisiana	Conner and Day (1992)
-	Field	Barataria Basin, Louisiana	Conner and Day (1992)
-	Field	Barataria Basin, Louisiana	Conner and Day (1992)
Seed	Field	Barataria Basin, Louisiana	Visser and Sasser (1995)
-			Keeland and Conner (1999)
Seed	Lab	SELU, lab	Souther and Shaffer (2000)

(continued next page)

APPENDIX 4 Summary of Studies Investigating Effects of Nutrients and Flooding on Regeneration and Seedling Growth for Baldcypress, Water Tupelo, and Red Maple, *continued*

Species	Treatment	Water properties	Regeneration rate (num. acre^{-1})	Growth rate (cm yr^{-1})	Diameter growth rate (cm yr^{-1})	Basal area growth rate (cm^2 yr^{-1})
Baldcypress	Hydrology	Variable flooding	-	80	-	-
Baldcypress	Hydrology	Variable flooding	-	25	-	-
Baldcypress	Hydrology	Variable flooding	-	60	-	-
Baldcypress	Hydrology	Variable flooding	-	-	2	-
Baldcypress	Hydrology	Variable flooding	-	-	8	-
Baldcypress	Hydrology	Variable flooding	-	-	5	-
Baldcypress	Hydrology	Variable flooding	-	-	12	-
Baldcypress	Hydrology	Not flooded	-	50	-	-
Baldcypress	Hydrology	Flooded-Aerated	-	25	-	-
Baldcypress	Hydrology	Flooded	-	30	-	-
Water Tupelo		-	-	-	0.186	-
Water Tupelo		-	-	-	0.108	-
Water Tupelo		-	-	-	0.132	-
Water Tupelo	Hydrology	Control	-	60	-	-
Water Tupelo	Hydrology	Flooded	-	40	-	-
Water Tupelo		-	-	-	-	4.9
Water Tupelo		-	-	-	-	13.03
Water Tupelo		-	-	-	0.208	9.8
Water Tupelo	Hydrology	Not flooded	-	80	-	-
Water Tupelo	Hydrology	Flooded-Aerated	-	60	-	-
Water Tupelo	Hydrology	Flooded	-	50	-	-
Red Maple		-	-	-	0.275	-
Red Maple	Hydrology	-	-	-	0.094	1.4

Sampling unit	Observation type (field/lab)	Location	Reference
Seed	Lab	SELU, lab	Souther and Shaffer (2000)
Seedling to 6 mo. old	Lab	SELU, lab	Souther and Shaffer (2000)
Seedling to 6 mo. old	Lab	SELU, lab	Souther and Shaffer (2000)
Seed	Lab	Southeastern Louisiana University	Souther and Shaffer (2000)
Seed	Lab	Southeastern Louisiana University	Souther and Shaffer (2000)
Seedling to 6 mo. old	Lab	Southeastern Louisiana University	Souther and Shaffer (2000)
Seedling to 6 mo. old	Lab	Southeastern Louisiana University	Souther and Shaffer (2000)
2 yr-old tree	Lab	Louisiana State University	Effler and Goyer (2006)
2 yr-old tree	Lab	Louisiana State University	Effler and Goyer (2006)
2 yr-old tree	Lab	Louisiana State University	Effler and Goyer (2006)
-	-	Great Dismal Swamp	Day (1985)
All diameter classes (5–55 cm dbh)	Field	Bayou Pigeon, Atchafalaya Basin	Dicke and Toliver (1990)
All diameter classes (5–55 cm dbh)	Field	Bayou Pigeon, Atchafalaya Basin	Dicke and Toliver (1990)
Seedling to 6 mo. old	Lab	Louisiana State University	Pezeshki (1990)
Seedling to 6 mo. old	Lab	Louisiana State University	Pezeshki (1990)
-	-	-	Conner and Day (1992)
-	-	-	Conner and Day (1992)
Seed	-	Barataria Basin, Louisiana	Visser and Sasser (1995)
2 yr-old tree	Lab	Louisiana State University	Effler and Goyer (2006)
2 yr-old tree	Lab	Louisiana State University	Effler and Goyer (2006)
2 yr-old tree	Lab	Louisiana State University	Effler and Goyer (2006)
-	-	Great Dismal Swamp	Day (1987)
Seed	-	Barataria Basin, Louisiana	Visser and Sasser (1995)

Most studies were conducted in Louisiana, in both field and laboratory settings.

LITERATURE CITED

Adams, C.E., Jr., J.T. Wells, and J.M. Coleman. 1982. Sediment transport on the continental shelf: implications for the developing Atchafalaya River Delta. Contributions in Marine Science 25: 133–148.

Adams S.R., M.B. Flinn, B.M. Burr, M.R. Whiles, and J.E. Garvey. 2006. Ecology of larval blue sucker (*Cycleptus elongatus*) in the Mississippi River. Ecology of Freshwater Fish 2006: 15: 291–300.

Aday, D.D., D.A. Rutherford, and W.E. Kelso. 2000. Field and laboratory determinations of hypoxic effects on RNA-DNA ratios of bluegill. American Midland Naturalist 143: 433–442.

Alexander, R.B., R.A. Smith, and G.E. Schwarz. 2000. Effect of stream channel size on the delivery of nitrogen to the Gulf of Mexico. Nature 403: 758–761.

Alexander, R.B., R.A. Smith, G.E. Schwarz, E.W. Boyer, J.V. Nolan, and J.W. Brakebill. 2008. Differences in phosphorus and nitrogen delivery to the Gulf of Mexico from the Mississippi River basin. Environmental Science and Technology 42: 822–830.

Alford, J.B. and M.R. Walker. 2011. Managing the flood pulse for optimal fisheries production in the Atchafalaya River Basin, Louisiana (USA). River Research and Applications DOI: 10.1002/rra.1610.

Allen, Y.C. 2010 A brief synopsis of land-water changes in the former extent of Grand Lake in the Atchafalaya Basin—early 1880s to late 2000s. Interim Report to Accompany Deliverables to the Louisiana Department of Natural Resources, Atchafalaya Basin Program, Baton Rouge, Louisiana.

Allen, Y.C., B.R. Couvillion, and J.A. Barras. 2011. Using multitemporal remote sensing imagery and inundation measures to improve land change. Estuaries and Coasts DOI: 10.1007/s12237-011-9437-z.

Allen, Y.C., G.C. Constant, and B.R. Couvillion. 2008. Preliminary classification of water areas within the Atchafalaya Basin Floodway System by using Landsat imagery. US Geological Survey Open-File Report 2008–1320.

Allison, M.A., G.C. Kineke, E.S. Gordon, and M.A. Goñi. 2000. Development and reworking of a seasonal flood deposit on the inner continental shelf off the Atchafalaya River. Continental Shelf Research 20: 2267–2294.

Allison, M.A., A. Sheremet, M.A. Goñi, and G.W. Stone. 2005. Storm layer deposition on the Mississippi-Atchafalaya subaqueous delta generated by Hurricane Lili in 2002. Continental Shelf Research 25: 2213–2232.

Allison, M.A., C.R. Demas, B.A. Ebersole, B.A. Kleiss, C.D. Little, E.A. Meselhe, N.J. Powell, T.C. Pratt, B.M. Vosburg. 2012. A water and sediment budget for the lower Mississippi-

Atchafalaya River in flood years 2008–2010: implications for sediment discharge to the oceans and coastal restoration in Louisiana. Journal of Hydrology 432–433: 84–97.

American Bird Conservancy. 2007. United States watchlist of birds of conservation concern. http://www.abcbirds.org/abcprograms/science/watchlist/watchlist.html.

Amos, J.B. 2006. Dendochronological analysis of productivity and hydrology in two Louisiana swamps [thesis]. Baton Rouge: Louisiana State University. 91 p.

Amos, J.B., R.F. Keim, and T.W. Doyle. 2005. Coastal forest productivity and wetland hydrology: a dendrochronological study. Eos Transactions, American Geophysical Union 86(18), Joint Assembly Supplement, Abstract NB33C-06.

Aslan, A., W.J. Austin, and M.D. Blum. 2005. Causes of river avulsion: insights from the late Holocene avulsion history of the Mississippi River, USA. Journal of Sedimentary Research 75: 650–664.

Atchafalaya Trace Commission. 2011. Atchafalaya National Heritage Area: management plan and environmental assessment. Louisiana Department of Culture, Recreation, and Tourism and the National Park Service.

Aust, W.M., S.H. Schoenholtz, T.W. Zaebst, and B.A. Szabo. 1997. Recovery status of a tupelo-cypress wetland seven years after disturbance: silvicultural implications. Forest Ecology and Management 90: 161–169.

Aust, W.M., T.C. Fristoe, P.A. Gellerstedt, L.A.B. Giese, and M. Miwa. 2006. Long-term effects of helicopter and ground-based skidding on site properties and stand growth in a tupelo-cypress wetland. Forest Ecology and Management 226: 72–79.

Baldwin, M.J. 2005. Winter bird use of the Chinese tallow tree in Louisiana [thesis]. Baton Rouge: Louisiana State University. 78 p.

Baltz, D.M. and R.F. Jones. 2003. Temporal and spatial patterns of microhabitat use by fishes and decapod crustaceans in a Louisiana estuary. Transactions of the American Fisheries Society 132: 662–678.

Barko, V.A. and R.A. Hrabik. 2004. Abundance of Ohio shrimp (*Macrobrachium ohione*) and glass shrimp (*Palaemonetes kadiakensis*) in the unimpounded upper Mississippi River. American Midland Naturalist 151: 265–273.

Barras, J.A. 2006. Land area change in coastal Louisiana after the 2005 hurricanes—a series of three maps. US Geological Survey Open-File Report 06–1274.

Barras, J.A. 2009. Land area change and overview of major hurricane impacts in coastal Louisiana, 2004–08. US Geological Survey Scientific Investigations Map 3080, scale 1:250,000, 6 p. pamphlet.

Barrow, W.C., Jr., L.A. Johnson-Randall, M.S. Woodrey, J. Cox, C.M. Riley, R.B. Hamilton, and C. Eberly. 2005. Coastal forests of the Gulf of Mexico: a description and some thoughts on their conservation. USDA Forest Service General Technical Report PSW-GTR-191:450–464.

Battle, J.M. and T.B. Mihuc. 2000. Decomposition dynamics of aquatic macrophytes in the lower Atchafalaya, a large floodplain river. Hydrobiologia 418: 123–136.

Bauer, R.T. 2011. Amphidromy and migrations of freshwater shrimps. I. Costs, benefits, evolutionary origins, and an unusual case of amphidromy. In: A. Asakura (ed.), New frontiers in crustacean biology. Leiden, Netherlands: BRILL. pp. 145–156.

Bauer, R.T. and J. Delahoussaye. 2008. Life history migrations of the amphidromous river shrimp *Macrobrachium ohione* from a continental large river system. Journal of Crustacean Biology 28: 622–632.

Baumann, R.H., J.W. Day, Jr., and C.A. Miller. 1984. Mississippi deltaic wetland survival: sedimentation versus coastal submergence. Science 224: 1093–1095.

Beck, M.B., R.D. Brumbaugh, L. Airoldi, A. Carranza, L.D. Coen, C. Crawford, O. Defeo, G.J. Edgar, B. Hancock, M. Kay, H. Lenihan, M.W. Lukenbach, C.L. Toropova, and G. Zhang.

2009. Shellfish reefs at risk: a global analysis of problems and solutions. Arlington, Virginia: The Nature Conservancy. 52 p.

Beck, M.B., R.D. Brumbaugh, L. Airoldi, A. Carranza, L.D. Coen, C. Crawford, O. Defeo, G.J. Edgar, B. Hancock, M. Kay, H. Lenihan, M.W. Lukenbach, C.L. Toropova, G. Zhang, and X. Guo. 2011. Oyster reefs at risk and recommendations for conservation, restoration, and management. BioScience 61: 107–116.

Bell, F.W. 1986. Competition from fish farming in influencing rent dissipation: the crawfish fishery. American Journal of Agricultural Economics 68: 95–101.

Benson, J.F. 2005. Ecology and conservation of Louisiana black bears in the Tensas River basin and reintroduced populations [thesis]. Baton Rouge: Louisiana State University. 114 p.

Bentley, S.J., Jr. 2002. Dispersal of fine sediments from river to shelf: process and product. Transactions of the Gulf Coast Association of Geological Societies 52: 1055–1067.

Bentley, S.J., H.H. Roberts, and K. Rotondo. 2003. The sedimentology of muddy coastal systems: the research legacy and new perspectives from the Coastal Studies Institute. GCACS/GCSSEPM Transactions 53: 52–63.

Bianchi, T.S., F. Garcia-Tigreros, S.A. Yvon-Lewis, M. Shields, H.J. Mills, D. Butman, C. Osburn, P. Raymond, G.C. Shank, S.F. DiMarco, N. Walker, B. Kiel Reese, R. Mullins-Perry, A. Quigg, G.R. Aiken, and E.L. Grossman. 2013. Enhanced transfer of terrestrially derived carbon to the atmosphere in a flooding event. Geophysical Research Letters 40: 116-122.

Black Bear Conservation Coalition. 2011. www.bbcc.org.

Blum, M.D. and H.H. Roberts. 2009. Drowning of the Mississippi Delta due to insufficient sediment supply and global sea-level rise. Nature Geoscience 2: 488–491.

Blum, M.D. and H.H. Roberts. 2012. The Mississippi Delta region: past, present, and future. Annual Review of Earth and Planetary Science 40: 655–683.

Bohora, S.B. 2012. Spatial variability in response of deltaic baldcypress forests to hydrology and climate [thesis]. Baton Rouge: Louisiana State University. 58 p.

Bonvillain, C.P. 2006. The use of a low-water refuge in the Atchafalaya River Basin by adult spotted gar *Lepisosteus oculatus* [thesis]. Thibodaux, Louisiana: Nicholls State University. 57 p.

Bonvillain, C.P. 2012. Red swamp crayfish *Procambarus clarkii* in the Atchafalaya River Basin: biotic and abiotic effects on population dynamics and physiological biomarkers of hypoxic stress [dissertation]. Baton Rouge: Louisiana State University. 141 p.

Bonvillain, C.P., B.T. Halloran, K.M. Boswell, W.E. Kelso, A.R. Harlan, and D.A. Rutherford. 2011. Acute physicochemical effects in a large river-floodplain system associated with the passage of Hurricane Gustav. Wetlands 31: 979–987.

Bonvillain, C.P., D.A. Rutherford, W.E. Kelso, and C.C. Green. 2012. Physiological biomarkers of hypoxic stress in the red swamp crawfish *Procambarus clarkii* from field and laboratory experiments. Comparative Biochemistry and Physiology, Part A 163: 15–21.

Brinson, M.M. 1977. Decomposition and nutrient exchange of litter in an alluvial swamp forest. Ecology 58: 601–609.

Brinson, M.M., H.D. Bradshaw, R.N. Holmes, and J.B. Elkins, Jr. 1980. Litterfall, stemflow, and throughfall nutrient fluxes in an alluvial swamp forest. Ecology 61: 827–835.

Brown, S. 1981. A comparison of the structure, primary productivity, and transpiration of cypress ecosystems in Florida. Ecological Monographs 51: 403–427.

Bruce, K.A., G.N. Cameron, P.A. Harcombe, and G. Jubinsky. 1997. Introduction, impact on native habitats, and management of a woody invader, the Chinese tallow tree, *Sapium sebiferum* (L.) Roxb. Natural Areas Journal 17: 255–260.

Brunet, L.A. 1997. Effects of environmental hypoxia on the reproductive processes of sunfishes from the Atchafalaya Basin [thesis]. Baton Rouge: Louisiana State University. 91 p.

Bruser, J.F. 1995. Ecology of Catahoula Lake plant communities in relation to an anthropogenic water regime [thesis]. Baton Rouge: Louisiana State University. 100 p.

Bryan, C.F. and D.S. Sabins. 1979. Management implications in water quality and fish standing stock information for the Atchafalaya River Basin, Louisiana. In: J.W. Day, Jr., D.D. Culley, Jr., R.E. Turner and A.J. Mumphrey Jr. (eds.), Proceedings of the third coastal marsh and estuary management symposium. Baton Rouge: Louisiana State University, Division of Continuing Education. pp. 293–316.

Bryan, C.F., W.E. Kelso, D.A. Rutherford, L.F. Hale, D.G. Kelly, B.W. Bryan, F. Monzyk, J. Noguera, and G. Constant. 1998. Atchafalaya River Basin hydrological management plan: research and development. Final Report. FEMA, LADWF Contract 512–3019. Baton Rouge: Louisiana State University, Louisiana Cooperative Fish and Wildlife Research Unit.

BryantMason, A. and Y.J. Xu. 2012. The record 2011 spring flood of the Mississippi River. How much nitrate was exported from its largest distributary, the Atchafalaya River, into the Gulf of Mexico? Journal of Hydrologic Engineering DOI: 10.1061/(ASCE) HE.1943-5584.0000660.

BryantMason, A., Y.J. Xu, and M. Altabet. 2012. Isotopic signature of nitrate in river waters of the lower Mississippi and its distributary, the Atchafalaya. Hydrological Processes DOI: 10.1002/hyp.9420.

BryantMason, A., Y.J. Xu, and M.A. Altabet. 2013. Limited capacity of river corridor wetlands to remove nitrate: a case study on the Atchafalaya River Basin during the 2011 Mississippi River flooding. Water Resources Research 49: 283-290.

Burns, A.C. 1980. Frank B. Williams: cypress lumber king. Journal of Forest History 24: 127–133.

Burr, B.M. and R.L. Mayden. 1999. A new species of *Cycleptus* (Cypriniformes: Catostomidae) from Gulf Slope drainages of Alabama, Mississippi, and Louisiana, with a review of the distribution, biology, and conservation status of the genus. Bulletin of the Alabama Museum of Natural History 20: 19–57.

Butman, D. and P.A. Raymond. 2011. Significant efflux of carbon dioxide from streams and rivers in the United States. Nature Geoscience 4: 839-842.

Byrne, M.E. and M.J. Chamberlain. 2011. Seasonal space use and habitat selection of adult raccoons (*Procyon lotor*) in a Louisiana bottomland hardwood forest. The American Midland Naturalist 166: 426-434.

Byrne, M.E., M.J. Chamberlain, and F.G. Kimmel. 2011. Seasonal space use and habitat selection of female wild turkeys in a Louisiana bottomland forest. Proceedings of the Southeastern Association of Fish and Wildlife Agencies 65: 8-14.

Calhoun, A.J.K. 1999. Forested wetlands. In: M.L. Hunter (ed.), Maintaining biodiversity in forest ecosystems. Cambridge, UK: Cambridge University Press. pp. 300–331

Campbell, D. 2005. The Congo River basin. In: L.H. Ferasier and P. Keddy, (eds.), The world's largest wetlands: ecology and conservation. Cambridge, UK: Cambridge University Press. pp. 149–165

Caraco, N.F. and J.J. Cole. 2002. Contrasting impacts of a native and alien macrophyte on dissolved oxygen on a large river. Ecological Applications 12: 1496–1509.

Cardoch, L. and J.W. Day, Jr. 2001. Energy analysis of nonmarket values of the Mississippi Delta. Environmental Management 28: 677–685.

Carle, M.V., C.E. Sasser, and H.H. Roberts. In Press. Accretion and vegetation community change in the Wax Lake subdelta following the historic 2011 Mississippi River flood. Journal of Coastal Research.

Carlson, D., M. Horn, T. Van Biersel, and D. Fruge. 2012. 2011 Atchafalaya Basin Inundation Data Collection and Damage Assessment Project. Report of Investigations No. 12-01. Louisiana Geological Survey, Baton Rouge, Louisiana.

Cary, L.R. 1906. The conditions for future oyster culture in the waters of the parishes of Vermilion and Iberia, Louisiana. Bulletin No. 4. Cameron, Lousiana: Gulf Biologic Station.

Castellanos, D.L. and L.P. Rozas. 2001. Nekton use of submerged aquatic vegetation, marsh, and shallow unvegetated bottom in the Atchafalaya River Delta, a Louisiana tidal freshwater ecosystem. Estuaries 24: 184–197.

Chabreck, R.H., J.E. Holcombe, R.G. Linscombe, and N.E. Kinler. 1982. Winter foods of river otters from saline and fresh environments in Louisiana. Proceedings of the Annual Conference of the Southeast Association of Fish and Wildlife Agencies 36: 473–483.

Chamberlain, M.J., M.E. Byrne, N.J. Stafford, III, K.L. Skow, and B.E. Collier. 2013. Wild turkey movements during flooding after opening of the Morganza Spillway, Louisiana. Southeastern Naturalist 12: 93-98.

Chambers, J.L., W.H. Conner, J.W. Day, Jr., S.P. Faulkner, E.S. Gardiner, M.S. Hughes, R.F. Keim, S.L. King, K.W. McLeod, C.A. Miller, J.A. Nyman, and G.P. Shaffer. 2005. Conservation, protection and utilization of Louisiana's coastal wetland forests. Coastal Wetland Forest Conservation and Use Science Working Group, Final Report to the Governor of Louisiana. 102 p.

Chambers, J.L., R.F. Keim, W.H. Conner, J.W. Day, Jr., S.P. Faulkner, E.S. Gardiner, M.S. Hughes, S.L. King, K.W. McLeod, C.A. Miller, J.A. Nyman, and G.P. Shaffer. 2006. Conservation of Louisiana's coastal wetland forests. In: T.F. Shupe and M.A. Dunn (ed.), Proceedings of Louisiana natural resources symposium. Baton Rouge, Louisiana. pp. 117–135.

Chick, J.H. and C.C. McIvor. 1994. Patterns in the abundance and composition of fishes among beds of different macrophytes: viewing a littoral zone as a landscape. Canadian Journal of Fisheries and Aquatic Sciences 51: 2873–2882.

Chick, J.H., R.J. Cosgriff, and L.S. Gittinger. 2003. Fish as potential dispersal agents for floodplain plants: first evidence in North America. Canadian Journal of Fisheries and Aquatic Sciences 60: 1437–1439.

Cobb, M., T.R. Keen, and N.D. Walker. 2008a. Modeling the circulation of the Atchafalaya Bay system during winter cold front events. Part 1: model description and validation. Journal of Coastal Research 24: 1036–1047.

Cobb, M., T.R. Keen, and N.D. Walker. 2008b. Modeling the circulation of the Atchafalaya Bay system during winter cold front events. Part 2: river plume dynamics during cold fronts. Journal of Coastal Research 24: 1048–1062.

Coleman, J.M. 1966. Ecological changes in a massive fresh-water clay sequence. Transactions of the Gulf Coast Association of Geological Societies 16: 159–174.

Coleman, J.M. 1988. Dynamic changes and processes in the Mississippi River delta. Geological Society of America Bulletin 100: 999–1015.

Coleman, J.M. and S.M. Gagliano. 1964. Cyclic sedimentation in the Mississippi River deltaic plain. Transactions of the Gulf Coast Association of Geological Societies 14: 67–80.

Coleman, J.M. and H.H. Roberts. 1991. The Mississippi River depositional system: a model for the Gulf Coast tertiary. In: D. Goldthwaite (ed.), An introduction to Gulf Coast geology. New Orleans, Louisiana: New Orleans Geological Society. pp. 99–121.

Coleman, J.M., H.H. Roberts, and G.W. Stone. 1998. Mississippi River delta: an overview. Journal of Coastal Research 14: 698–716.

Coleman, J., O. Huh, D. Braud, Jr., and H. Roberts. 2005. Major world delta variability and wetland loss. Transactions of the Gulf Coast Association of Geological Societies 55: 102–130.

Colon-Gaud, J.C., W.E. Kelso, and D.A. Rutherford. 2004. Spatial distribution of macroinvertebrates inhabiting hydrilla and coontail beds in the Atchafalaya Basin, Louisiana. Journal of Aquatic Plant Management 42: 85–91.

Conaway, L.K. and R.A. Hrabik. 1997. The Ohio shrimp *Macrobrachium ohione* in the upper Mississippi River. Transactions of the Missouri Academy of Sciences 31: 44–46.

Condrey, R.E, P. Hoffman, D.E. Evers, J. Anderson, and D. Morgan. 2010. The last natural delta of the Mississippi River. 1519–1880: Discovery, demise, and coastal-restoration implications. 2010 State of the Coast Conference, Coalition to Restore Coastal Louisiana. Baton Rouge, Louisiana.

Conner, W.H. 1988. Natural and artificial regeneration of baldcypress (*Taxodium distichum* (L.) Rich.) in the Barataria basins of Louisiana [dissertation]. Baton Rouge: Louisiana State University. 129 p.

Conner, W.H. and M.A. Buford. 1998. Southern deepwater swamps. In: M.G. Messina and W.H. Conner (eds.), Southern forested wetlands: ecology and management. Boca Raton, Florida: Lewis Publishers. pp. 261–287.

Conner, W.H. and J.W. Day, Jr. 1976. Productivity and composition of a baldcypress-water tupelo site and a bottomland hardwood forest site in a Louisiana swamp. American Journal of Botany 63: 1354–1364.

Conner, W.H. and J.W. Day, Jr. 1988. Rising water levels in coastal Louisiana: implications for two coastal forested wetlands in Louisiana. Journal of Coastal Research 4: 589–596.

Conner, W.H. and J.W. Day, Jr. 1992a. Water level variability and litterfall productivity of forested freshwater wetlands in Louisiana. American Midland Naturalist 128: 237–245.

Conner, W.H. and J.W. Day, Jr. 1992b. Diameter growth of *Taxodium distichum* (L.) Rich. and *Nyssa aquatica* L. from 1979–1985 in four Louisiana swamp stands. American Midland Naturalist 127: 290–299.

Conner, W.H., J.W. Day, Jr., and W.R. Slater. 1993. Bottomland hardwood productivity: case study in a rapidly subsiding, Louisiana, USA, watershed. Wetlands Ecology and Management 2: 189–197.

Conner, W.H., J.G. Gosselink, and R.T. Parrondo. 1981. Comparison of the vegetation of three Louisiana swamp sites with different flooding regimes. American Journal of Botany 68: 320–331.

Conner, W.H., I. Mihalia, and J. Wolfe. 2002. Tree community structure and changes from 1987 to 1999 in three Louisiana and three South Carolina forested wetlands. Wetlands 22: 58–70.

Conner, W.H. and J.R. Toliver. 1990. Long-term trends in the bald-cypress (*Taxodium distichum*) resource in Louisiana (USA.). Forest Ecology and Management 333/34: 543–557.

Conner, W.H., J.R. Toliver, and F.H. Sklar. 1986. Natural regeneration of baldcypress (*Taxodium distichum* (L.) Rich.) in a Louisiana swamp. Forest Ecology and Management. 14: 305–317.

Consensus Building Institute 2010a. Final stakeholder assessment: the views of stakeholders on management of the Atchafalaya Basin. A report to the National Audubon Society by the Consensus Building Institute and the MIT Science Impact Collaborative, Massachusetts Institute of Technology, Cambridge, Massachusetts, USA.

Consensus Building Institute 2010b. Consensus building opportunities in the Atchafalaya Basin: two potential strategies. A report to the National Audubon Society by the Consensus Building Institute and the MIT Science Impact Collaborative, Massachusetts Institute of Technology, Cambridge, Massachusetts, USA.

Constant, G.C., W.E. Kelso, and D.A. Rutherford. 1999. Effects of variation in river stage on the water quality and biota in the Buffalo Cove Management Unit of the Atchafalaya Basin, Louisiana. MIPR W42HEM 8343-1433. Final Report to the US Army Corps of Engineers, New Orleans District.

Constant, G.C., W.E. Kelso, D.A. Rutherford, and C.F. Bryan. 1997. Habitat, movement, and reproductive status of pallid sturgeon (*Scaphirhynchus albus*) in the Mississippi and Atchafalaya rivers. Report prepared for U.S Army Corp of Engineers, New Orleans District. 78 p.

Conway, W C., L.M. Smith, and J.F. Bergan. 2002a. Potential allelopathic interference by the exotic Chinese tallow tree (*Sapium sebiferum*). The American Midland Naturalist 148: 43–53.

Conway, W.C., L.M. Smith, and J.F. Bergan. 2002b. Avian use of Chinese tallow seeds in coastal Texas. The Southwestern Naturalist 37: 550–556.

Couvillion, B.R., J.A. Barras, G.D. Steyer, W. Sleavin, M. Fischer, H. Beck, N. Trahan, B. Griffin, and D. Heckman. 2011. Land area change in coastal Louisiana from 1932 to 2010. US Geological Survey Scientific Investigations Map 3164, scale 1:265,000, 12 p. pamphlet.

Cratsley, D.W. 1975. Recent deltaic sedimentation in Atchafalaya Bay, Louisiana [thesis]. Baton Rouge: Louisiana State University. 142 p.

Creasman, L., N.J. Craig, and M. Swan. 1992. The Forested Wetlands of the Mississippi River: An Ecosystem in Crisis. Baton Rouge, Louisiana: The Nature Conservancy.

Crook, A. 2008. A multi-scale assessment of den selection in Louisiana black bears (*Ursus americanus luteolus*) in northern and central Louisiana [thesis]. Baton Rouge: Louisiana State University. 63 p.

Dale, V.H., C.L. Kling, J.L. Meyer, J. Sanders, H. Stallworth, T.H. Armitage, D. Wangsness, T.H. Bianchi, A. Blumberg, W. Boynton, D.J. Conley, W. Crumpton, M. David, D. Gilbert, R.W. Howarth, R. Lowrance, K. Mankin, J. Opaluch, H. Paerl, K. Reckhow, A.N. Sharpley, T.H.W. Simpson, C.S. Snyder, and D. Wright. 2010. Hypoxia in the Northern Gulf of Mexico. New York: Springer. 284 p.

Davidson, N.L., Jr., W.E. Kelso, and D.A. Rutherford. 1998. Relationships between environmental variables and the abundance of cladocerans and copepods in the Atchafalaya River basin. Hydrobiologia 379: 175–181.

Davidson, N.L., Jr., W.E. Kelso, and D.A. Rutherford. 2000. Characteristics of cladoceran and copepod communities in floodplain habitats of the Atchafalaya River Basin. Hydrobiologia 435: 99–107.

Davis, J.C. 1975. Minimal dissolved oxygen requirements of aquatic life with emphasis on Canadian species: a review. Journal of the Fisheries Research Board of Canada 32: 2295–2332.

Day. F.P., Jr. 1985. Tree growth rates in the periodically flooded Great Dismal Swamp. Castanea 50: 89–95.

Day, F.P., Jr. 1987. Effects of flooding and nutrient enrichment on biomass allocation in *Acer rubrum* seedlings. American Journal of Botany 74: 1541–1554.

DeBell, D.S. and A.W. Naylor. 1972. Some factors affecting germination of swamp tupelo seeds. Ecology 53: 504–506.

de Evía, J.A. 1785. José de Evia y sus reconocimientos del Golfo de México: [diarios, cartas, explicaciones, descripciones, planos y mapas]. Reprint Madrid: Ediciones J. Porrúa Turanzas, 1968.

Dellenbarger, L.E. and E.J. Luzar. 1988. The economics associated with crawfish production from Louisiana's Atchafalaya Basin. Journal of the World Aquaculture Society 19: 41–46.

Demaree, D. 1932. Submerging experiments with *Taxodium*. Ecology 13: 258–262.

Demas, C.R., S.T. Brazelton, and N.J. Powell. 2001. The Atchafalaya Basin—river of trees. US Geological Survey Fact Sheet 021–02.

DeMaster, D.J., B.A. McKee, C.A. Nittrouer, Q. Jiangchu, and C. Quodong. 1985. Rates of sediment accumulation and particle reworking based on radiochemical measurements from continental shelf deposits in the East China Sea. Continental Shelf Research 4: 143–158.

Denes, T.A. and J.M Caffrey. 1988. Changes in seasonal water transport in a Louisiana estuary, Fourleague Bay, Louisiana. Estuaries 11: 184–191.

Department of Conservation of Louisiana. 1920. Coast line and oyster bottoms of Louisiana. Map compiled under the direction of the Department of Conservation of Louisiana.

Dicke, S.G. and J.R. Toliver. 1990. Growth and development of bald-cypress/water tupelo stands under continuous versus seasonal flooding. Forest Ecology and Management 33/34: 523–530.

Donoghue, J.F. 2011. Sea level history of the northern Gulf of Mexico coast and sea level rise scenarios for the near future. Climatic Change 107: 17–33.

Doyle, T.W., B.D. Keeland, L.E. Gorham, and D.J. Johnson. 1995. Structural impact of Hurricane Andrew on the forested wetlands of the Atchafalaya Basin in south Louisiana. Journal of Coastal Research 21: 354–364.

Draut, A.E., G.C. Kineke, O.K. Huh, J.M. Grymes, III, K.A. Westphal, and C.C. Moeller. 2005a. Coastal mudflat accretion under energetic conditions, Louisiana chenier-plain coast. Marine Geology 214: 27–47.

Draut, A.E., G.C. Kineke, D.W. Velasco, M.A. Allison, and R.J. Prime. 2005b. Influence of the Atchafalaya River on recent evolution of the chenier-plain inner continental shelf, northern Gulf of Mexico. Continental Shelf Research 25: 91–112.

DuBarry, A.P., Jr. 1963. Germination of bottomland tree seed while immersed in water. Journal of Forestry 61: 225–226.

Dubois, K.D., D. Lee and J. Veizer. 2010. Isotopic constraints on alkalinity, dissolved organic carbon, and atmospheric carbon dioxide fluxes in the Mississippi River. Journal of Geophysical Research 115: G02018.

Dundee, H.A. and D.A. Rossman. 1989. The Amphibians and Reptiles of Louisiana. Baton Rouge: Louisiana State University Press. 300 p.

Eadie, B.J., B.A. McKee, M.B. Lansing, J.A. Robbins, S. Metz, and J.H. Trefry. 1994. Records of nutrient-enhanced coastal ocean productivity in sediments from the continental shelf. Estuaries 17: 754–765.

Effler, R.S. and R.A. Goyer. 2006. Baldcypress and water tupelo sapling response to multiple stress agents and reforestation implications for Louisiana swamps. Forest Ecology and Management 226: 330–340.

Elsey, R.M. and A.R. Woodward. 2010. American alligator *Alligator mississippiensis*. In S.C. Manolis and C. Stevenson (eds.), Crocodiles: status survey and conservation action plan, 3rd ed. Darwin, Australia: IUCN Crocodile Specialist Group. 4 p. Available http://www.iucncsg.org/pages/Publications.html.

Engel, M.A. 2003. Physicochemical effects on the abundance and distribution of larval fishes in the Atchafalaya River Basin, Louisiana [thesis]. Baton Rouge: Louisiana State University. 133 p.

Eubanks, T. 1994. The status and distribution of the Piping Plover in Texas. Bulletin of the Texas Ornithological Society 27: 19–25.

Evers, D.E., C.E. Sasser, J.G. Gosselink, D.A. Fuller, and J.M. Visser. 1998. The impact of vertebrate herbivores on wetland vegetation in Atchafalaya Bay, Louisiana. Estuaries 21: 1–13.

Ewel, K.C. 1996. Sprouting by pondcypress (*Taxodium distichum* var. *nutans*) after logging. Southern Journal of Applied Forestry 20: 209–213.

Falcini, F., N.S. Khan, L. Macelloni, B.P. Horton, C.B. Lutken, K.L. McKee, R. Santoleri, S. Colella, C. Li, G. Volpe, M. D'Emidio, A. Salusti, and D.J. Jerolmack. 2012. Linking the historic 2011 Mississippi River flood to coastal wetland sedimentation. Nature Geoscience DOI: 10.1038/NGE01615.

Faulkner, S.P., J.L. Chambers, W.H. Conner, R.F. Keim, J.W. Day, Jr., E.S. Gardiner, M.S. Hughes, S.L. King, K.W. McLeod, C.A. Miller, J.A. Nyman, and G.P. Shaffer. 2007. Conservation and use of coastal wetland forests in Louisiana. In: W.H. Conner, T.W. Doyle, and K.W. Krauss (eds.), Ecology of tidal freshwater forested wetlands of the southeastern United States. pp. 447–460.

Faulkner, S.P., P. Bhattarai, Y. Allen, J. Barras, and G. Constant. 2009. Identifying baldcypress-water tupelo regeneration classes in forested wetlands of the Atchafalaya Basin, Louisiana. Wetlands 29: 809–817.

Fisk, H.N. 1944. Geological investigation of the alluvial valley of the lower Mississippi River. US Army Corps of Engineers, Mississippi River Commission. 78 p.

Fisk, H.N. 1952. Geologic investigation of the Atchafalaya Basin and the problem of Mississippi River diversion. US Army Corps of Engineers, Mississippi River Commission, Vicksburg, Mississippi. 145 p.

Fitzpatrick, J.W., M. Lammertink, M.D. Juneau, Jr., T.W. Gallagher, B.R. Harrison, G.M. Sparling, K.V. Rosenbergy, R.W. Rohrbaugh, E.C.H. Swarthout, P.H. Wrege, S.B. Swarthout, M.S. Dantzker, R.A. Charif, T.R. Barksdale, J.V. Remsen, Jr., S.D. Simon, and D. Zollner. 2005. Ivory-billed woodpecker (*Campephilus principalis*) persists in continental North America. Science 308: 1460–1462.

Fontenot, Q.C., D.A. Rutherford, and W.E. Kelso. 2001. Effects of environmental hypoxia associated with the annual flood pulse on the distribution of larval sunfish and shad in the Atchafalaya River Basin, Louisiana. Transactions of the American Fisheries Society 130: 107–116.

Fontenot, W.R. 2004. Atchafalaya Basin Program bird survey final report. Louisiana Department of Natural Resources, Atchafalaya Basin Program, Baton Rouge, Louisiana. 15 p.

Ford, M. and J.A. Nyman. 2011. Preface: an overview of the Atchafalaya River. Hydrobiologia 658: 1–5.

Frazier, D.E. 1967. Recent deltaic deposits of the Mississippi River, their development and chronology. Transactions of the Gulf Coast Association of Geological Societies 17: 287–315.

Freeman, A.M. and H.H. Roberts. 2012. Storm layer deposition on a coastal Louisiana lake bed. Journal of Coastal Research DOI: 10.2112/JCOASTRES-D-12–00026.1.

Fremling, C.R., J.L. Rasmussen, R.E. Sparks, S.P. Cobb, C.F. Bryan, and T.O. Claflin. 1989. Mississippi River fisheries: a case history. Proceedings of the International Large Rivers Symposium, Canadian Special Publication of Fisheries and Aquatic Sciences 106: 309–351.

Froese, R. and D. Pauly (eds). 2011. FishBase. www.fishbase.org, version (12/2011).

Fuller D.A., C.E. Sasser, W.B. Johnson, and J.G. Gosselink. 1985. The effects of herbivory on vegetation on islands in Atchafalaya Bay, LA. 4: 105–113.

Gagliano, S.M. and J.L. van Beek. 1975. Environmental base and management study Atchafalaya Basin, Louisiana. EPA Socioeconomic Environmental Studies Series, EPA-600/5–75–006.

Gagnon, P.F. 2006. Population biology and disturbance ecology of a native North American bamboo (*Arundinaria gigantea*) [dissertation]. Baton Rouge: Louisiana State University. 71 p.

Gastaldo, R.A., D.P. Douglass, and S.M. McCarroll. 1987. Origin, characteristics, and provenance of plant macrodetritus in a Holocene crevasse splay, Mobile Delta, Alabama. Palaios 2: 229–240.

Gelwicks, K.R. 1996. Physicochemical influences on the distribution and abundance of fishes of the Atchafalaya River Basin, Louisiana [thesis]. Baton Rouge: Louisiana State University. 59 p.

Goñi, M.A., Y. Alleau, R. Corbett, J.P. Walsh, D. Mallinson, M.A. Allison, E. Gordon, S. Petsch, and T.M. Dellapenna. 2007. The effects of Hurricanes Katrina and Rita on the seabed of the Louisiana shelf. The Sedimentary Record 5: 4–9.

Goñi, M.A., E.S. Gordon, N.M. Monacci, R. Clinton, R. Gisewhite, M.A. Allison, and G. Kineke. 2006. The effect of Hurricane Lili on the distribution of organic matter along the inner Louisiana shelf (Gulf of Mexico, USA). Continental Shelf Research 26: 2260–2280.

Gonzáles, J.L. and T.E. Törnqvist. 2006. Coastal Louisiana in crisis: subsidence or sea level rise? Eos 87: 493–508.

Goolsby, D.A., W.A. Battaglin, G.B. Lawrence, R.S. Artz, B.T. Aulenbach, R.P. Hooper, D.R. Keeney, and G.J. Stensland. 1999. Flux and sources of nutrients in the Mississippi—Atchafalaya River Basin. Topic 3 Report for the Integrated Assessment on Hypoxia in the Gulf of

Mexico. US Department of Commerce, National Oceanic and Atmospheric Administration, National Ocean Service, Coastal Ocean Program.

Green, M.C., M.C. Luent, T.C. Michot, C.W. Jeske, and P.L. Leberg. 2006. Statewide wading bird and seabird nesting colony inventory, 2004–2005. Louisiana Department of Wildlife and Fisheries, Louisiana Natural Heritage Program Report. Baton Rouge, Louisiana. 158 p.

Green, M.C., P. Leberg, and M. Luent. 2010. Evaluation of aerial sampling methods for the detection of waterbird colonies. Journal of Field Ornithology 81: 411–419.

Gregory, R., T. McDaniels, and D. Fields. 2001. Decision aiding, not dispute resolution: creating insights through structured environmental decisions. Journal of Policy Analysis and Management 20: 415–432.

Grinson, M.M. and A.I. Malvarez. 2002. Temperate freshwater wetlands: types, status, and threats. Environmental Conservation 29: 115–133.

Grisham, B.A., M.J. Chamberlain, and F.G. Kimmel. 2008. Spatial ecology and survival of male wild turkeys in a bottomland hardwood forest. Proceedings of the Southeastern Association of Fish and Wildlife Agencies 62: 70-76.

Gu, B., Y. Bian, C.L. Miller, W. Dong, X. Jiang, and L. Liang. 2011. Mercury reduction and complexation by natural organic matter in anoxic environments. Proceedings of the National Academy of Sciences DOI: 10.1073/pnas.1008747108.

Haase, C. S. 2010. Analysis of flow alteration associated with the Old River Control Structure near Simmesport, LA. Unpublished Project Report, The Nature Conservancy of Louisiana, Baton Rouge, Louisiana.

Hale, L. 1995. A historical view of events surrounding the controversy of the Old River Control Structures. Final Report to Louisiana Cooperative Fish and Wildlife Research Unit, Louisiana State University, Baton Rouge, Lousiana. 41 p.

Halloran, B.T. 2010. Early life history dynamics of the fish community in the Atchafalaya River Basin [dissertation]. Baton Rouge: Louisiana State University. 113 p.

Harms, W.R., H.T. Schreuder, D.D. Hook, and C.L. Brown. 1980. The effects of flooding on the swamp forest in Lake Ocklawaha, Florida. Ecology 61: 1412–1421.

Hart, W.O. 1913. The oyster and fish industry of Louisiana. Transactions of the American Fisheries Society 42: 151–156.

Hebert, K.L. 1966. The flood control capacities of the Atchafalaya Basin Floodway [thesis]. Baton Rouge: Louisiana State University. 97 p.

Hern, S.C., W.D. Taylor, L.R. Williams, V.W. Lambou, M.K. Morris, F.A. Morris, and J.W. Hilgert. 1978. Distribution and importance of phytoplankton in the Atchafalaya Basin. US Environmental Protection Agency, Ecological Research Series, EPA-600/3-78-001.

Hern, S.C., V.W. Lambou, and J.R. Butch. 1980. Descriptive water quality for the Atchafalaya Basin, Louisiana. US Environmental Protection Agency, EPA-600/4-80-014.

Hightower, D.A. 2003. Fine-scaled movements and habitat use of black bears in south central Louisiana [thesis]. Baton Rouge: Louisiana State University. 79 p.

Hightower, D.A., R.O. Wagner, and R.M. Pace, III. 2002. Denning ecology of female American black bears in south central Louisiana. Ursus 13: 11–17.

Hoese, H.D. 1976. Study of sport and commercial fishes of the Atchafalaya Bay region. US Fish and Wildlife Service, Ecological Services. Lafayette, Louisiana.

Holbrook, R.S., F.C. Rohwer, and W.P. Johnson. 2000. Habitat use and productivity of mottled ducks on the Atchafalaya River Delta, Louisiana. Proceedings of the Annual Conference of the Southeastern Association of Fish and Wildlife Agencies 54: 292–303.

Holland, L.E., C.F. Bryan, and J.P. Newman, Jr. 1983. Water quality and the rotifer population in the Atchafalaya River Basin. Hydrobiologia 98: 55–69.

Holloway, H.A., D.R. Lavergne, and B.E. McManus. 1998. Characteristics and attitudes of Louisiana freshwater anglers. Louisiana Department of Wildlife and Fisheries, Baton Rouge, Louisiana.

Holm, G.O., Jr. and C.E. Sasser. 2001. Differential salinity response between two Mississippi River subdeltas: implications for changes in plant composition. Estuaries 24: 78–89.

Hook, D.D. 1984. Adaptations to flooding with fresh water. In: T.T. Kozlowski (ed.), Flooding and plant growth. Orlando, Florida: Academic Press. pp. 265–294.

Hook, D.D. and D.S. DeBell. 1970. Factors influencing stump sprouting of swamp and water tupelo seedlings. US Department of Agriculture Forest Service, Research Paper SE-57. Southeastern Forest Experiment Station, Asheville, North Carolina. 9 p.

Hook, D.D., W.P. Le Grande, and W.P. Langdon. 1967. Stump sprouts on water tupelo. Southern Lumberman 215: 111–112.

Huh, O.K., N.D. Walker, and C. Moeller. 2001. Sedimentation along the eastern Chenier plain coast: down drift impact of a delta complex shift. Journal of Coastal Research 17: 72–81.

Hunter, R.G., J.W. Day, Jr., R.R. Lane, J. Lindsey, J.N. Day, and M.G. Hunter. 2009. Impacts of secondarily treated municipal effluent on a freshwater forested wetland after 60 years of discharge. Wetlands 29: 363–371.

Hupp, C.R. 2000. Hydrology, geomorphology, and vegetation of coastal plain rivers in the southeastern United States. Hydrological Processes 14: 2991–3010.

Hupp, C.R., C.R. Demas, R.H. Day, and D.E. Kroes. 2002. Sediment trapping and carbon sequestration in theAtchafalaya River Basin, Louisiana. Bulletin of the Society of Wetland Scientists, Supplement. 19: S-40.

Hupp, C.R., C.R. Demas, D.E. Kroes, R.H. Day, and T.W. Doyle. 2008. Recent sedimentation patterns within the central Atchafalaya Basin, Louisiana. Wetlands 28: 125–140.

Hupp, C.R. and G.B. Noe. 2006. Sediment and nutrient accumulation within lowland bottomland ecosystems: an example from the Atchafalaya River Basin, Louisiana. In: Proceedings of the international conference on hydrology and management of forested wetlands. ASABE 701P0406. St. Joseph, Michigan: American Society of Agricultural and Biological Engineers. pp. 175–187.

Isaacs, J.C. and D. Lavergne. 2010. Louisiana commercial crawfish harvesters survey report. Louisiana Department of Wildlife and Fisheries, Socioeconomic Research and Development Section, Baton Rouge, Louisiana.

Irwin, E.R. and M.C. Freeman. 2002. Proposal for adaptive management to conserve biotic integrity in a regulated segment of the Tallapoosa River, Alabama, USA. Conservation Biology 16: 1212–1222.

Isphording, W.C., F.D. Imsand, and G.C. Flowers. 1989. Physical characteristics and aging of Gulf Coast estuaries. Transactions of the Gulf Coast Association of Geological Societies 29: 387–401.

Isphording, W.C., F.D. Imsand, and R.B. Jackson. 1996. Fluvial sediment characteristics of the Mobile River delta. Transactions of the Gulf Coast Association of Geological Societies 46: 185–191.

Jaramillo, S., A. Sheremet, M.A. Allison, A.H. Reed, and K.T. Holland. 2009. Wave-mud interactions over a muddy Atchafalaya subaqueous clinoform, Louisiana, United States: wave-supported sediment transport. Journal of Geophysical Research 114: C04002, doi:10.1029/2008JC004821.

Jennings C.A. and S.J. Zigler. 2000. Ecology and biology of paddlefish in North America: historical perspectives, management approaches, and research priorities. Reviews in Fish Biology and Fisheries 10: 167–181.

Joanen, T. and L. McNease. 1989. The management of alligators in Louisiana, USA. In: G.J.W. Webb, S.C. Manolis and P.J. Whitehead (eds.), Wildlife management: crocodiles and alligators. Sydney, Australia: Surrey Beatty and Sons. pp. 33–42.

Johnson, W.B, C.E. Sasser, and J.G. Gosselink. 1985. Succession of vegetation in an evolving river delta, Atchafalaya Bay, Louisiana. Journal of Ecology 73: 973–986.

Johnson, W.P., R.S. Holbrook, and F.C. Rohwer. 2002. Nesting chronology, clutch size, and egg size in the mottled duck. Wildfowl 53: 155–166.

Johnson, W.P. and F.C. Rohwer. 2000. Foraging behavior of green-winged teal and mallards on tidal mudflats in Louisiana. Wetlands 20: 184–188.

Johnson, W.P. and F.C. Rohwer. 2007. Early nesting by blue-winged teal in coastal Louisiana. The Journal of Louisiana Ornithology 7: 89–93.

Johnson, W.P., F.C. Rohwer, and M. Carloss. 1996. Evidence of nest parasitism in mottled ducks. Wilson Bulletin 108: 187–189.

Jubinsky, G. 1994. A review of the literature: Chinese tallow (*Sapium sebiferum*). Florida Department of Environmental Protection, Bureau of Aquatic Plant Management. Technical Services Section, TSS-93–03. 12 pp.

Juneau, C.L., Jr. 1984. Shell dredging in Louisiana 1914–1984. Louisiana Department of Wildlife and Fisheries, Baton Rouge, Louisiana.

Junk, W.J., P.G. Bayley, and R.E. Sparks. 1989. The flood-pulse concept in river-floodplain systems. In: D.P. Dodge (ed.), Proceedings of the international large river symposium. Canadian Special Publication of Fisheries and Aquatic Sciences 106. pp. 110–127.

Justic, D., V.J. Bierman, Jr., D. Scavia, and R.D. Hetland. 2007. Forecasting Gulf's hypoxia: the next 50 years? Estuaries and Coasts 30: 791–801.

Kadlec, R.H. and R.L. Knight. 1996. Treatment wetlands. Boca Raton, Florida. 893 p.

Kaller, M.D., W.E. Kelso, B.T. Halloran, and D.A. Rutherford. 2011. Effects of spatial scale on assessment of dissolved oxygen dynamics in the Atchafalaya River Basin, Louisiana. Hydrobiologia 658: 7–15.

Kammerer, J.C. 1990. Largest rivers in the United States. US Geological Survey, Department of the Interior Water Fact Sheet, Open-file Report 87–242.

Kanouse, S.C. 2003. Nekton use and growth in three brackish marsh pond microhabitats [thesis]. Baton Rouge: Louisiana State University. 67 p.

Kanouse, S., M.K. LaPeyre, and J.A. Nyman. 2006. Nekton use of *Ruppia maritima* and non-vegetated bottom habitat types within brackish marsh ponds. Marine Ecology Progress Series 327: 61–69.

Keeland, B.D. and L.E. Gorham. 2009. Delayed tree mortality in the Atchafalaya Basin of southern Louisiana following Hurricane Andrew. Wetlands 29: 101–111.

Keeland, B.D. and W.H. Conner. 1999. Natural regeneration and growth of *Taxodium distichum* (L.) Rich in Lake Chicot, Louisiana, after 44 years of flooding. Wetlands 19: 149–155.

Keeland, B.D., W.H. Conner, and R.R. Sharitz. 1997. A comparison of wetland tree growth response to hydrologic regime in Louisiana and South Carolina. Forest Ecology and Management 90: 237–250.

Keim, R.F. and J.B. Amos. 2012. Dendochronological analysis of baldcypress (*Taxodium distichum*) responses to climate and contrasting flood regimes. Canadian Journal of Forest Research 42: 423–436.

Keim, R.F. and S.L. King. 2006. Spatial assessment of coastal forest conditions. Louisiana Governor's Office of Coastal Activities, Governor's Applied Coastal Research and Development Program, Final Report. 24 p.

Keim, R.F., J.L. Chambers, M.S. Hughes, J.A. Nyman, C.A. Miller, and J.B. Amos. 2006a. Ecological consequences of changing hydrological conditions in wetland forests of coastal

Louisiana. In: Y.J. Xu and V.P. Singh (eds.), Coastal environment and water quality. Highlands Ranch, Colorado: Water Resources Publications, LLC. pp. 383–396.

Keim, R.F., J.L. Chambers, M.S. Hughes, L.D. Dimov, W.H. Conner, G.P. Shaffer, E.S. Gardiner, and J.W. Day, Jr. 2006b. Long-term success of stump sprouts in high-degraded baldcypress-water tupelo swamps in the Mississippi Delta. Forest Ecology and Management 234: 24–33.

Keim, R.F., T.J. Dean, and J.L. Chambers. 2013. Flooding effects on stand development in cypress-tupelo. In: Proceedings of the 15th biennial southern silvicultural research conference. USDA Forest Service General Technical Report SRS-175. pp. 431–437.

Keim, R.F., C.W. Izdepski, and J.W. Day, Jr. 2012. Growth responses of baldcypress to wastewater nutrient additions and changing hydrologic regime. Wetlands 32: 95–103.

Kemp, G.P. 1986. Mud deposition at the shoreface: wave and sediment dynamics on the chenier plain of Louisiana [dissertation]. Baton Rouge: Louisiana State University. 148 p.

Kemp, G.P., J.T. Wells, and I.L. van Heerden. 1980. Frontal passages affect delta development in Louisiana. Coastal Oceanography and Climatology News 3: 4–5.

Kennedy, H.E., Jr. 1982. Growth and survival of water tupelo coppice regeneration after six growing seasons. Southern Journal of Applied Forestry 6: 133–135.

Keown, M.P., E.A. Dardeau, Jr., and E.M. Causey. 1986. Historic trends in the sediment flow regime of the Mississippi River. Water Resources Research 22: 1555–1564.

Khan, N.S, B.P. Horton, K.L. McKee, D. Jerolmack, F. Falcini, M.D. Enache, and C.H. Vane. 2013. Tracking sedimentation from the historic A.D. 2011 Mississippi River flood in the deltaic wetlands of Louisiana. Geology 41: 391-394.

Killgore, K.J., J.J. Hoover, S.G. George, B.R. Lewis, C.E. Murphy and W.E. Lancaster. 2007. Distribution, relative abundance and movements of pallid sturgeon in the free-flowing Mississippi River. Journal of Applied Ichthyology 23: 476–483.

Kim, W. D. Mohrig, R. Twilley, C. Paola, and G. Parker. 2009. Land building in the delta of the Mississippi River: is it feasible? Eos Transactions. AGU, 90(42) DOI:10.1029/2009E0420001.

Kineke, G.C., E.E. Higgins, K. Hart, and D. Velasco. 2006. Fine-sediment transport associated with cold-front passages on the shallow shelf, Gulf of Mexico. Continental Shelf Research 26: 2073–2091.

King, S.L. 1995. Effects of flooding regimes on two impounded bottomland hardwood stands. Wetlands 15: 272–284.

Kneib R.T. 1997. Early life stages of resident nekton in intertidal marshes. Estuaries 20: 214–230.

Kolb, C.R. and J.R. Van Lopik, 1958, Geology of the Mississippi River deltaic plain, southeastern Louisiana. Technical Report 3–483, US Army Engineer Waterways Experiment Station, Vicksburg, Mississississippi.

Kolker, A.S., M.A. Allison, and S. Hameed. 2011. An evaluation of subsidence rates and sea-level variability in the northern Gulf of Mexico. Geophysical Research Letters 38, L21404. DOI: 10.1029/2011GL049458.

Kozak, J. and B.P. Piazza. In Review. A proposed process for applying structured decision making to restoration planning in the Atchafalaya River Basin, Louisiana, USA. Restoration Ecology.

Kozlowski, T.T. 1997. Responses of Woody Plants to Flooding and Salinity. Tree Physiology Monograph No. 1. Victoria, Canada: Heron Publishing.

Kozlowski, T.T. 2002. Physiological-ecological impacts of flooding on riparian forest ecosystems. Wetlands 22: 550–561.

Kroes, D.E. and C.R. Hupp. 2010. The effect of channelization on floodplain sediment deposition and subsidence along the Pocomoke River, Maryland. Journal of the American Water Resources Association 1–14, DOI: 10.1111/j.1752–1688.2010.00440.x.

Kroes, D.E. and Y.C. Allen 2012. Flow pattern development in the Atchafalaya River Basin. Proceedings, American Water Resources Association, Spring Specialty Conference GIS and Water Resources. http://www.awra.org/proceedings/Spring2012/doc/abs/Daniel%20 E.%20Kroes_d2f47aa7_7856.pdf

Kroes. D.E. and T.F. Kraemer. 2013. Human-induced storm channel abandonment/capture and filling of floodplain channels within the Atchafalaya River Basin, Louisiana. Geomorphology 201: 148–156.

Kuehl, S.A., D.J. DeMaster, and C.A. Nittrouer. 1986. Nature of sediment accumulation on the Amazon continental shelf. Continental Shelf Research 6: 209–225.

Kuehl, S.A., T.D. Pacioni, and J.M. Rine. 1995. Seabed dynamics of the inner Amazon continental shelf: temporal and spatial variability of surficial strata. Marine Geology 125: 283–302.

Kynard, B., E. Henyey., and M. Horgan. 2002. Ontogenetic behavior, migration, and social behavior of pallid sturgeon, *Scaphirhynchus albus*, and shovelnose sturgeon, *S. platorynchus*, with notes on the adaptive significance of body color. Environmental Biology of Fishes 63: 389–403.

Lafon B. 1806. Carte générale du territoire d'Orléans comprenant aussi la Floride Occidentale et une portion du territoire du Mississipi. Dresée d'après les observations les plus récentes par Bmi. Lafon, ingénieur géographe à la N[ouve]lle. Orléans.

Lambou, V.W. 1959. Fish populations of backwater lakes in Louisiana. Transactions of the American Fisheries Society 88: 7–15.

Lambou, V.W. 1961. Utilization of macrocrustaceans for food by fresh-water fishes in Louisiana and its effects on the determination of predator-prey relations. The Progressive Fish Culturist 23: 18–25.

Lambou, V.W. 1963. The commercial and sport fisheries of the Atchafalaya Basin Floodway. Proceedings of the 17th Annual Conference of the Southeastern Association of Game and Fish Commissioners 17: 256–281.

Lambou, V.W. 1990. Importance of bottomland hardwood forest zones to fishes and fisheries: the Atchafalaya Basin, a case history. In: J.G. Gosslink, L.C. Lee, and T.A. Muir (eds.), Ecological processes and cumulative impacts: illustrated by bottomland hardwood wetland ecosystems. Chelsea, Michigan: Lewis Publishers. pp. 125–193.

Lambou, V.W. and S.C. Hern. 1983. Transport of organic carbon in the Atchafalaya Basin, Louisiana. Hydrobiologia 98: 25–34.

Lane, R.R., J.W. Day, B. Marx, E. Reyes, and G.P. Kemp. 2002. Seasonal and spatial water quality changes in the outflow plume of the Atchafalaya River, Louisiana, USA. Estuaries 25: 30–42.

Lane, R.R., J.W. Day, and B. Thibodeaux. 1999. Water quality analysis of a freshwater diversion at Caernarvon, Louisiana. Estuaries 22: 327–336.

Lane, R.R., C.J. Madden, J.W. Day, Jr. and D.J. Solet. 2011. Hydrologic and nutrient dynamics of a coastal bay and wetland receiving discharge from the Atchafalaya River. Hydrobiologia 658: 55–66.

Latch, E.K., D.G. Scognamillo, J.A. Fike, M.J. Chamberlain, and O.E. Rhodes, Jr. 2008. Deciphering ecological barriers to North American river otter (*Lontra canadensis*) gene flow in the Louisiana landscape. Journal of Heredity DOI: 10.1093/jhered/esn009.

Leberg, P.L., P. Deshotels, S.M. Pius, and M. Carloss. 1995. Nest sites of seabirds on spoil islands in coastal Louisiana. Proceedings of the Southeastern Association of Fish and Wildlife Agencies 49: 356–366.

Leberg, P.L, M.R. Carloss, L.J. Dugas, K.L. Pilgrim, L.S. Mills, M.C. Green, and D. Scognamillo. 2004. Recent record of a cougar (*Puma concolor*) in Louisiana, with notes on diet, based on analysis of fecal materials. Southeastern Naturalist 3: 653–658.

Lemlich, S.K. and K.C. Ewel. 1984. Effects of wastewater disposal on growth rates of cypress trees. Journal of Environmental Quality 13: 602–604.

Lester, G.D., S.G. Sorenson, and P.L. Faulkner. 2005. Louisiana comprehensive wildlife conservation strategy. Louisiana Department of Wildlife and Fisheries. Baton Rouge, Louisiana.

Levine, S.J. 1977. Food and feeding habits of juvenile and adults of selected forage, commercial, and sport fishes in the Atchafalaya Basin, Louisiana [thesis]. Baton Rouge: Louisiana State University. 127 p.

Li, C., H. Roberts, G.W. Stone, E. Weeks, and Y. Luo. 2011. Wind surge and saltwater intrusion in Atchafalaya Bay during onshore winds prior to cold front passage. Hydrobiologia 658: 27–39.

Lindau, C.W., R.D. Delaune, A.E. Scaroni, and J.A. Nyman. 2008. Denitrification in cypress swamp within the Atchafalaya River Basin, Louisiana. Chemosphere 70: 886–894.

Lohrenz, S.E., G.L. Fahnenstiel, D.G. Redalje, G.A. Lang, M.J. Dagg, T.E. Whitledge, and Q. Dortch. 1999. Nutrients, irradiance, and mixing as factors regulating primary production in coastal waters impacted by the Mississippi River plume. Continental Shelf Research 19: 1113-1141.

Louisiana Department of Agriculture and Forestry. 2011. Louisiana forestry reports. http://www.ldaf.state.la.us/portal/Offices/Forestry/ForestryReports/tabid/449/Default.aspx).

Louisiana Department of Natural Resources. 1998. Atchafalaya Basin Floodway System, Louisiana State Master Plan. Louisiana Department of Natural Resources, Baton Rouge, Louisiana.

Louisiana Department of Natural Resources. 2002. Water in the basin committee, recommendations to the Governor. Final Report presented to Governor M.J. "Mike" Foster. Louisiana Department of Natural Resources, Atchafalaya Basin Program, Baton Rouge, Louisiana.

Louisiana Department of Natural Resources. 2010. Atchafalaya Basin, FY 2010 annual plan. Atchafalaya Basin Program.

Louisiana Department of Wildlife and Fisheries. 2009. Louisiana's alligator management program, 2008–2009 annual report. Louisiana Department of Wildlife and Fisheries, Office of Wildlife, Coastal and Nongame Resources Division. Baton Rouge, Louisiana.

Louisiana Department of Wildlife and Fisheries. 2010. Oyster stock assessment report of the public oyster areas in Louisiana, seed grounds and seed reservations. Oyster Data Report Series No. 16. July 2010.

Louisiana Department of Wildlife and Fisheries. 2011. Louisiana alligator program. http://www.wlf.louisiana.gov/wildlife/alligator-program.

Louisiana Natural Heritage Program. 2009. The natural communities of Louisiana. Louisiana Department of Wildlife and Fisheries. Baton Rouge, Louisiana.

Louisiana Natural Heritage Program. 2011. Rare animals of Louisiana: Louisiana black bear *Ursus americanus luteolus*. http://www.wlf.louisiana.gov/sites/default/files/pdf/fact_sheet _animal/32305-Ursus%20americanus%20luteolus/ursus_americanus_luteolus.pdf.

Louisiana State University Agricultural Center. 2010. Louisiana summary of agriculture and natural resources—1978 through 2010. Louisiana State University Agricultural Center, Baton Rouge, Louisiana.

Louisiana State University Agricultural Center. 2011. Crawfish. http://www.lsuagcenter.com /en/crops_livestock/aquaculture/crawfish/.

Louisiana Wildlife and Fisheries Commission. 1968. The history and regulation of the shell dredging industry in Louisiana. Baton Rouge, Louisiana.

Lowery, G. H., Jr. 1974a. Louisiana Birds. Baton Rouge: Louisiana State University Press. 681 p.

Lowery, G.H. Jr., 1974b. The Mammals of Louisiana and Its Adjacent Waters. Baton Rouge: Louisiana State University Press. 565 p.

Loyola University Center for Environmental Communication. 2011. Louisiana Old River Control Structure and Mississippi River flood protection. America's Wetland Resource Center. http://www.americaswetlandresources.com/background_facts/detailedstory/Louisiana RiverControl.html.

Luyssaert, S., E. Detlef Schulze, A. Börner, A. Knohl, D. Hessenmöller, B.E. Law, P. Ciais, and J. Grace. 2008. Old-growth forests as global carbon sinks. Nature 455: 213–215.

Majersky-FitzGerald, S. 1998. The development and sand body geometry of the Wax Lake Outlet delta, Atchafalaya Bay, Louisiana [dissertation]. Baton Rouge: Louisiana State University. 130 pp.

Mallach, T.J. and P.L. Leberg. 1999. Use of dredged material substrates by nesting terns and black skimmers. Journal of Wildlife Management 63: 137–146.

Martin, R.P. and G.D. Lester. 1990. Atlas and census of wading bird and seabird nesting colonies in Louisiana. Louisiana Natural Heritage Special Publication No. 3. Louisiana Department of Wildlife and Fisheries, Baton Rouge, Louisiana. 182 p.

Martinez J.D. and W.G. Haag. 1987. The Atchafalaya River and its basin—a field trip. Guidebook Series No.4, Louisiana Geological Survey, Baton Rouge. 22 p.

Marvin-Dipasquale, M., J. Agee, C. McGowan, R.S. Oremland, M. Thomas, D. Krabbenhoft, and C.C. Gilmour. 2000. Methyl-mercury degradation pathways: a comparison among three mercury-impacted ecosystems. Environmental Science and Technology 34: 4908–4916.

Mason, A. and Y.J. Xu. 2006. Spatiotemporal relations of carbon to nitrogen ratio in Atchafalaya River Basin. In: Y.J. Xu and V.P. Singh (eds.) Coastal environment and water quality. Highlands Ranch, Colorado: Water Resources Publications, LLC. pp. 227–236.

Mason, T.D. 2002. The influence of hydrilla infestation and drawdown on the food habits and growth of age-0 largemouth bass in the Atchafalaya River Basin, Louisiana [thesis]. Baton Rouge: Louisiana State University. 101 p.

Mattoon, W.R. 1915. The southern cypress. USDA Forest Service Bulletin No. 272. 74 p.

McClain, W.R. and R.P. Romaire. 2007. Procambarid crawfish: life history and biology. USDA Southern Regional Aquaculture Center Publication No. 2403.

McClain, W.R., R.P. Romaire, C.G. Lutz, and M.G. Shirley. 2007. Louisiana crawfish production manual. Louisiana State University Agricultural Center Publication 2637, Baton Rouge, Louisiana.

McClurkin, D.C. 1965. Diameter growth and phenology of trees on sites with high water tables. US Forest Service Research Note SO-22. Southern Forest Experiment Station, New Orleans, Louisiana.

McIlhenny, E.A. 1941. The passing of the Ivory-Billed Woodpecker. The Auk 58: 582–584.

McKee, B.A., C.A. Nittrouer, and D.J. DeMaster. 1983. Concepts of sediment deposition and accumulation applied to the continental shelf near the mouth of the Yangtze River. Geology 11: 631–633.

McKee, B.A., D.J. DeMaster, and C.A. Nittrouer. 1984. The use of ^{234}Th/^{238}U disequilibrium to examine the fate of particle-reactive species on the Yangtze continental shelf. Earth and Planetary Science Letters 68: 431–442.

McPhee, J. 1989. The Control of Nature. New York: Farrar, Straus Giroux.

Meade, R.H. and J.A. Moody. 2010. Causes for the decline of suspended-sediment discharge in the Mississippi River system 1940–2007. Hydrological Processes 24: 35–49.

Megonigal, J.P., W.H. Conner, S. Kroeger, and R.R. Sharitz. 1997. Aboveground production in southeastern floodplain forests: a test of the subsidy-stress hypothesis. Ecology 78: 370–384.

Messina, M.G. and W.H. Conner (eds.). 1998. Southern Forested Wetlands Ecology and Management. Boca Raton, Florida: CRC Press. 640 p.

Messina, M.G., D.J. Frederick, and A. Clark. 1986. Nutrient content distribution in southern coastal plain hardwoods. Biomass 10: 59–79.

Middleton, B.A. and K.L. McKee. 2004. Use of a latitudinal gradient in baldcypress production to examine physiological controls on biotic boundaries and potential responses to environmental change. Global Ecology and Biology 13: 247–258.

Middleton, B.A. and K.L. McKee. 2005. Primary production in an impounded baldcypress swamp (*Taxodium distichum*) at the northern limit of its range. Wetlands Ecology and Management 13: 15–24.

Mississippi River/Gulf of Mexico Watershed Nutrient Task Force. 2008. Gulf hypoxia action plan 2008 for reducing, mitigating, and controlling hypoxia in the northern Gulf of Mexico and improving water quality in the Mississippi River Basin. Washington, D.C.

Mitchell, D.S. 1976. The growth and management of *Eichhornia crassipes* and *Salvinia* spp. in their native environment and in alien situations. In C.K. Varshney and J. Rzoska (eds.), Aquatic weeds in Southeast Asia. The Hague, Netherlands: Dr. W. Junk, Publisher. pp. 167–175.

Mitsch, W.J., J.W. Day, J.W. Gilliam, P.M. Groffman, D.L. Hey, G.W. Randall, and N. Wang. 2001. Reducing nutrient loading to the Gulf of Mexico from the Mississippi River Basin: strategies to counter a persistent ecological problem. Bioscience 51: 373–388.

Mitsch, W.J. and K.C. Ewel. 1979. Comparative biomass and growth of cypress in Florida wetlands. American Midland Naturalist 101: 417–426.

Mitsch, W.J. and J.G. Gosselink. 1993. Wetlands, 2nd ed. New York: Van Nostrand Reinhold. 721 p.

Momot, W.T., H. Gowing, and P.D. Jones. 1978. The dynamics of crawfish and their role in ecosystems. American Midland Naturalist 99: 10–35.

Moorman, A.M., T.E. Moorman, G.A. Baldassarre, and D.M. Richard. 1991. Effects of saline water on growth and survival of mottled duck ducklings in Louisiana. Journal of Wildlife Management 55: 471–476.

Moorman, T.E. and P.N. Gray. 1994. Mottled duck (*Anas fulvigula*). In: A. Poole and F. Gill (eds.).The birds of North America, No. 81. Philadelphia: The Birds of North America, Inc.

Morey, S.L., P.J. Martin, J.J. O'Brien, A.A. Wallcraft, and J. Zavala-Hidalgo. 2003. Export pathways for river discharged fresh water in the northern Gulf of Mexico. Journal of Geophysical Research 108: C10, 3303, DOI: 10.1029/2002JC001674.

Morgan, J.P. and P.B. Larimore. 1957. Change in the Louisiana shoreline. Transactions of the Gulf Coast Association of Geological Societies 7: 303–310.

Morgan, J.P., J.R. Van Lopik, and J.G. Nichols. 1953. Occurrence and development of mudflats along the western Louisiana coast. Louisiana State University, Coastal Studies Institute Technical Report 2.

Moss, R.E., J.W. Scanlan, and C.S. Anderson. 1983. Observations on the Natural history of the blue sucker (*Cycleptus elongatus* Le Sueur) in the Neosho River. American Midland Naturalist 109: 15–22.

Mossa, J. 1996. Sediment dynamics in the lowermost Mississippi River. Engineering Geology 45: 457–479.

Mossa, J. and H.H. Roberts. 1990. Synergism of riverine and winter storm-related sediment transport processes in Louisiana's coastal wetlands. Transactions of the Gulf Coast Association of Geological Societies 40: 635–642.

National Audubon Society. 2011a. Important bird area site report for Atchafalaya Basin. http://iba.audubon.org/iba/profileReport.do?siteId=3015.

National Audubon Society. 2011b. Important bird area site report for Atchafalaya Delta. http://iba.audubon.org/iba/profileReport.do?siteId=3272.

NatureServe. 2011. NatureServe Explorer: An online encyclopedia of life [web application]. Version 7.1. NatureServe, Arlington, Virginia. Available http://www.natureserve.org/explorer.

Neill, C.F. and M.A. Allison. 2005. Subaqueous deltaic formation on the Atchafalaya shelf, Louisiana. Marine Geology 214: 411–430.

Nessel, J.K. and S.E. Bayley. 1984. Distribution and dynamics of organic matter and phosphorus in a sewage enriched cypress swamp. In: K.C. Ewel and H.T. Odum (eds.), Cypress swamps. Gainesville, Florida: University of Florida Press. pp. 262–278.

Nichols, M.M. 1989. Sediment accumulation rates and relative sea-level rise in lagoons. Marine Geology 88: 201–219.

Nyland, P.D. 1995. Black bear habitat relationships in coastal Louisiana [thesis]. Baton Rouge: Louisiana State University.

Odum, E.P. 2000. Tidal marshes as outwelling/pulsing systems. In: M.P. Weinstein and D.A. Kreeger (eds.), Concepts and controversies in tidal marsh ecology. Dordrecht, The Netherlands: Kluwer Academic Publishers. pp. 3–7.

Odum, W.E., E.P. Odum, and H.T. Odum. 1995. Nature's pulsing paradigm. Estuaries 18: 547–555.

Olivier, T.J. and R.T. Bauer. 2011. Female downstream-hatching migration of the river shrimp *Macrobrachium ohione* in the lower Mississippi River and the Atchafalaya River. American Midland Naturalist 166: 379–393.

Olivier, T.J., S.L. Conner, and R.T. Bauer. 2012. Evidence of extended marine planktonic larval development in far-upstream populations of the river shrimp *Macrobrachium ohione* (Smith, 1874) from the Mississippi River. Journal of Crustacean Biology 32: 899-905.

Orem, W.H., R.J. Rosenbauer, P.W. Swarzenski, H.E. Lerch, M.D. Corum, and A.L. Bates. 2007. Organic geochemistry of sediments in nearshore areas of the Mississippi and Atchafalaya Rivers: I. general organic characterization. US Geological Survey Open-File Report 2207–1180.

Orlando, S.P., Jr., L.P. Rozas, G.H. Ward, and C.J. Klein. 1993. Salinity characteristics of Gulf of Mexico estuaries. National Oceanic and Atmospheric Administration, Office of Ocean Resources Conservation and Assessment, Silver Spring, Maryland.

Pace, R.M., III. 2000. Winter survival rates of American Woodcock in south central Louisiana. Journal of Wildlife Management 64: 933–939.

Penn, G.H., Jr. 1943. A study of the life history of the Louisiana red-crawfish, *Cambarus clarkii* Girard. Ecology 24: 1–18.

Penn, G.H., Jr. 1956. The genus *Procambarus* in Louisiana (Decapoda, Astacidae). American Midland Naturalist 56: 406–422.

Perez, B.C., J.W. Day, Jr., L.J. Rouse, R.F. Shaw, and M. Wang. 2000. Influence of Atchafalaya River discharge and winter frontal passage on suspended sediment concentration and flux in Fourleague Bay, Louisiana. Estuarine, Coastal and Shelf Science 50: 271–290.

Perez, B.C., J.W. Day, Jr., D. Justic, R.R. Lane, and R.R. Twilley. 2011. Nutrient stoichiometry, freshwater residence time, and nutrient retention in a river-dominated estuary in the Mississippi Delta. Hydrobiologia 658: 41–54.

Perez, B.C., J.W. Day, Jr., D. Justic, and R.R. Twilley. 2003. Nitrogen and phosphorus transport between Fourleague Bay, LA, and the Gulf of Mexico: the role of winter cold fronts and Atchafalaya River discharge. Estuarine, Coastal and Shelf Science 57: 1065–1078.

Perret, A.J., M.D. Kaller, W.E. Kelso, and D.A. Rutherford. 2010. Effects of hurricanes Katrina and Rita on sport fish community abundance in the eastern Atchafalaya Basin, Louisiana. North American Journal of Fisheries Management 30: 511–517.

Perret, W.S. and C.W. Caillouet. 1974. Abundance and size of fishes taken by trawling in Vermilion Bay, Louisiana. Bulletin of Marine Science 24: 52–75.

Peterson, G.W., R.F. Shaw, B.J. Milan, and B.A. Thompson. 2008. Comparing fish assemblages between natural (Wax Lake) and altered (Atchafalaya River) deltas as a means of evaluating aquatic habitat quality in coastal reconstruction/restoration projects. Final Report, Louisiana Department of Wildlife and Fisheries, Sport Fish Restoration Program, Baton Rouge, Louisiana.

Pezeshki, S.R. 1990. A comparative study of the response of *Taxodium distichum* and *Nyssa aquatica* seedlings to soil anaerobisis and salinity. Forest Ecology and Management 33/34: 531–541.

Pezeshki, S.R., R.D. Delaune, and W.H. Patrick, Jr. 1990. Flooding and saltwater intrusion: potential effects on survival and productivity of wetland forests along the US Gulf Coast. Forest Ecology and Management 33/34: 287–301.

Piazza, B.P. 1997. Effects of within-season nesting dynamics on estimation of rookery size in wading bird rookeries in the Atchafalaya Delta, Louisiana [thesis]. Baton Rouge: Louisiana State University. 119 p.

Piazza, B.P., Y.C. Allen, R. Martin, J.F. Bergan, K. King, and R. Jacob. In Review. Floodplain conservation in the Mississippi River Valley—combining spatial analysis, landowner outreach, and market assessment to enhance land protection for the Atchafalaya River Basin, Louisiana, USA. Restoration Ecology.

Piazza, B.P., P.D. Banks, and M.K. La Peyre. 2005. The potential for created oyster reefs as a sustainable shoreline protection strategy in Louisiana. Restoration Ecology 13: 1–8.

Piazza, B.P. and M.K. La Peyre. 2007. Restoration of the annual flood pulse in Breton Sound, Louisiana, USA: habitat change and nekton community response. Aquatic Biology 1: 109–119.

Piazza, B.P. and M.K. La Peyre. 2009. The effect of Hurricane Katrina on nekton communities in the tidal freshwater marshes of Breton Sound, Louisiana, USA. Estuarine, Coastal and Shelf Science 83: 97–104.

Piazza, B.P. and M.K. La Peyre. 2010. Using *Gambusia affinis* growth and condition to assess estuarine habitat quality: a comparison of indices. Marine Ecology Progress Series 412: 231–245.

Piazza, B.P. and M.K. La Peyre. 2011. Nekton community response to a large-scale Mississippi River discharge: examining spatial and temporal response to river management. Estuarine, Coastal and Shelf Science 91: 379–387.

Piazza B.P. and M.K. La Peyre. 2012. Measuring changes in consumer resource availability to riverine pulsing in Breton Sound, Louisiana, USA. PLoS ONE 7(5): e37536. doi:10.1371/journal.pone.0037536.

Piazza, B.P. and V.L. Wright. 2004. Within-season nest persistence in large wading bird rookeries. Waterbirds 27: 362–367.

Pollard, J.E., S.M. Melancon and L.S. Blakey. 1983. Importance of bottomland hardwoods to crawfish and fish in the Henderson Lake area, Atchafalaya Basin, Louisiana. Wetlands 3: 1–25.

Pulliam, W.M. 1992. Methane emissions from cypress knees in a southeastern floodplain swamp. Oecologia 91: 126–128.

Rabalais, N.N. and R.E. Turner. 2001. Hypoxia in the northern Gulf of Mexico: description, causes, and change. In N.N. Rabalais, and R.E. Turner (eds.), Coastal Hypoxia: Consequences for Living Resources and Ecosystems. Coastal and Estuarine Studies, No. 58. Washington, D.C.: American Geophysical Union. pp. 1–36.

Rabalais, N.N., R.E. Turner, and W.J. Wiseman, Jr. 2002. Gulf of Mexico hypoxia, a.k.a. "The Dead Zone." Annual Review of Ecology and Systematics. 33: 235–263.

Rayner, D.A. 1976. A monograph concerning the water elm *Planera aquatica* (Walt.) J.F. Gmelin (Ulmaceae) [dissertation]. Columbia: University of South Carolina. 172 p.

Reed, B.C., W.E. Kelso, D.A. Rutherford. 1992. Growth, fecundity, and mortality of paddlefish in Louisiana. Transactions of the American Fisheries Society 121: 378–384.

Remsen, J.V., Jr. 1986. Was Bachman's warbler a bamboo specialist? The Auk 103: 216–219.

Reuss, M. 1982. The Army Corps of Engineers and flood-control politics on the lower Mississippi. Louisiana History 23: 131–148.

Reuss, M. 2004. Designing the Bayous: The Control of water in the Atchafalaya Basin 1800–1995. College Station: Texas A&M University Press. 496 p.

Rice, G.E., D.B. Senn, and J.P. Shine. 2009. Relative importance of atmospheric and riverine mercury sources to the northern Gulf of Mexico. Environmental Science and Technology 43: 415–422.

Roberts, H.H. 1997. Dynamic changes of the Holocene Mississippi River delta plain: the delta cycle. Journal of Coastal Research 13: 605–627.

Roberts, H.H. 1998. Delta switching: early responses of the Atchafalaya diversion. Journal of Coastal Research 14: 882.

Roberts, H.H., R.D. Adams, and R.H.W. Cunningham. 1980. Evolution of sand-dominant subaerial phase, Atchafalaya Delta, Louisiana. The American Association of Petroleum Geologists Bulletin 64: 264–279.

Roberts, H.H., S. Bentley, J.M. Coleman, S.A. Hsu, O.K. Huh, K. Rotondo, M. Inoue, L.J. Rouse, Jr., A. Sheremet, G. Stone, N. Walker, S. Welsh, and W.J. Wiseman, Jr. 2002. Geological framework and sedimentology of recent mud deposition on the eastern chenier plain coast and adjacent inner shelf, western Louisiana. Transactions of the Gulf Association of Geological Societies 52: 849–859.

Roberts, H.H. and J.M. Coleman. 1996. Holocene evolution of the deltaic plain: a perspective—from Fisk to present. Engineering Geology 45: 113–138.

Roberts, H.H., J.M. Coleman, S.J. Bentley, and N. Walker. 2003. An embryonic major delta lobe: a new generation of delta studies in the Atchafalaya-Wax Lake Delta system. GCAGS/GCSSEPM Transactions 53: 690–703.

Roberts, H.H., O.K. Huh, S.A. Hsu, L.J. Rouse, Jr., and D.A. Rickman. 1989. Winter storm impacts on the Chenier Plain coast of southwestern Louisiana: Transactions of the Gulf Coast Association of Geological Societies 39: 512–522.

Roberts, H.H., N. Walker, R. Cunningham, G.P. Kemp, and S. Majersky. 1997. Evolution of sedimentary architecture and surface morphology: Atchafalaya and Wax Lake deltas, Louisiana (1973–1984). Transactions of the Gulf Coast Association of Geological Societies 47: 477–484.

Rodriguez, A.B., A.R. Simms, and J.B. Anderson. 2010. Bay-head deltas across the northern Gulf of Mexico back step in response to the 8.2 ka cooling event. Quaternary Science Reviews 29: 3983–3993.

Rome, N.E., S.L. Conner, and R.T. Bauer. 2009. Delivery of hatching larvae to estuaries by an amphidromous river shrimp: tests of hypotheses based on larval moulting and distribution. Freshwater Biology 54: 1924–1932.

Rosen, T. and Y.J. Xu. 2013. Recent decadal growth of the Atchafalaya River Delta complex: effects of variable riverine sediment input and vegetation succession. Geomorphology 194: 108-120.

Rosson, J.F., Jr., W.H. McWilliams, and P.D. Frey. 1988. Forest resources of Louisiana. US Department of Agriculture, Forest Service, Southern Forest Experiment Station Resource Bulletin SO-130. New Orleans, Louisiana.

Rouse, L.J., H.H. Roberts, R.D. Adams, and R.H.W. Cunningham. 1978. Satellite observations of the subaerial growth of the Atchafalaya Delta, Louisiana. Geology 6: 405–408.

Rozas, L.P. and W.E. Odum. 1988. Occupation of submerged aquatic vegetation by fishes: testing the roles of food and refuge. Oecologia 77: 101–106.

Rutherford, D.A., K.R. Gelwicks, and W.E. Kelso. 2001. Physicochemical effects of the flood pulse on fishes in the Atchafalaya River Basin, Louisiana. Transactions of the American Fisheries Society 130: 276–288.

Sabo, M.J., C.F. Bryan, W.E. Kelso, and D.A. Rutherford. 1999a. Hydrology and aquatic habitat characteristics of a riverine swamp: I. influence of flow on water temperature and chemistry. Regulated Rivers: Research and Management 15: 505–523.

Sabo, M.J., C.F. Bryan, W.E. Kelso, and D.A. Rutherford. 1999b. Hydrology and aquatic habitat characteristics of a riverine swamp: II. hydrology and the occurrence of chronic hypoxia. Regulated Rivers: Research and Management 15: 525–542.

Sager, D.R. and C.F. Bryan. 1981. Temporal and spatial distribution of phytoplankton in the lower Atchafalaya River Basin, Louisiana. In: L.A. Krumholz, C.F. Bryan, G.E. Hall, and G.T. Pardue (eds.), The warmwater streams symposium. Bethesda, Maryland: American Fisheries Society. pp. 91–101.

Sager, W.W., W.W. Schroeder, J.S. Laswell, K.S. Davis, R. Rezak, and S.R. Gittings. 1992. Mississippi—Alabama outer continental shelf topographic features formed during the late Pleistocene—Holocene transgression. Geo-Marine Letters 12: 41–48.

Sasser, C.E. and D.A. Fuller. 1988. Vegetation and waterfowl use of islands in Atchafalaya Bay. Final report submitted to Louisiana Board of Regents, Baton Rouge, Louisiana. Contract No. 86-LBR/018-B04.

Sasser, C.E., J.G. Gosselink, G.O. Holm, Jr., and J.M. Visser. 2009. Tidal freshwater wetlands of the Mississippi River deltas. In: A. Barendregt, D.F. Whigham, and A.H. Baldwin (eds.), Tidal freshwater wetlands. Leiden, The Netherlands: Backhuys Publishers. pp. 167–178.

Scaroni, A.E. 2011. The effect of habitat change on nutrient removal in the Atchafalaya River Basin, Louisiana [dissertation]. Baton Rouge: Louisiana State University. 122 p.

Scaroni, A.E., C.W. Lindau, and J.A. Nyman. 2010. Spatial variability of sediment denitrification across the Atchafalaya River Basin, Louisiana, USA. Wetlands: 30: 949–955.

Scaroni, A.E., J.A. Nyman, and C.W. Lindau. 2011. Comparison of denitrification characteristics among three habitat types of a large river floodplain: Atchafalaya River Basin, Louisiana. Hydrobiologia 658: 17–25.

Scyphers, S.B., S.P. Powers, K.L. Heck, Jr., and D. Byron. 2011. Oyster reefs as natural breakwaters mitigate shoreline loss and facilitate fisheries. PLoS ONE 6(8): e22396. DOI: 10.1371/journal.pone.0022396.

Shaffer, G.P., C.E. Sasser, J.G. Gosselink, and M. Rejmánek. 1992. Vegetation dynamics in the emerging Atchafalaya Delta, Louisiana, USA. Journal of Ecology 80: 677–687.

Shaffer, G.P., W.B. Wood, S.S. Hoeppner, T.E. Perkins, J. Zoller, and D. Kandalepas. 2009. Degradation of baldcypress-water tupelo swamp to marsh and open water in southeastern Louisiana, USA.: an irreversible trajectory? Journal of Coastal Research 54: 152–465.

Shen, Y., C.G. Fichot, and R. Benner. 2012. Floodplain influence on dissolved organic matter composition and export from the Mississippi–Atchafalaya river system. Limnology and Oceanography 57: 1149–1160.

Sheremet, A. and G.W. Stone. 2003. Observations of nearshore wave dissipation over muddy sea beds. Journal of Geophysical Research 108: C11, 3357, DOI: 10.1029/2003JC001885.

Shoch, D.T., G. Kaster, A. Hohl, and R. Souter. 2009. Carbon storage of bottomland hardwood afforestation in the lower Mississippi Valley, USA. Wetlands 29: 535–542.

Sibley, D.A., L.R. Bevier, M.A. Patten and C.S. Elphick. 2006. Comment on "Ivory-billed Woodpecker (*Campephilus principalis*) persists in continental North America." Science 311: 1555.

Smith, N.A. 2004. Feeding ecology and morphometric analysis of paddlefish *Polydon spathula* in the Mermentau River, Louisiana [thesis]. Baton Rouge: Louisiana State University. 76 p.

Smith, S.M. and G.L. Hitchcock. 1994. Nutrient enrichments and phytoplankton growth in the surface waters of the Louisiana Bight. Estuaries 17: 740-753.

Snedden, G.A., W.E. Kelso, and D.A. Rutherford. 1999. Diel and seasonal patterns of spotted gar movement and habitat use in the lower Atchafalaya River Basin, Louisiana. Transactions of the American Fisheries Society 128: 144–154.

Sommerfield, C.K. and C.A. Nittrouer. 1999. Modern accumulation rates and a sediment budget for the Eel shelf: a flood-dominated depositional environment, Marine Geology 154: 227–241.

Sommerfield, C.K., C.A. Nittrouer, and C.R. Alexander. 1999. [7]Be as a tracer of flood sedimentation on the northern California continental margin. Continental Shelf Research 19: 335–361.

Souther, R.F. and G.P. Shaffer. 2000. The effects of submergence and light on two age classes of baldcypress (*Taxodium distichum* (L.) Richard) seedlings. Wetlands 20: 697–706.

Sparks, R.E. 1992. The Atchafalaya Basin. In: S. Maurizi and F. Poillon (eds.), Restoration of aquatic ecosystems. Washington, D.C.: National Academy Press. pp. 398–405.

Sprague, L.A., R.M. Hirsch, and B.T. Aulenbach. 2011. Nitrate in the Mississippi River and its tributaries, 1980 to 2008: are we making progress? Environmental Science and Technology 45: 7209–7216.

Stanley, D.J. and A.G. Warne. 1994. Worldwide initiation of Holocene marine deltas by deceleration of sea-level rise. Science 265: 228–231.

State of Louisiana. 2012. Louisiana's comprehensive master plan for a sustainable coast: committed to our coast. Baton Rouge.

State of the Birds. 2009. The state of the birds: United States of America. www.stateofthebirds.org.

State of the Birds. 2010. The state of the birds: 2010 report on climate change. www.stateofthebirds.org.

Stone, G.W., X. Zhang, and A. Sheremet. 2005. The role of barrier islands, muddy shelf and reefs in mitigating the wave field along coastal Louisiana. Journal of Coastal Research 44: 40–55.

Straw, J.A., Jr., D.G. Krementz, M.W. Olinde, and G.F. Sepik. 1984. American woodcock. In: T.C. Tacha and C.E. Braun (eds.), Migratory shore and upland game bird management in North America. Washington, D.C.: The International Association of Fish and Wildlife Agencies. pp. 97–114.

Sunderland, E.M., D.P. Krabbenhoft, J.W. Moreau, S.A. Strode, and W.M. Landing. 2009. Mercury sources, distribution, and bioavailability in the North Pacific Ocean: insights from data and models. Global Biogeochemical Cycles 23, GB2010, DOI: 10.1029/2008GB003425.

Sutton, D.L., T.K. Van, and K.M. Portier. 1992. Growth of a dioecious and monoecious hydrilla from single tubers. Journal of Aquatic Plant Management 30: 15–20.

Swarzenski, C.M. 2003. Surface-water hydrology of the Gulf Intracoastal Waterway in south-central Louisiana, 1996–99. US Geological Survey Professional Paper 1672.

Swarzenski, P.W. 2001. Evaluating basin/shelf effects in the delivery of sediment-hosted contaminants in the Atchafalaya and Mississippi River Deltas—a new US Geological Survey coastal and marine geology project. USGS Open-File Report 01-215.

Sylvan, J.B. Q. Dortch, D.M. Nelson, A.F. Maier Brown, W. Morrison, and J.W. Ammerman. 2006. Phosphorus limits phytoplankton growth on the Louisiana shelf during the period of hypoxia formation. Environmental Science and Technology 40: 7548–7553.

Tanner, J.T. 1942. The ivory-billed woodpecker. National Audubon Society Research Report No. 1, National Audubon Society, New York.

The Nature Conservancy. 2002. Conservation planning in the Mississippi River alluvial plain. The Nature Conservancy, Baton Rouge, Louisiana.

Thompson, B.A. and L.A. Deegan. 1983. The Atchafalaya River Delta: A "new" fishery nursery, with recommendations for management. In F.J. Webb (ed.), Proceedings of the tenth annual conference on wetland restoration and creation. Tampa, Florida: Hillsborough Community College. pp. 217–239.

Thompson, B.A. and G.W. Peterson. 2001. Examining artificial wetlands: are we building suitable habitat for Fundulus jenkinsi, a federal candidate species? Final Report National Fish and Wildlife Foundation, Washington, D.C.

Thompson, B.A. and G.W. Peterson. 2003. Evaluating sportfish use of habitat created by coastal restoration projects. Louisiana Department of Wildlife and Fisheries, Wallop-Breaux Program, Final Report.

Thompson, B.A. and G.W. Peterson. 2006. Evaluating sportfish use of habitat created by coastal restoration projects: II. Can we make post-construction alterations to these habitats to make them more suitable for fish life cycles? Louisiana Department of Wildlife and Fisheries, Wallop-Breaux Program, Final Report.

Thomson, D.A., G.P. Shaffer, and J.A. McCorquodale. 2002. A potential interaction between sea level rise and global warming: implications for coastal stability on the Mississippi River deltaic plain. Global Planetary Change 32: 49–59.

Thompson, W.C. 1951. Oceanographic analysis of Atchafalaya Bay, Louisiana and adjacent continental shelf area marine pipeline problems. Texas A&M Foundation, Department of Oceanography, Sec. 2, Project 25. 31 pp.

Thompson, W.C. 1955. Sandless coastal terrain of the Atchafalaya Bay area, Louisiana. SEPM Special Publication No. 3: 52–76.

Tockner, K. and J.A. Stanford. 2002. Riverine flood plains: present state and future trends. Environmental Conservation 29: 308–330.

Tri-Parish Partnership. 2009. Atchafalaya East Watershed (upper Terrebonne Basin), phase 2A, detailed problem identification and technical evaluation, Iberville Parish, Pointe Coupee Parish, West Baton Rouge Parish. 58050 Meriam Street, P.O. Box 389, Plaquemine, Louisiana 70765.

Troutman, J.P., D.A. Rutherford, and W.E. Kelso. 2007. Patterns of habitat use among vegetation-dwelling littoral fishes in the Atchafalaya River Basin, Louisiana. Transactions of the American Fisheries Society 136: 1063–1075.

Truesdale, F.M. and W.J. Mermilliod. 1979. The river shrimp *Macrobrachium ohione* (Smith) (Decapoda, Paleaemonidae): its abundance, reproduction, and growth in the Atchafalaya Basin of Louisiana. Crustaceana 36: 61–73.

Turner, R.E., J.J. Baustian, E.M. Swenson, J.S. Spicer. 2006. Wetland sedimentation from hurricanes Katrina and Rita. Science 10.1126/science.1129116.

Turner, R.E., N.N. Rabalais, R.B. Alexander, G. McIsaac, and R.W. Howarth. 2007. Characterization of nutrient, organic carbon, and sediment loads and concentrations from the Mississippi River into the northern Gulf of Mexico. Estuaries and Coasts 30: 773–790.

Tweel, A.W. and R.E. Turner. 2011. Watershed land use and river engineering drive wetland formation and loss in the Mississippi River birdfoot delta. Limnology and Oceanography 57: 18–28.

Tweel, A.W. and R.E. Turner. 2012. Landscape-scale analysis of wetland sediment deposition from four tropical cyclone events. PLoS ONE 7(11): e50528.DOI: 10.1371/journal.pone.0050528.

Tye, R.S. and J.M. Coleman. 1989a. Evolution of Atchafalaya lacustrine deltas, south-central Louisiana. Sedimentary Geology 65: 95–112.

Tye, R.S. and J.M. Coleman. 1989b. Depositional processes and stratigraphy of fluvially dominated lacustrine deltas: Mississippi Delta plain. Journal of Sedimentary Petrology 59: 973–996.

Urbatsch, L. 2000. Chinese tallow tree *Triadica sebifera* (L.) Small. USDA Natural Resources Conservation Service Plant Guide.

US Army Corps of Engineers 1982. Atchafalaya Basin Floodway System final report and environmental impact statement. US Army Corps of Engineers Report submitted to the Mississippi River Commission.

US Army Corps of Engineers. 1987. Oyster shell dredging in Atchafalaya Bay and adjacent waters, Louisiana: final environmental impact statement and appendixes. US Army Corps of Engineers, New Orleans District.

US Army Corps of Engineers. 1993. Oyster shell dredging in Gulf of Mexico waters, St. Mary and Terrebonne parishes, Louisiana. Draft Environmental Impact Statement and Appendices. US Army Corps of Engineers, New Orleans District.

US Army Corps of Engineers. 1998. Atchafalaya Basin. In: M.H. Saucier (ed.), Water resources development in Louisiana. New Orleans: US Army Corps of Engineers. pp. 36–61.

US Army Corps of Engineers. 2000. Atchafalaya Basin Floodway System project, Louisiana master plan. US Army Corps of Engineers, New Orleans District.

US Army Corps of Engineers. 2001. Atchafalaya Basin Floodway System: a project of the US Army Corps of Engineers in Louisiana. US Army Corps of Engineers Fact Sheet, New Orleans District.

US Army Corps of Engineers. 2009. Old River control. US Army Corps of Engineers, New Orleans District.

US Army Corps of Engineers. 2010. Dredging schedule. http://operations.usace.army.mil /nav/090ctweda/FY10MVNDredgingSlideshow.pdf.

US Army Corps of Engineers. 2011. The Atchafalaya Basin Project. US Army Corps of Engineers Fact Sheet.

US Environmental Protection Agency. 2007. Status and life history of the pallid sturgeon. US Environmental Protection Agency, Office of Pesticide Programs. 16 pp. http://www.epa .gov/espp/litstatus/effects/appendix_c_life_history_sturgeon.pdf

US Fish and Wildlife Service. 1993. Pallid sturgeon recovery plan. US Fish and Wildlife Service, Bismarck, North Dakota.

US Fish and Wildlife Service. 1995. Louisiana Black Bear Recovery Plan. Jackson, Mississippi.

US Fish and Wildlife Service. 2006. Atchafalaya National Wildlife Refuge bird list. US Fish and Wildlife Service, Southeast Louisiana Refuges, Lacombe, Louisiana.

US Geological Survey. 2001. The Atchafalaya Basin—river of trees. USGS Fact Sheet 021–02.

US Geological Survey. 2004. Latitudinal variation in carbon storage can help predict changes in swamps affected by global warming. US Department of the Interior, US Geological Survey Fact Sheet 2004–3019.

US Geological Survey. 2011. Upper Midwest Environmental Sciences Center paddlefish study project. http://www.umesc.usgs.gov/aquatic/fish/paddlefish/main.html.

van Beek, J.L., A.L. Harmon, C.L. Wax, and K.M. Wicker. 1979. Operation of the Old River Control Project, Atchafalaya Basin: an evaluation from a multiuse management standpoint. US Environmental Protection Agency, Office of Research and Development, Research Report EPA-600/4–79–073.

van Beek, J.L., W.G. Smith, J.W. Smith, and P. Light. 1977. Plan and concepts for multi-use management of the Atchafalaya Basin. US Environmental Protection Agency, Office of Research and Development Report. EPA-600/3–77–062.

van Heerden, I.L. 1983. Deltaic sedimentation in eastern Atchafalaya Bay, Louisiana [dissertation]. Baton Rouge: Louisiana State University. 117 p.

van Heerden, I.L. 2001. The dynamic Atchafalaya—an essay. Report prepared for The Atchafalaya Trace Program, Louisiana Department of Culture, Recreation and Tourism, Baton Rouge, Louisiana.

van Heerden, I.L. and H.H. Roberts. 1980. The Atchafalaya Delta—Louisiana's new prograding coast. Transactions of the Gulf Coast Association of Geological Societies 30: 497–506.

van Heerden, I.L. and H.H. Roberts. 1988. Facies development of Atchafalaya Delta, Louisiana: a modern bayhead delta. The American Association of Petroleum Geologists Bulletin 72: 439–453.

van Maasakkers, M.J. 2009. Environmental restoration in the Atchafalaya Basin: boundaries and interventions [thesis]. Cambridge: Massachusetts Institute of Technology. 72 p.

Viosca, P., Jr. 1927. Flood control in the Mississippi valley in its relation to Louisiana fisheries. Transactions of the American Fisheries Society 57: 49–64.

Viosca, P., Jr. 1928. Louisiana wetlands and the value of their wildlife and fishery resources. Ecology 9: 216–229.

Visser, J.M. and C.E. Sasser. 1995. Changes in tree-species composition, structure and growth in a baldcypress-water tupelo swamp forest, 1980–1990. Forest Ecology and Management 72: 119–129.

Wagner, R.O. 1995. Movement patterns of black bears in south central Louisiana [thesis]. Baton Rouge: Louisiana State University. 57 p.

Wagner, R.O. 2003. Developing landscape-scaled habitat selection functions for forest wildlife from Landsat data: judging black bear habitat quality in Louisiana [dissertation]. Baton Rouge: Louisiana State University. 164 p.

Walker, N.D. 2001. Tropical storm and hurricane wind effects on water level, salinity, and sediment transport in the river-influenced Atchafalaya-Vermilion system, Louisiana, USA. Estuaries 24: 498–508.

Walker, N.D. and A.B. Hammack. 1999. Satellite observations of circulation, sediment distribution and transport in the Atchafalaya-Vermilion Bay system. US Army Corps of Engineers, Waterways Experiment Station, Vicksburg, Mississippi.

Walker, N.D. and A.B. Hammack. 2000. Impacts of winter storms on circulation and sediment transport: Atchafalaya-Vermilion Bay region, Louisiana, USA. Journal of Coastal Research 16: 996–1010.

Walley, R.C. 2007. Environmental factors affecting the distribution of native and invasive aquatic plants in the Atchafalaya River Basin, Louisiana, USA [thesis]. Baton Rouge: Louisiana State University. 106 pp.

Walls, L.G. 2009. Crawfishes of Louisiana. Baton Rouge: Louisiana State University Press.

Walls, S.C., J.H. Waddle, and R.M. Dorazio. 2011. Estimating occupancy dynamics in an anuran assemblage from Louisiana, USA. Journal of Wildlife Management 75: 751–761.

Watson, J. T., S.C. Reed, R.H. Kadlec, R.L. Knight, and A.E. Whitehouse. 1989. Performance expectations and loading rates for constructed wetlands. In: D.A. Hammer (ed.), Constructed wetlands for wastewater treatment. Chelsea, Michigan: Lewis Publishers. pp. 319–351.

Weaver, J.E. and L.F. Holloway. 1974. Community structure of fishes and macrocrustaceans in ponds of a Louisiana tidal marsh influenced by weirs. Contributions in Marine Science 18: 57–69.

Wellner, R., R. Beaubouef, J. Van Wagoner, H. Roberts, and S. Tao. 2005. Jet-plume depositional bodies—the primary building blocks of Wax Lake Delta. Transactions of the Gulf Coast Association of Geological Societies 55: 867–909.

Wells, J.T. and G.P. Kemp. 1981. Atchafalaya mud stream and recent mudflat progradation: Louisiana chenier plain. Transactions of the Gulf Coast Association of Geological Societies 31: 409–416.

Wells, J.T., Chinburg, S.J. and J.M. Coleman. 1982. Development of the Atchafalaya River Deltas: generic analysis. US Army Corps of Engineers Contract Report, Louisiana State University, Coastal Studies Institute, Center for Wetland Resources, Baton Rouge, Louisiana.

Wiedenfeld, D.A. and M.M. Swan. 2000. Louisiana breeding bird atlas. Louisiana Sea Grant College Program, Louisiana State University, Baton Rouge, Louisiana.

Williston, H.L., F.W. Shropshire, and W.E. Balmer. 1980. Cypress management: a forgotten opportunity. USDA Forest Service Forestry Report SA-FR 8.

Wilson, W.B., M.J. Chamberlain, and F.G. Kimmel. 2005a. Seasonal space use and habitat selection of female wild turkeys in a Louisiana bottomland hardwood forest. Proceedings of the Southeastern Association of Fish and Wildlife Agencies 59: 114-125.

Wilson, W.B., M.J. Chamberlain, and F.G. Kimmel. 2005b. Survival and nest success of female wild turkeys in a Louisiana bottomland hardwood forest. Proceedings of the Southeastern Association of Fish and Wildlife Agencies 59: 126-134.

Withers, K. 2002. Shorebird use of coastal wetland and barrier island habitat in the Gulf of Mexico. The Scientific World JOURNAL 2: 514–526.

Xu, Y.J. 2006a. Total nitrogen inflow and outflow from a large river swamp basin to the Gulf of Mexico. Hydrological Sciences Journal 51: 531–542.

Xu, Y.J. 2006b. Organic nitrogen retention in the Atchafalaya River swamp. Hydrobiologia 506: 133–143.

Xu, Y.J. 2010. Long-term sediment transport and delivery of the largest distributary of the Mississippi River, the Atchafalaya, USA. In: Sediment dynamics for a changing future. International Association of Hydrological Sciences Publ. 337. Wallingford, UK: IAHS Press. pp. 282–290.

Xu, Y.J. 2013. Transport and retention of nitrogen, phosphorus and carbon in North America's largest river swamp basin, the Atchafalaya River Basin. Water 5: 379-393.

Xu, Y.J. and A. Patil. 2006. Organic carbon fluxes from the Atchafalaya River into the Gulf of Mexico. In: Y.J. Xu and V.P. Singh (eds.), Coastal environment and water quality. Highlands Ranch, Colorado: Water Resources Publications. pp. 217–226.

Xu, Y.J. and A. BryantMason. 2011. Determining the nitrate contribution of the Red River to the Atchafalaya River in the northern Gulf of Mexico under changing climate. In: J. Peters, (ed.), Water quality: current trends and expected climate change impacts. International Association of Hydrological Sciences Publ. 348. Wallingford UK: IAHS Press. pp. 95–100.

Zhang, X., S. Feagley, J. Day, W. Conner, I. Hesse, J. Rybczyk, and W. Hudnall. 2000. A water chemistry assessment of wastewater remediation in a natural swamp. Journal of Environmental Quality 29: 1960–1968.

INDEX

The letter *f* following a page number denotes a figure, and the letter *t* following a page number denotes a table.

agricultural production: atrazine runoff from, 150; chloride runoff, 150; nitrogen runoff from, 138, 145; nutrient input from, 59–60; phosphorus runoff from, 145; runoff and hypoxic events, 64; species threatened by, 132, 135; water drainage canals, 165

agricultural resources, 83–86, 84f

Amazon River, 48t

American alligator (*Alligator mississippiensis*), 121, 132, 136, 136f

American coot (*Fulica americana*), 131

American Pass, 167, 181, 183

American sweetgum (*Liquidambar styraciflua*), 52

American woodcock (*Scolopax minor*), 121, 126f

aquatic animals, 63, 65f, 69, 189f, 196f

aquatic vegetation, 74, 92–94, 193–97

ash (*Fraxinus* spp.), 229

Atchafalaya Basin Floodway System, 127f, 157, 160, 163, 170f, 186f, 211f

Atchafalaya Basin Floodway System Project, 169

Atchafalaya Basin lakes, 38, 40f

Atchafalaya Basin Program, 169

Atchafalaya Basin Program Natural Resource Inventory and Assessment Tool, 169

Atchafalaya Basin Project, 162

Atchafalaya Bay (AB): AR flow into, 22, 26f, 27, 69, 70f; bird populations, 121; characteristics, 78–79; engineering for flood control, 69; geological transformation, 38–44, 39f, 41f; nitrogen concentration reduction, 141; oyster reefs, 107, 109, 113, 114f; sedimentation, 48, 141, 144, 184; sediment redistribution, 41, 44–49, 45f, 193; tides and events overcoming, 78; WLO flow into, 69, 70f

Atchafalaya Delta (AD): AR flow received, 26f, 141; bayhead delta phase, 38, 40–41, 41f, 43f; bayhead deltas, comparisons with other, 42f; birds of the, 76, 124, 126, 129, 131–32, 131f; boundaries, 3f, 70f; deltaic formation, 18f; emergence of, 39f, 40; future growth, possibility of, 40–41, 44, 45; geological transformation, 40–41, 43f; land growth rates, 191–92; land ownership, 153; management and restoration models, 237; marsh habitat, 69, 70f; mercury deposition, 149; navigation channel bisecting, 69; nekton in the, 92–93, 222–223; nitrate transportation, 141; nutrient storage, 147; recreational fishing, 101; sedimentation, 27, 27f, 40, 40f, 41f, 43f, 44, 75–76, 184, 193; viability, requirements for, 237; water and salinity levels, 77, 78f; WLD compared, 72, 75, 77, 77t, 78f

Atchafalaya Delta (AD) islands: birds of the, 126, 128–29, 129f, 130f; creation of, 71–74, 72f, 74f; habitat, 73–75; shape, 73, 74; vegetation, 128–29, 130f, 131–32; wetland environments on, 71–72, 72f

Atchafalaya Delta complex, 27f

Atchafalaya Delta plain, 190–93

Atchafalaya East Watershed, 183–184

Atchafalaya Flood Protection Levees, 1, 8f

Atchafalaya Mud Stream, 47–48, 48t

Atchafalaya National Wildlife Refuge, 153

Atchafalaya River (AR): bayhead delta phase, 42f; carbon dioxide (CO_2) flux rates, 146; characteristics, 15, 16f, 17t, 19; discharge, 17t, 22, 25, 26f, 69, 70f, 199, 200f; drainage basin, 15, 17t; elemental mercury delivery rates, 149; flow capture/split, 15–16, 19–22, 20f, 23f, 27, 31, 69, 171–172; formation of the, 17–21, 19f, 20f; nutrients, 137, 140–41, 143, 145; restoration, role in, 237; sediment load/discharge, 25, 27–28, 27f, 28t, 192, 192f; wildlife habitat, 118–21

Atchafalaya River (AR) channel: bank stabilization levees, 69–70; development and distributary closure, effects of, 172, 174; dredging, purpose and results of, 69, 71f, 72, 166f; engineering of the, 69, 71f, 166–69, 166f, 168f; salinity incursions, effect on, 77

Atchafalaya River (AR) plume, 43f, 47, 141, 145

Terrebonne basin, 28, 228

Thistlewaite, 153

threatened species, 89

Three Rivers region, 17, 134

tidal marsh habitat, 92, 222

timber harvest: bird habitats impacted, 223; current status of, 212–13; economic value, 85–86; historically, 83, 84f; species protected from, 229; species threatened by, 135; statistics, 84f, 85–86, 85f, 246. *See also* forest management

timber resources: hurricane damage to, 54–55, 61; landowner income, 249; pulpwood, 85–86, 85f; sawtimber, 85–86, 85f. *See also* forest management

total Kjeldahl nitrogen (TKN), 138, 139f

tree species, invasive, 55

tricolored heron (*E. tricolor*), 126

Trinity Delta, 42f

Tropical Storm Frances, 46

tropical storms, 46–49, 64, 77, 193

trotline fishing, 100

Turnbull's Bend, 17–18, 18f, 19f, 22

Twenty One Inch Canal, 181

Upper Belle River, 170f, 177

US Army Corps of Engineers, 151, 153, 157, 158f, 159f, 169, 225, 228

vegetated wetlands, 94f, 137

vegetation occurrence records, 88t

Vermilion/Atchafalaya Bays Public Oyster Seed Ground, 113–14, 115f

Vermilion Bay: AR flow received, 22; boundaries, 3f; commercial fishing, 104, 104f; fish habitats, 93, 95; land cover, 5f; nitrogen reduction, 141; oysters and oyster reefs, 105–109, 105f, 106f, 113–15, 115f; recreational fishing, 102, 102f; suspended sediment during cold front passage, 45f

Verret Swamps, 1, 3f, 53f, 57, 59–61, 183, 207

wading birds, 124

warmouth (*Lepomis gulosus*), 220

wastewater treatment, 150

waterbird colonies, active, 127f

water-elm (*Planera aquatica*), 60–61

waterfowl, 76–77, 121, 124, 124f, 251–52

water habitats, 61–69, 68f, 89, 237

water hickory (*Carya aquatic*), 52

water hyacinth (*Eichhornia crassipes*), 194, 195f, 196, 236

water management: ARB floodway, 169–74, 171f; bird habitats, impact on, 223; crawfish, impact on, 217–18; for cypress-tupelo forest sustainability, 207; fish, impacts on, 219–21, 220f; ideal model, 235–37; present-day model, 233–37; scales of control, 171; stakeholders position, 228

water moccasin, 122f

water oak (*Quercus nigra*), 52

water quality: ARB, present-day, 177–83; crawfish, impact on, 217–19, 218f; cypress-tupelo forest regeneration, 208–209; East and West Atchafalaya Protection Levees, outside the, 183–84; levee construction and, 60; water management model for, 233–34. *See also* hypoxia

water stargrass (*Heteranthera dubia*), 74

water temperature: in backswamps, 57; hypoxia and, 67, 69

water tupelo (*Nyssa aquatica*): bear dens in, 134f, 135, 223; carbon dioxide (CO_2) sequestration, 147; dominance of, 2; flooding, studies on effects of, 266–69; flood tolerance, 57, 200–209, 205f, 212–13, 233, 237; harvesting the, 83, 84f, 134f, 135, 213, 229; hurricane tolerance, 61; importance of, 233; management and restoration models, 233, 237; nutrients, studies on effects of, 266–69; regeneration ability, 84f, 202f, 233, 237, 266–69; stump regeneration, 205–206, 206f. *See also* cypress-tupelo swamp forest

Wax Lake Delta (WLD): AD compared, 76–77, 77t, 78f; aquatic vegetation, 94f; AR flow received, 22, 26f, 141; bayhead deltas, comparisons with other, 42f; boundaries, 3f, 53f, 70f; emergence, 39f, 40, 41f; growth of, 43f, 191, 191f, 193; land gain to land loss shift, hurricane influence, 46; land ownership, 153; management strategies, 237; marsh erosion, 113; nutrients, 141, 147; oyster reef removal, 113; recreational fishing, 101; sedimentation, 40, 41f, 42f, 43f, 44, 47f; size and shape, 72; spring flood pulse, 72f; subaerial extent, 43f;

viability, requirements for, 237; water and salinity levels, 77, 78f

Wax Lake Delta (WLD) habitat: AD compared, 75; area of, 1; biomass in relation to grazing pressure, 76–77, 77t; bird populations, 121, 124, 127f, 129, 131, 131f; delta marshes, 69; nekton in the, 92–93, 94f, 222–23; nonnative species invasion, 76f; vegetation patterns, 75; wintering waterfowl, 76

Wax Lake Delta (WLD) islands, 72–77, 72f, 73f, 75f, 77t

Wax Lake Outlet (WLO): AR flow received, 22, 26f, 69; bayhead deltas, 40, 43f; boundaries, 3f; channelization, 168; construction purpose of, 1, 69, 71; fresh water pulses, effect on oyster habitat, 113–15, 114f; photograph of, 72f; river discharge to Atchafalaya Bay, 69, 70f; sediment transport, 43f; Wax Lake Delta, effect on, 72

West Atchafalaya Floodway, 153, 160, 162, 163

West Atchafalaya Floodway management, 157

West Atchafalaya Protection Levee, 34, 133, 165, 166–67

West Cote Blanche Bay, 25

western Terrebonne marshes: AR flow received, 22, 26f, 141; boundaries, 3f; floodway management, 165; levees preventing freshwater flow, 163; nitrate nitrogen removal, 141; recreational fishing, 102, 102f; research recommendations, 240; suspended sediment during cold front passage, 45f, 49

wetland forest, 199. *See also* forested wetlands

wetlands habitat: birds of the, 123–24, 129; creation of, 16f; food webs, 95; game species, 121, 122f; sediment supply, 192–93; sustainability, 189, 234

Whiskey Bay Pilot Channel, 166–69

white crappie (*Pomoxis annularis*), 91, 101

white-faced ibis (*Plegadis chihi*), 126

white ibis (*Eudocimus albus*), 121, 124, 125t, 126, 216f

white river crawfish (*Procambarus zonangulus*), 97

white-tailed deer (*Odocoileus virginianus*), 121, 122f